Stem Cell Biology and Regenerative Medicine

Series Editor:

Kursad Turksen
Ottawa Hospital Research Institute, Ottawa, ON, Canada

Our understanding of stem cells has grown rapidly over the last decade. While the apparently tremendous therapeutic potential of stem cells has not yet been realized, their routine use in regeneration and restoration of tissue and organ function is greatly anticipated. To this end, many investigators continue to push the boundaries in areas such as the reprogramming, the stem cell niche, nanotechnology, biomimetics and 3D bioprinting, to name just a few. The objective of the volumes in the Stem Cell Biology and Regenerative Medicine series is to capture and consolidate these developments in a timely way. Each volume is thought-provoking in identifying problems, offering solutions, and providing ideas to excite further innovation in the stem cell and regenerative medicine fields.

More information about this series at http://www.springer.com/series/7896

Ayse Begum Tekinay
Editor

Nanomaterials for Regenerative Medicine

Humana Press

Editor
Ayse Begum Tekinay
Eryigit Research & Development Center
Ivedik OSB, Ankara, Turkey

ISSN 2196-8985 ISSN 2196-8993 (electronic)
Stem Cell Biology and Regenerative Medicine
ISBN 978-3-030-31201-5 ISBN 978-3-030-31202-2 (eBook)
https://doi.org/10.1007/978-3-030-31202-2

This Humana imprint is published by the registered company Springer Nature Switzerland AG
The registered company address is: Gewerbestrasse 11, 6330 Cham, Switzerland

Preface

Regenerative medicine studies offer promising therapeutic approaches for damaged tissues through tissue engineering, stem cell use, or utilization of biomaterials that can replace, regrow, or repair these tissues. Due to the difficulties associated with organ transplantation such as donor availability and immune problems, regenerative medicine studies have attracted considerable attention for finding alternative therapeutic solutions for many degenerative diseases and organ or tissue damages caused by other reasons such as accidents. Developing suitable materials for regenerative medicine is an important part of delivering such therapeutic approaches from bench to bedside.

In the recent decades, plenty of research efforts in materials science have been focused on medical applications. Many researchers working in chemistry, materials science, physics, etc. have been trying to learn more biology to apply their research in developing a more value-added product such as medical products since there are several important requirements for using materials for medical applications. This book aims to provide knowledge on the requirements for the design of the optimal nanomaterials for regenerative medicine and cover the most recent approaches in nanomaterial design.

Nanotechnology is the use, control, and characterization of materials at nanoscale, which is usually defined between 0.1 and 100 nm. We now know that the materials behave differently at nanoscale than they do at macroscale and materials gain superior properties that can be effective for many therapeutic purposes at nanoscale. In this book, we discussed the properties of nanomaterials and important information for providing an efficient therapy for designing better materials for regenerative medicine applications. We tried to cover a broad range of specific regenerative medicine applications while focusing on the most widely studies fields of regenerative medicine such as neuroregeneration and cartilage regeneration to provide more in-depth knowledge on the most widely used areas of nanomaterials for regenerative medicine.

The topics that are covered will enable the reader to have a broad knowledge on the state-of-the-art nanomaterials systems that are used for regenerative medicine applications, providing guidelines for clinicians. In addition, the sections regarding the requirements of nanomaterials to be used for potential clinical applications will guide chemists and materials scientists to make the most effective designs for specific regenerative purposes.

It should be kept in mind that the research that has been done in this field are only the beginning steps of a vast area of developments and future research on novel nanomaterials and more clinical research that will be pursued by larger pharmaceutical companies can enable faster application of the newly found therapeutic methods from the bench to the bedside.

Ankara, Turkey Ayse Begum Tekinay

Contents

1 **Nanomaterials for Regenerative Medicine** 1
Ayse Begum Tekinay

2 **Design and Development of Electrospun Nanofibers in Regenerative Medicine** 47
Brabu Balusamy, Anitha Senthamizhan, and Tamer Uyar

3 **Nanomaterials for Wound Healing** 81
Gozde Uzunalli

4 **Immunomodulatory Nanomaterials** 119
Turgay Tekinay

5 **Neuroregenerative Nanotherapeutics** 143
Ayse Begum Tekinay, Brabu Balusamy, Anitha Senthamizhan, and Tamer Uyar

6 **Nanomaterials for Cartilage Regeneration** 183
Huseyin Ozkan and Orhan Yanmis

7 **Nanotechnology in Dentistry: Past, Present, and Future** 197
Metin Calisir

Index 217

Contributors

Brabu Balusamy Fondazione Istituto Italiano di Tecnologia, Genoa, Italy

Metin Calisir Department of Periodontology, Faculty of Dentistry, Adiyaman University, Adiyaman, Turkey

Huseyin Ozkan University of North Carolina, Chapel Hill, NC, USA

Anitha Senthamizhan Fondazione Istituto Italiano di Tecnologia, Genoa, Italy

Ayse Begum Tekinay Eryigit Research & Development Center, Ivedik OSB, Ankara, Turkey

Turgay Tekinay Algitek Biotechnology, Ankara, Turkey

Tamer Uyar Department of Fiber Science & Apparel Design, College of Human Ecology, Cornell University, Ithaca, NY, USA

Gozde Uzunalli Department of Comparative Pathobiology, College of Veterinary Medicine, Purdue University, West Lafayette, IN, USA

Orhan Yanmis Department of Orthopaedics Surgery and Rehabilitation, Wake Forest University School of Medicine, Winston-Salem, NC, USA

Chapter 1
Nanomaterials for Regenerative Medicine

Ayse Begum Tekinay

Abstract Organ or tissue loss due to accidents, diseases, or aging has been a major problem in medicine which until recently could have only been alleviated with organ or tissue transplantation. Tremendous efforts of scientists working in the fields of tissue engineering and regenerative medicine have resulted in several successful cases where engineered scaffold systems were used to replace several major organs such as trachea (Hofstetter, Journal of Thoracic and Cardiovascular Surgery, 156(3):1273–1274, 2018). In addition, there are also many studies that have been performed on animal models (Annesini et al., Artificial organ engineering, Springer, Cham, 2017; Brohem et al., Pigment Cell and Melanoma Research, 24:35–50, 2011; Malchesky, Artificial Organs, 33:273–295, 2009), which are expected to pave the way for novel treatment options for humans who need organ transplantation. Extracellular matrix (ECM) is an especially important factor to consider when engineering artificial organs or tissue replacements. In this chapter, we will overview the major challenges and crucial design principles for engineering new materials for regenerative medicine. To better explain these principles, we will also provide a brief introduction on what regenerative medicine is and its major components and the types of nanomaterials that are currently being used for regenerative medicine applications.

1.1 Introduction

1.1.1 What Is Regenerative Medicine?

Tissue regeneration is the renewal or regrowth of cells and their ECM to replace the lost tissue through an orchestrated chain of events that require cell division, cell proliferation, cell migration, and production of ECM elements. This chain of events

A. B. Tekinay (✉)
Eryigit Research & Development Center, Ivedik OSB, Ankara, Turkey
e-mail: abt@eryigit.com.tr

A. B. Tekinay (ed.), *Nanomaterials for Regenerative Medicine*, Stem Cell Biology and Regenerative Medicine, https://doi.org/10.1007/978-3-030-31202-2_1

changes depending on the tissue and the type and the age of the organism in which they take place. For many species, tissue regeneration results in the complete reconstruction of the lost tissue, such as the tissue regeneration that is observed in planaria, hydra, or different amphibians. In other species which have more developed and complex structures, however, there is only limited regeneration of the tissues. For example, although a Hawaiian starfish can regenerate even its whole body from a limb, in amphibians, the organ or tissue regeneration is more limited and the ability for complete regeneration decreases with the age of the animal. In mammals, the extent of tissue regeneration is even poorer and, like amphibians, older mammals usually have poorer regeneration capability compared to younger ones. In humans, most tissue regeneration takes place in epithelial cells in the skin, lungs, and intestines. The difference in the regeneration capabilities of different species and the change in regeneration capability through aging have attracted much attention form scientists who work in the field of regenerative medicine.

Regenerative medicine is the study of tissue regeneration and spans a wide range of scientific disciplines as it requires knowledge from different fields of science. These fields include molecular biology, materials science, nanotechnology, pharmaceutical sciences, medicine, chemistry, physics, and engineering. Tissue engineering, biomaterials and stem cell biology are the major subdisciplines of regenerative medicine and these fields are also cross-disciplinary. Although most of the previous regenerative medicine studies have focused on decellularization or acellularization of organs and their recellularization with the patient's own cells, recent advances in the fields of nanotechnology and engineering have enabled bottom-up approaches such as reconstruction of the ECM elements with synthetic or natural alternatives through self-assembly or layer-by-layer modelling and 3D-printing attempts for reformation of the ECM structure together with the cellular components. Although we are still far from the type of tissue regeneration that were depicted in sci-fi movies, the pace of advances in the regenerative medicine field shows strong promise that the clinical trials of the bottom-up approaches with tailored treatment options will soon enable therapies that will enable regeneration of most tissues.

1.1.2 Importance of Extracellular Matrix and Stem Cells

ECM is a combination of fibrous proteins, non-fibrous proteins, proteoglycans, glycosaminoglycans (GAGs) and other secreted materials that surround the cells in the body. It has many functions including providing structural support, allowing access to nutrition and oxygen, guidance for cell migration and providing the signals for cell division, differentiation, adhesion, and apoptosis. In humans, the composition of the ECM molecules varies vastly among different tissues. Not only the ratio of the fibrous vs. non-fibrous proteins or GAGs change, but also the types of the proteins and GAGs vary enormously depending on the maturity of the tissue, its inflammation state, the age of the person, and the type and the function of the tissue. The complexity of the ECM molecules has fascinated the scientists working in this field, and analyzing the components of the ECM according to their regenerative potential has paved the way for developing novel materials that can mimic these components. Since developing and mature tissues have different ECM elements, the comparison of these elements and the analysis of specific functions of each domain of proteins or GAGs have also been providing clues for developing better materials for regenerative medicine applications.

In addition to ECM components, stem cells also are important for tissue regeneration studies. Essentially, stem cells are the cells that can differentiate into more than one cell type and they have been the major targets of regenerative medicine studies due to their differentiation potential (Madl, Heilshorn, & Blau, 2018). The differentiation potential of the stem cells varies depending on their types and the age of the donor. In addition, the differentiation potential also gives clues about the teratogenic potential of the stem cells. The specific clues that can aid in navigating the stem cell differentiation pathways have been the major focus of nanomaterial development studies for regenerative medicine. Nanomaterials can also provide a suitable niche for stem cell transplantation in order to increase the viability of the stem cells after transplantation. In this section, the major players of the ECM and the fundamental characteristics of the stem cells will be described and how this information can be used for developing better nanomaterials for regenerative medicine will be discussed.

1.1.2.1 Extracellular Matrix

The major functions of the ECM include providing mechanical support to cells and guiding cell migration, viability, differentiation, and proliferation. The ECM maintains these functions through the physical, chemical, and biological signals that it harbors. The major structural characteristics of the ECM include porosity and the self-assembly ability of its components. Porosity is a vital characteristic for any ECM mimetic material and is crucial for the transfer and diffusion of nutrients within the matrix, the discarding of waste materials, the transfer of signaling molecules from one cell to another and cellular growth, proliferation, and migration. Several types of porous materials have previously been developed as ECM mimetic materials and have been successfully used in in vitro and in vivo studies. These include electrospun polymeric nanofibers (Pham, Sharma, & Mikos, 2006), electrospun polysaccharides and other natural biomaterials (K. Y. Lee, Jeong, Kang, Lee, & Park, 2009), cryogels (Hixon, Lu, & Sell, 2017), nanocomposites (Pina, Oliveira, & Reis, 2015), natural biomimetic molecules such as collagen fibers, chitosan, and alginate. (Hinderer, Layland, & Schenke-Layland, 2016), self-assembling synthetic amyloid-like peptide molecules that form nanofibers (Feyzizarnagh, Yoon, Goltz, & Kim, 2016), and self-assembling peptide amphiphile nanofibers (Cui, Webber, & Stupp, 2010).

The size of the pores and the flexibility and biodegradability of the material are the important factors in determining whether a porous material is suitable for a regenerative medicine application. The type of the tissue that is aimed to be regenerated is also important when determining the pore structure. In humans, the cell size mostly varies from 5 μm in diameter to 150 μm, although there are cells that are much longer than these. Thus, the sizes of the pores should be in line with the diameters of the cells; that is, they should not be much larger or much smaller that the cell sizes. The larger pores will result in poor mechanical support while stiff materials that have smaller pores might prevent the migration or proliferation of cells.

The flexibility and biodegradability of the material is also of utmost importance when designing nanomaterials for regenerative medicine. While materials that totally biodegrade in 2–4 weeks is desirable for applications such as soft tissue regeneration in organs with highly proliferative cells such as skin, a faster biodegradation rate would be more suitable for mesenchymal stem cell (MSC) transplantation and a much slower biodegradation rate is necessary for bone regeneration since the mechanical support should be maintained until the bone cells have the time to form natural stiff bone ECM to replace the synthetic one.

As explained above, the constituents of the natural ECM can be grouped depending on their structure under the following categories: fibrous components and the non-fibrous components. The non-fibrous components include not only laminins and GAGs which are important structural elements of the ECM, but also growth factors, enzymes such as matrix metalloproteases or cytokines which are only present in the ECM at certain times for specific functions. The fibrous constituents of the ECM mostly include collagen and fibronectin. Depending on the type and the age of the tissue, the collagen types that are prevalent in the ECM change. There are 16 different types of collagens; however, most of the collagens that are found in our tissues are collagen type I, II, and III (Bornstein & Sage, 2003). Collagen type I forms ordered fibrillar structures which are structurally important in many tissues including bone or cornea (Cen, Liu, Cui, Zhang, & Cao, 2008; Viguet-Carrin, Garnero, & Delmas, 2006). The ordering of collagen I determines the well-organized structures of these tissues that are crucial for their basic functions such as strength or transparency. On the other hand, cartilage, which is a less organized tissue, where the flexibility and load-bearing capabilities are more important, has mostly collagen type II, which forms a network-like structure (Lees & Partington, 2016).

The morphology of the fibrous proteins of the ECM is not the only characteristics that should be mimicked while designing ECM like materials. There are many biological signals that are present on these proteins as well. For example, GFOGER is a short peptide sequence which has been derived from collagen type I and have been used for various purposes from cell attachment to enhancing biomineralization (Knight et al., 2000; S. Q. Liu, Tian, et al., 2010; Mhanna et al., 2013; Shekaran et al., 2014). Another short peptide sequence, RGD, was originally derived from fibronectin, but was later found in other ECM proteins as well. These short peptide sequences interact with various integrins on the cell surface and modulate cellular activities such as adhesion and differentiation (Bellis, 2011; Ruoslahti, 2002).

The non-fibrous proteins of the ECM have also attracted much attention for modulating cellular behavior. Laminins are a major example to such proteins and are crucial components of basal lamina in addition to other ECM types. The peptide sequences that were derived from laminins, such as IKVAV and YIGSR have also been major players in nanomaterial design, from neuroregeneration applications to cardiac tissue regeneration and implant functionalization (T. Y. Cheng, Chen, Chang, Huang, & Wang, 2013; Farrukh et al., 2017; Graf et al., 1987; Hamsici, Cinar, et al., 2017; G. A. Silva et al., 2004; W. Sun, Incitti, et al., 2017; Y. Sun et al., 2016; Yoon et al., 2012; Zhu et al., 2017).

Growth factors that are secreted into the ECM by different cell types are major targets for biomaterial design due to their ability to significantly alter cell fate. Growth factors are proteins that mostly function in cell division and differentiation, and several growth factors have been approved to be used for therapeutic applications such as PDGF and NGF. Although the importance of growth factors for regeneration and cell proliferation has been known for more than 20 years, the scientists have been cautious in using these proteins as therapeutic agents due to their tumorigenic properties. The growth factor-based drugs that have been approved by European Medicines Agency (EMA) or U.S. Food and Drug Administration (FDA) carry warning signs about risk of tumor formation as a side effect. Thus, the initial attempts and aims of using growth factors for regenerative medicine focused on controlled delivery of these molecules in a specific tissue limiting their tumorigenic potential while utilizing their regenerative capabilities (F. M. Chen, Zhang, & Wu, 2010; Cinar et al., 2017; Tekinay, Guler, Mumcuoglu, & Ustun, 2013). However, like many other proteins, growth factors also have several different domains for various functions and engineering these domains to obtain the optimal regenerative effect while eliminating the side effects is the major challenge for their use (Mitchell, Briquez, Hubbell, & Cochran, 2016). In addition, the size of a protein can be a critical factor for eliciting immune response, and for most regenerative medicine applications, immunogenicity of the material is not a desired quality. Thus, shorter peptides derived from the growth factors have been tailored for mimicking the growth factor activity (Cai, Dinh, & Heilshorn, 2014; Chan, Stahl, & Yu, 2011; Colangelo et al., 2008; Lin et al., 2006; Perugini, Varelias, Sadlon, & D'Andrea, 2009; Zouani, Chollet, Guillotin, & Durrieu, 2010).

In addition to the proteinaceous components of the ECM, GAGs have also been widely studied in order to understand the functions of the ECM and its molecular characteristics. However, the molecular, genetic and biochemical techniques that enable the study of the proteins are much more advanced and diverse compared to the techniques that are currently available for the study of GAGs. Peptides are polymers of amino acids which are bound to each other through peptide bonds. Although there are 20 different amino acids, combination of these amino acids and the diversity of the sequences is enormous, this diversity is not even comparable to the diversity of the carbohydrates which are made of many different monomers, and can also form branches at various points of the polymers with each different branch being able to end up in a different GAG. The major classes of GAGs found in the ECM are heparan sulfate, chondroitin sulfate, dermatan sulfate, keratan sulfate, heparin, and hyaluronate. Among these GAGs, hyaluronate is the only type that is not sulfated. The ratios of different GAGs and their sulfation degrees also vary in different tissues. Since GAGs also carry out important functions such as growth factor binding, water encapsulation and providing mechanical strength, different types of GAGs or GAG-mimetic molecules have been utilized for regenerative medicine applications (Arslan, Guler, & Tekinay, 2016; Rufaihah et al., 2017; Tansik et al., 2016; Yaylaci et al., 2016).

MMPs have also been important targets for designing ECM mimetic nanomaterials for regenerative medicine applications. MMPs normally function in keeping the homeostasis of the ECM by constantly reorganizing the structural protein components such as collagens, aggrecan or fibronectin. There are various types of MMPs, which can be present either by themselves or in combinations within different tissues depending on the age and the type of the tissue. MMPs are especially important during wound healing since they are key players in ECM remodeling. Thus, many nanomaterial studies for regenerative medicine or drug delivery applications have focused on MMP digest sites when designing new materials (D. S. Ferreira et al., 2015; Olson et al., 2009; Salinas & Anseth, 2008).

Overall, the natural ECM is a complex combination of different types of biomacromolecules and this complexity aids in maintaining the homeostasis of the tissues. This complexity also provides a wide array of opportunities for finding the best design for producing nanomaterials tailored for specific regenerative medicine applications.

1.1.2.2 Stem Cells

Stem cells are the cells that have the ability to renew themselves and differentiate into more specialized cell types. They are the origins of all of the tissues in our body and thus can potentially be used to regenerate any of those tissues. However, as explained before, the niche in which the stem cells reside in, the ECM, and the epigenetic regulations inside the stem cells limit the potential of these cells for regenerative medicine applications.

The stem cells are classified into several groups depending on their differentiation capabilities. Among these, the most widely known ones are MSCs, embryonic stem cells, induced pluripotent stem cells (iPSCs) and tissue specific stem cells. The potency of the stem cells to differentiate into different cell types is also used to classify them as totipotent stem cells, pluripotent stem cells and multipotent stem cells. Totipotent stem cells are the cells that can differentiate into all of the cells of the body and the placental cells, such as the embryonic cells that are in the first divisions of the embryonic development. The multipotent cells, such as embryonic stem cells, can differentiate into all of the cell types in the body; whereas multipotent stem cells can only differentiate into several cell types. In this section, we will briefly describe their characteristics and give examples of their uses with different types of nanomaterials for regenerative medicine applications.

1.1.2.2.1 Mesenchymal Stem Cells

MSCs are the most widely studied and used stem cell type for regenerative medicine applications. They are multipotent stem cells that reside in the bone marrow and provide a niche for the other bone marrow stem cells, hematopoietic stem cells. Hematopoietic stem cells are the stem cells that form the blood cells. MSCs on the other hand can differentiate into osteoblasts, chondrocytes, adipocytes, and fibroblasts which can form the tissues that surround the bone marrow such as bone, cartilage, and fat. Many controversial therapies have been developed that utilize the MSCs for tissue regeneration and they are currently used in clinics in many countries around the world. Most of these treatments are focused on orthopedic applications; however, there are also applications ranging from antiaging cosmetic applications to neuroregeneration.

1.1.2.2.1.1 The Variety Among MSCs

The reason that we have used the term "controversial" for the MSC-based treatments is that many scientists who work in this field are concerned about several aspects of these treatments. One major hurdle in stem cell transplantation in general is the difficulty of keeping cells alive during and after the transplantation procedure. Another major concern is regarding the heterogeneity of the MSCs (Kfoury & Scadden, 2015; Lv, Tuan, Cheung, & Leung, 2014; Uccelli, Moretta, & Pistoia, 2008). This is a rather significant problem not only for MSC-based treatments in clinics, but also for scientific studies which use MSCs as tools to study different molecular processes. The differentiation potential of the MSCs vary depending on the genetic and epigenetic makeup of the cells which is also affected by the age, gender, overall health and lifestyle of the donor, in addition to the passage number and the culture media components of the cells. So, depending on all these complex variables the rates of proliferation and differentiation of MSCs to specific lineages such as osteogenic or chondrogenic lineage can vary significantly. In order to further understand the molecular basis under this complexity, several markers have been identified to select only a certain group of MSCs for scientific research; however, the clinical studies are still far behind such developments.

Another major variability among MSCs comes from the tissue origin of the MSCs. Although most of the current research on MSCs have been performed on bone marrow derived cells, several other tissue sources have also been suggested and were investigated (Al-Nbaheen et al., 2013; Hass, Kasper, Böhm, & Jacobs, 2011; Kern, Eichler, Stoeve, Klüter, & Bieback, 2006; Wagner et al., 2005). Adipose and umbilical vein are the most widely studied tissues among them.

1.1.2.2.1.2 Bone Marrow Derived MSCs

Bone marrow derived MSCs are the most widely used MSC type both for scientific studies and for clinical use. MSCs were first identified in the bone marrow and were shown to differentiate into several specialized cells in their natural niche such as osteoblasts, chondrocytes and adipocytes (Friedenstein, Gorskaja, & Kulagina, 1976). This differentiation capability into all three of these cell types later became a trademark for MSCs and was checked after isolation before being able to classify them as MSCs. MSC differentiation can be induced by addition of specific growth factors or changing the chemical composition of the culture media. Osteogenic differentiation of the MSCs can be easily induced by using growth factors such as bone morphogenetic proteins (BMPs) or simply by the addition of ascorbic acid, β-glycerophosphate, and dexamethasone into the culture media after they are grown to more than 85% confluency. Moreover, addition of other ions, nanoparticles or the morphological characteristics of the culture surface have also been shown to induce the osteogenic differentiation of MSCs to a great extent (Abagnale et al., 2015; Dalby et al., 2007; D. Liu, Yi, Zhang, Zhang, & Yang, 2010; Yi, Liu, Fong, Zhang, & Yang, 2010). Among the physicochemical characteristics of the culture surface, stiffness has been shown to be one of the most effective methods for tuning the differentiation potential of MSCs towards osteogenic lineage (Pek, Wan, & Ying, 2010; Shih, Tseng, Lai, Lin, & Lee, 2011). Mechanotransduction through integrins are thought to modulate the transcription factor activity altering the differentiation pathway and also change the ECM composition to further advance the differentiation pathway, while changing the mechanical characteristics of the cells themselves as well (Topal et al., 2017). The osteogenic differentiation may also be induced through using specific peptide or chemical ligands. For example, tenascin-C and collagen derived, alkaline phosphatase mimetic or BMP-mimetic peptides have been shown to mediate the osteogenic differentiation of MSCs, while the spacings between fibronectin derived RGD peptides were shown to alter their fate (Bilem et al., 2016; Gulseren et al., 2015; Horii, Wang, Gelain, & Zhang, 2007; Mehta, Madl, Lee, Duda, & Mooney, 2015; Sever, Mammadov, Guler, & Tekinay, 2014; Tansik et al., 2016; X. Wang et al., 2013). The peptide ligands presented on nanomaterials can not only induce differentiation through binding specific cell surface receptors, but also bind to the growth factors that are released by the cells themselves and induce their differentiation through a nanomaterial based feedback mechanism (Arslan et al., 2016; Tansik et al., 2016). When the MSCs start differentiating towards osteogenic or chondrogenic lineage, the cells aggregate to form nodular structures. As they express more osteogenic proteins such as alkaline phosphatase (ALP), they also increase calcium accumulation which can be stained with Alizarin red and von Kossa staining techniques. For assessing chondrogenic differentiation of MSCs, it is customary to stain the GAGs that are deposited by the differentiated cells with a method such as saffranin-O staining. Adipogenic differentiation can also be assessed by measuring ALP activity, since adipogenic cells also have elevated ALP activity. Adipogenic activity may also be analyzed by cytostaining methods such as Oil-Red-O staining. All staining methods should be accompanied by qRT-PCR and Western blot or ELISA analysis to make sure that the differentiation markers are expressed. Due to the heterogenous nature of the MSCs, the MSC cultures usually differentiate into multiple cell types depending on their epigenetic state and the strength of the external cues that drive differentiation.

1.1.2.2.1.3 Adipose Derived MSCs

Adipose tissue derived MSCs are the second most studied MSC type in the literature. The ease of obtaining MSCs from discarded adipose tissue compared to isolating them from bone marrow has made this MSC type an attractive target for clinical studies. Similar to bone marrow derived MSCs, adipose derived MSCs are also heterogenous as a cell population and can differentiate into osteogenic, chondrogenic and adipogenic lineages. Adipose derived MSCs have been used for regenerative medicine applications for various purposes including cardiomyocyte differentiation after cardiac infarct, bone tissue regeneration and others (Konno et al., 2013; Miyahara et al., 2006).

Similar to bone marrow derived MSCs, material stiffness is an important factor for inducing differentiation of the adipose-derived MSCs into specific lineages (Wen et al., 2014). Nanofibers made from natural biomaterials and synthetic polymers have been utilized as ECM-mimetic materials to induce differentiation of adipose-derived MSCs for neuroregeneration and bone regeneration (Fesharaki et al., 2018; Francis, 2010; Hu et al., 2017; C. C. Lin & Fu, 2016; Ravichandran, Venugopal, Sundarrajan, Mukherjee, & Ramakrishna, 2012).

1.1.2.2.1.4 Other MSC Types

MSCs have been located in tissues other than bone marrow and adipose tissue as well, and since then, they have been studied for regenerative medicine applications alone and together with different types of nanomaterials. ECM-mimetic nanofibers have been extensively used for such purposes. Umbilical cord-derived MSCs, for example, have been shown to differentiate into hepatocyte-like cells or used for regeneration of tissues including cardiac tissue and bone tissue by using ECM-mimetic scaffolds such as poly(ε-caprolactone) nanofiber scaffolds, self-assembling peptide nanofibers or fibronectin-immobilized polycaprolactone nanofibers (W. Chen et al., 2013; M. S. V. Ferreira & Mousavi, 2018; Tahlawi et al., 2019). Further studies should be performed in order to better understand the heterogeneity and molecular mechanisms underpinning the differences between different MSC types. Methods to better sort MSCs from different tissue sources in order to have more homogenous cell populations are also necessary in order to enhance the efficiency of MSC treatments. In addition, use of synthetic, chemically well defined scaffolds as carrier systems for MSC transplantation would also increase the efficacy of MSC-based therapies, which have shown promising results as not only cellular sources for more differentiated cells but also as cytokine and growth factor sources.

1.1.2.2.2 Embryonic Stem Cells

Embryonic stem cells are the cells that are isolated from the inner cell mass of blastocysts, and can give rise to all the specialized cells in our body, that is, they are pluripotent. Most of the human embryonic stem cells have been derived from blastocysts which were produced for in vitro fertilization purposes, but have been deemed unnecessary afterwards. Due to their potential to develop into a complete human being, their use for research purposes has been highly debated for a long time and is still illegal in several countries. However, due to their pluripotency, they also provide an invaluable source for studying developmental processes and tissue regeneration. On the other hand, their pluripotency also makes these cells tumorigenic when transplanted into an organism (Ben-David & Benvenisty, 2011).

Several different types of nanomaterials have been utilized for both extended culture and directed differentiation of embryonic stem cells into specific cell lineages. Neural differentiation of embryonic stem cells, for example, have been induced by culturing them on three-dimensional electrospun poly ε-caprolactone nanofibers, peptide functionalized nanofibers, and graphene oxide or even culturing with functionalized gold nanoparticles (Abbasi et al., 2016; Philip, Silantyeva, Becker, & Willits, 2019; Wei, Li, Yang, Zheng, & Le, 2017; Yang et al., 2014). In addition to neuronal lineage, nanomaterials have been used for directed differentiation of embryonic stem cells into different cell fates including bone cells, chondrocytes or hepatocytes (Dehdilani et al., 2016; Farrukh et al., 2017; Gerecht et al., 2007; Hwang, Varghese, Zhang, & Elisseeff, 2006; Smith, Liu, Hu, & Ma, 2010). Even though embryonic stem cells are quite attractive for tissue regeneration due to their diverse potential for differentiation, the ethical concerns behind their generation, their teratogenic potential and the limitedness of the source material have encouraged the scientists to find alternative pluripotent cell sources such as iPSCs.

1.1.2.2.3 Induced Pluripotent Stem Cells

Induced pluripotent stem cells (iPSCs) are pluripotent cells that have been engineered from differentiated cells in the laboratory. With this technology, we can generate stem cells that behave like embryonic stem cells by using skin cells as the starting material. Although this technology is fairly recent, due to the scarcity of embryonic stem cells, iPSCs have been adopted for the molecular biology studies quite rapidly and have been widely used ever since. iPSCs provide important tools not only for studying normal development and pathophysiology of diseases but also the high-throughput screening of novel drugs and therapeutics. A major advantage of iPSCs over embryonic stem cells is that they can carry the same genetic information as the host, which can make them ideal candidates for tissue regeneration purposes in personalized medicine. In addition, they can also be used to study an individual's disease pathophysiology within the context of that person's own genetic makeup, which makes iPSCs invaluable tools for studying genetic diseases as well.

While they are similar to embryonic stem cells with respect to their differentiation potential, embryonic stem cells and iPSCs are different cell types, since iPSCs are generated through addition of genetic material by using viruses as tools, and have different epigenetic characteristics. Thus, more comprehensive molecular studies are needed before they can be used as alternatives of embryonic stem cells for therapeutic purposes. Nevertheless, iPSCs have been used for many regenerative medicine studies together with nanomaterials. Similarly to embryonic stem cells, tissue regeneration studies of iPSCs have mostly focused on neural regeneration. Polymeric nanofibers such as poly(γ-caprolactone) and poly-L-lactic acid are among the nanomaterials that have been shown to selectively induce neural differentiation of iPSCs (C. Lin et al., 2018; Mohtaram et al., 2015; Pang et al., 2016). In addition to neural differentiation studies, iPSCs have also been shown to differentiate into cardiomyocytes, hepatocytes and osteogenic cells in the presence of nanomaterials (Khan, Johnson, Xu, Belevych, & Gyorke, 2015; Mahmoodinia Maymand et al., 2018; J. Xie et al., 2016).

1.1.2.2.4 Tissue Specific Stem Cells

Tissue specific stem cells are found in almost all tissues and have the ability to self-renew and differentiate into very specific cell types. Satellite cells that can renew themselves and differentiate into skeletal muscle cells are a widely studied example (Kuang, Kuroda, Le Grand, & Rudnicki, 2007). Recently different nanomaterials with specific morphological characteristics have been used to induce cell proliferation and differentiation of satellite cells (Allur Subramaniyan et al., 2018; Cha, Lee, & Koh, 2017). Peptide nanofibers that display laminin-mimetic epitopes which are abundantly found in the natural basal lamina have also been used for the treatment of acute muscle defects by inducing satellite cell proliferation and differentiation (Eren Cimenci et al., 2017).

Hematopoietic stem cells are another widely studied tissue specific stem cell type and have utmost importance for the treatment of various diseases that are either hematological in origin or ones that may be treated by specific immune cells. Nanomaterials, especially in the form of nanofibers, have been used for selective proliferation, enrichment and culture of hematopoietic stem cells and their differentiation into specific cell types. Electrospun nanofibers have been used for efficient culture of hematopoietic stem cells (Chua et al., 2006; M. S. V. Ferreira & Mousavi, 2018; K. Ma et al., 2008). Nanofibers with specific functional groups have also been used for selective differentiation of hematopoietic stem cells into distinct cell types such as lymphocytes or megakaryocytes (Sabaghi et al., 2016; Zakrzewski, Tuckett, & van den Brink, 2009).

There are many other types of tissue specific stem cells that are found in other organs as well, and as their roles in maintenance of the tissue structure are studied at a molecular level, their interaction with the natural ECM will continue to give clues on how to design better nanomaterials for tissue specific regeneration.

1.2 Nanoscale Materials: Nanomaterials

Nanoscale materials, also called nanomaterials, are materials that were synthesized to have at least one dimension between 0.1 and 100 nm. For the synthesis of nanomaterials, two different synthesis approaches can be used: top-down and bottom-up methods. In top-town methods, the fabrication process starts with larger pieces of materials and these materials are reduced to nanoscale by using different types of finely tuned nanofabrication tools. The most widely used nanofabrication method for top-down approach is photolithography, where optical radiation is used to generate micro-nano sized patterns through the use of photoresist and a mask. This method has also been used for the production of computer chips that are currently being used today. However, compared to bottom-up methods, top-down methods are more expensive, require heavy machinery and harsh chemicals and are slower. The bottom-up approaches start with atoms or molecules and form larger systems through self-assembly either due to physicochemical interaction between the smaller groups or by the use of externally applied driving forces. Self-assembly is also a widely used method by the biological organisms for the production of many biomacromolecules such as cytoskeletal components and ECM elements. Thus, it has been widely adapted for producing materials for regenerative medicine. Within the next subsections, we will provide a brief description of the most widely used materials for regenerative medicine with several examples and explain their advantages and disadvantages.

1.2.1 Nonbiological Nanomaterials for Regenerative Medicine

Due to their inherent biocompatibility, biological materials are frequently employed for biomedical applications. However, the difficulties in extracting, purifying, and processing biological materials and the exciting characteristics, ease of manufacture and processing, and lack of immunogenicity have paved the way for wide usage of nonbiological nanomaterials for regenerative applications.

1.2.1.1 Carbon-Based Nanomaterials

Due to their extraordinary electrical and mechanical characteristics, carbon-based nanomaterials have attracted much attention from scientists who develop novel materials for regenerative medicine. Especially for applications where electrical conductivity might be a determinant factor, such as neurological applications or implantable-sensors, carbon nanotubes have been extensively studied. The most widely studied carbon-based materials for this purpose are zero-dimensional fullerenes, one-dimensional carbon nanotubes, and two-dimensional graphene. Fullerenes have been mostly proposed as drug delivery vehicles or theranostic agents (J. H. Liu, Cao, et al., 2010; Montellano, Da Ros, Bianco, & Prato, 2011; Partha, Mitchell, Lyon, Joshi, & Conyers, 2008; Shi, Yu, et al., 2013; Shi, Zhang, et al., 2013); however, there are also a limited number of studies where the chemically modified fullerenes were used for inducing cellular differentiation for tissue regeneration purposes (J. Li et al., 2013; Minami et al., 2015; Xiao, Aoshima, Saitoh, & Miwa, 2010). General biomedical application areas of the carbon-based nanomaterials include cancer therapy, targeted drug delivery, biosensors, bioimaging, and regenerative medicine. However, while functionalization of carbon-based materials enables their water-solubility which is a most desirable characteristic for biomedical applications, it also brings the risk of high toxicity which has also been studied for several different types of carbon-based nanomaterials.

1.2.1.1.1 Carbon Nanotubes

One-dimensional carbon-based nanomaterials, known as carbon nanotubes, have been one of the most controversial nanomaterials used for regenerative medicine applications. The striking electrical characteristics of the carbon nanotubes make them ideal candidates for biomedical applications such as neural prostheses, ECM-mimetic scaffolds, and in-body sensors. Different post-processing methods enable manufacturing of nanoribbon or nanoyarn forms of carbon nanotubes which can be used for unique biomedical applications. Carbon nanotubes have been suggested to be used for retinal applications where they can be used as both sensors and as part of the electrical circuit in a wire-free way (Bareket et al., 2014; Shoval, 2009). In addition, there are many studies where carbon nanotubes or composite carbon nanotubes have been suggested to be used for neuroregeneration. Peptide–carbon nanotube composites are one example of these composite materials (Antonucci, Kupis-Rozmysłowicz, & Boghossian, 2017; Erol et al., 2018; Shrestha et al., 2019). Carbon nanotubes have been used by themselves or together with polymeric materials for neural applications as well (Ahn et al., 2015; Fabbro et al., 2013; Fan et al., 2012; Guo et al., 2017; P. Gupta, Sharan, Roy, & Lahiri, 2015; Massobrio, Massobrio, & Martinoia, 2016; N. Singh et al., 2016; X. Xu et al., 2018). In addition to neural regeneration, carbon nanotubes have also been proposed to be used for bone, cardiac and muscle regeneration (Ahadian et al., 2017, 2016; E. E. da Silva et al., 2009; A. Gupta, Liberati, et al., 2015; Im, Li, Wang, Zhang, & Keidar, 2012; Mukherjee et al., 2016; Patel et al., 2016; Raucci, Alvarez-Perez, Giugliano, Zeppetelli, & Ambrosio, 2016).

The toxicity of carbon nanotubes have been a major concern for their use in biomedical applications. The degree of toxicity of these materials depends on their type (single or multiwalled), their size and the functional groups that are present on the nanotubes (Lanone, Andujar, Kermanizadeh, & Boczkowski, 2013). The high-aspect ratios and the fibrous nature of the carbon nanotubes, nanofibers, and nanorods in addition to their persistence in biological systems might be several of the reasons for their toxicity. The dosage and the mechanism of exposure are also main determinants of their cytotoxic effects. Pulmonary exposure, for example, has been one of the most widely studied factor since it also creates an occupational safety risk for manufacturers of carbon nanotubes and devices made of carbon nanotube-including composites (Bhattacharya, Andón, El-Sayed, & Fadeel, 2013; Fenoglio et al., 2012; Lam, James, McCluskey, & Hunter, 2004; Magrez et al., 2006; Muller et al., 2005; Warheit et al., 2004). Their toxicity on keratocytes, due to skin exposure, and immune system has also been investigated (A. J. Andersen, Wibroe, & Moghimi, 2012; Bottini et al., 2006; Shvedova et al., 2003). Although minimal or no toxicity was shown for several types of the carbon nanotubes and nanorods, it is crucial to investigate the cytotoxicity and genotoxicity of these materials in detail prior to proposing them for biomedical use since the toxicity and carcinogenicity of multiple types and sizes of carbon nanotubes have been shown both in vitro and in vivo.

1.2.1.1.2 Graphene

Graphene is a two-dimensional form of carbon-based materials. It is a single layer of carbon atoms which are tightly bound in a hexagonal lattice with a molecular bond length of 0.142 nm. Graphene is one of the most promising semimetals, and its biocompatibility, excellent mechanical characteristics, ease and flexibility of functionalization make it a suitable nanomaterial for regenerative medicine applications. Several types of cells have been grown on graphene sheets in order to understand their suitability for cell culture, including fibroblasts and osteoblasts (Mehrali et al., 2014; Ryoo, Kim, Kim, & Min, 2010). Embryonic stem cells and other types of stem cells have also been cultured on functionalized and non-functionalized graphene sheets and were shown to differentiate into distinct lineages such as osteoblasts and neuronal cells (Kenry, Lee, Loh, & Lim, 2018; W. C. Lee et al., 2011; Nayak et al., 2011; Y. Wang et al., 2012; Yang et al., 2014).

1.2.1.2 Metal Nanoparticles

As one-dimensional nanomaterials, metal nanoparticles have been extensively studied for tissue regeneration and drug delivery applications in addition to theranostic purposes. Metal nanoparticles have been utilized either by themselves or as parts of scaffold systems and have been shown to extensively alter cell fate in both cases. While their in vivo effects are based on the dosage of usage, several different types of metal nanoparticles have been shown to be biocompatible. Similar to many one-dimensional and two-dimensional nanomaterials, liver and spleen are the major organs where nanoparticles accumulate as observed in in vivo studies when they are administered intravenously. Various metal nanoparticles such as gold, silver, iron oxide and zinc nanoparticles were proposed to be used for biomedical applications and their characteristics differ depending on their size, shape and morphology. In the next subsections, the most widely studied metal nanoparticles and their advantages and disadvantages will be discussed.

1.2.1.2.1 Gold Nanoparticles

Gold nanoparticles can be produced in various sizes and shapes including nanocubes, nanorods, nanoclusters, nanoshells, nanospheres, or even nanostars. The size and shape of the gold nanoparticles change their electronic and optical properties dramatically, which increases the benefits of their use for biosensing and bioimaging applications. Due to their biocompatibility and the ease of functionalization with targeting agents, gold nanoparticles have also been investigated to be used for drug delivery applications and as theranostic agents (Brown et al., 2010; Y. Cheng et al., 2008; H. Zhang et al., 2017). Various sizes and shapes of gold nanoparticles and other engineered nanomaterials that contain gold nanoparticles such as their combinations of carbon nanotubes have been used as theranostic agents (Bishop, Tzeng, & Green, 2015; Curry, Kopelman, Shilo, & Popovtzer, 2014; Song et al., 2016).

Gold nanoparticle containing scaffolds have also been developed for use in regenerative medicine. For example, functionalized gold nanoparticles have been used to increase the bioactivity of polymeric scaffold systems and for enhancing the mechanical strength and conductivity of silk nanofibers (Cohen-Karni et al., 2012; Jung et al., 2012). Gold nanoparticles have also been incorporated into hydrogels in order to enhance the regeneration of various tissues including bone and cardiac tissue (Baei et al., 2016; Heo et al., 2014). Incorporation of these nanoparticles enabled modulation of the conductivity and mechanical properties of these hydrogels. Gold nanoparticle-hydrogel combinations have also been investigated for drug delivery and immunotherapy purposes (Baei et al., 2016; Yata et al., 2017).

1.2.1.2.2 Silver Nanoparticles

Silver nanoparticles are 1–100 nm sized nanoparticles of silver or silver oxide and have been utilized for biomedical applications due to their optical and thermal characteristics, high electrical conductivity and antibacterial effects. Especially, the antibacterial activity of silver nanoparticles make them attractive targets for various purposes where bacterial infection is an important problem such as dermal wound healing or dental applications. The antibacterial activity of the silver nanoparticles is dependent on the shape and size of the particles in addition to the functional groups that are present on their surfaces (Martínez-Castañón, Niño-Martínez, Martínez-Gutierrez, Martínez-Mendoza, & Ruiz, 2008; Pal, Tak, & Song, 2007). In addition to their antibacterial activity, silver nanoparticles were also shown to have antifungal and antiviral characteristics (Jo, Kim, & Jung, 2009; J. Lee, Kim, Sang, Guk, & Gun, 2012; Milovanovic, Arsenijevic, Milovanovic, Kanjevac, & Arsenijevic, 2017; Mori et al., 2013). Thus, scaffold systems with silver nanoparticles have been widely utilized for regenerative medicine. Electrospun nanofibers with different silver nanoparticle formulations have been used as wound dressing materials and for bone regeneration and dental applications (Annur, Wang, Der Liao, & Kuo, 2015; Bapat et al., 2018; Brennan et al., 2015; Corrêa et al., 2015; L. Lu et al., 2013; Thomas, Soumya, Mathew, & Radhakrishnan, 2015; S. Xu et al., 2013). Silver nanoparticle incorporated hydrogels have also been developed for regenerative medicine purposes by using natural and synthetic materials as hydrogels such as chitosan, alginate and synthetic peptides and polymers. Several research areas of these materials include dermal wound healing and orthopedics (Brennan et al., 2015; Das, Kumar, Patil, Viswanathan, & Ghosh, 2015; González-Sánchez et al., 2015; Varaprasad, Mohan, Vimala, & Mohana Raju, 2011).

1.2.1.2.3 Other Metal Nanoparticles

Although gold and silver nanoparticles are the most widely used metal nanoparticles that have been investigated for biomedical purposes, other types of metal-based nanomaterials have also been shown to be biocompatible and bioactive in several studies. For example, copper containing mesoporous glass nanoparticles have been shown to enhance bone regeneration and modulate angiogenesis (Bari et al., 2017; Romero-Sánchez, Marí-Beffa, Carrillo, Medina, & Díaz-Cuenca, 2018). Although iron oxide nanoparticles have been mostly utilized as MRI contrast agents and theranostic purposes, they have also been used for tissue regeneration in several studies (Fang et al., 2014; C. Li, Armstrong, et al., 2018; H. Mok & Zhang, 2013; Ozdemir, Ekiz, Dilli, Guler, & Tekinay, 2016; Shi, Yu, et al., 2013; Z. Q. Zhang & Song, 2016). Zinc oxide nanoparticles are already currently used in many industrial products such as paints, coating, and cosmetics. Depending on their dose, zinc oxide nanoparticles are also biocompatible and they have been proposed to be used for cancer treatment due to their potent ability to trigger reactive oxygen species production. For tissue regeneration, their antibacterial and anti-inflammatory activities are attractive characteristics and have paved the way for their use for purposes such as wound dressings (Rath, Hussain, Chauhan, Garg, & Goyal, 2016; Sudheesh Kumar et al., 2012, 2013).

1.2.1.3 Synthetic Nonbiological Polymers

Biodegradable synthetic polymers are the largest and most versatile class of biomaterials and their use in tissue engineering and regenerative medicine has been studied for several decades. They can be designed and synthesized in various structures with distinct physicochemical characteristics that are suitable for a variety of biomedical applications. Especially over the last 20 years, the polymer synthesis and processing methods have developed dramatically, paving the way for designing novel nanomaterials for biological purposes. Within the next subsections, we will briefly describe some of these methods and their use for regenerative medicine.

1.2.1.3.1 Preparation Techniques

The preparation technique of the polymeric materials should be determined according to the tissue where the material will be used. The main purpose is to provide an environment where cells can easily live and thrive; and the main challenge regarding this issue is to have a porous structure with interconnected channels. A high surface-to-volume ratio is advantageous since it enables more cells to attach and enhances their proliferation and migration. It also permits the diffusion of nutrients, oxygen and waste to and from the cells to the microenvironment. Polymeric materials can be manufactured through a variety of methods. For any fabrication method, the important parameters are to have a suitable pore size, pore density and interconnectivity of the pores. Another important factor is the functionalization of the surface to have the desired bioactivity. Mechanical properties of the materials are also important depending on the application and since increased porosity results in decreased mechanical strength, these two values should be optimized with respect to the final purpose. The fabrication methods for polymeric materials can be classified depending on the requirement for solvents during the process.

1.2.1.3.1.1 Wet Methods

For wet methods, polymers are dissolved in solvents to form polymer solutions, which are further processed to manufacture porous ECM like structures. Wet methods include electrospinning, thermally induced phase separation and solvent casting/salt leaching. One of the main concerns regarding wet preparation methods is the remains of solvents after the fabrication due to the damaging effects of the organic solvents on cells. Thus, all remaining solvents should be carefully removed after using these methods. In addition, the scaffolds should be appropriately sterilized before further use for biological applications.

1.2.1.3.1.2 Cryogels

Cryogels are macroporous gels that are made by using different types of synthetic polymers in addition to natural biomaterials such as DNA and silk fibroin. The gels exhibit high levels of toughness and fast responsivity which makes them ideal candidates for tissue engineering areas such as orthopedic applications. They can withstand high degrees of mechanical forces such as elongation and torsion. The cryogel preparation technique is based on the insolubility of salts in ice compared to water. In cryogelation processes, the monomer and initiator solution is cooled below freezing point and their combinations are localized into dispersed zoned between crystallized solvents where the monomers form polymers through the action of initiators. After the removal of the ice, which acts as a porogen, by increasing the temperature above the freezing point of the solvent, a macroporous scaffold is produced with pore sizes of approximately 100 μm and average wall thicknesses of several micrometers. Cryogels have been used for various purposes in regenerative medicine studies including cartilage tissue regeneration, hepatocyte or lung epithelial cell growth, and wound healing of the skin (Bölgen et al., 2011; Damania et al., 2018; Hixon et al., 2017; Humpolíček et al., 2018; Odabas et al., 2013; Priya et al., 2016; D. Singh, Zo, Kumar, & Han, 2013).

1.2.1.3.1.3 Electrospinning

Although electrospinning is a relatively old technique that dates back to the sixteenth century, it is still one of the most desirable methods for preparing ECM-mimetic scaffolds for regenerative medicine purposes. Electrospinning enables production of fibers with diameters between nanometers to milimeters with an inexpensive and simple setup. The electrospinning equipment includes a syringe pump, a collector, and a high-voltage supply. Polymer fibers are formed through electrostatic repulsion forces and surface tension. High-voltage is used to burst the polymer droplets from the needle to the collector as fibers and the solvent evaporates during the process leaving the jet to solidify into fibers. There are various parameters that should be optimized for producing the perfect nanofibers for producing ECM-mimetic scaffolds: solution parameters, processing parameters and environmental parameters. Solution parameters include the types and characteristics of the solvent and the polymers, while processing parameters are the voltage, the distance between the needle and the collector and the flow rate of the solution. Environmental parameters are the temperature and humidity of the environment. Although most nanofibers are nonwoven and randomly dispersed, the collector plates can also be oriented to have woven or aligned nanofibers.

Electrospun nanofibers can be fabricated as single polymers, blends of polymers and natural biomaterials, and core–shell fibers, which enhances their bioactivity through addition of bioactive signals or enables precise modulation of their mechanical characteristics. Briefly, the most widely used application areas for electrospun nanofibers in regenerative medicine are neural tissue engineering, dermal wound healing, dentistry, skeletal muscle regeneration, and bone tissue regeneration (Aviss, Gough, & Downes, 2010; Jang, Castano, & Kim, 2009; Norouzi, Boroujeni, Omidvarkordshouli, & Soleimani, 2015; Panseri et al., 2008; J. Xie, MacEwan, Schwartz, & Xia, 2010; Zafar et al., 2016; Y. Zhou et al., 2008). Electrospun nanofibers are also utilized for manipulating stem cell fate in cell culture and in vivo (Chua et al., 2006; Lim & Mao, 2009; C. C. Lin & Fu, 2016; J. Xie et al., 2016).

1.2.1.3.1.4 Solvent Free Methods

Unlike wet processing methods, solvent free methods are not based on dissolving the monomers in solvents. Gas foaming, solid freeform fabrication, including 3D printing, and blending and particle leaching are among these methods.

1.2.1.3.1.5 3D Printing

3D printing is a computer-guided printing method which enables to print any structure that you can design in 3D. Since its first invention in 1990s, it has become much cheaper and more easily manipulated and more accessible to all researchers working in this field. In addition to printing three-dimensional scaffolds, 3D printers that have been designed for biomedical applications can be used for printing cells and hydrogels for the fabrication of three-dimensional artificial tissues (Gopinathan & Noh, 2018; Murphy & Atala, 2014; Ratheesh et al., 2017). Although it will be a while before several challenges such as hypoxia due to restricted diffusion of nutrients and oxygen and correct alignment of the specialized cell types with respect to each other will be overcome, 3D printing is still the best manufacturing option for producing implant materials with specific shapes and sizes (S. W. Mok et al., 2016; Oropallo & Piegl, 2016).

1.2.1.3.2 Advantages and Disadvantages of Polymeric Nanomaterials

Polymeric materials are easy to synthesize and handle, and their chemistry has been extensively investigated for several decades. They can be manufactured in bulk and are more economically feasible than many of the other nanomaterial types such as DNA or peptide-based nanomaterials or metal nanoparticles. In addition, since their chemistry is well known, their functionalization with specific chemical groups is relatively easier. In addition to small bioactive chemical groups, they can also be readily functionalized with biomaterials such as peptides or DNA, which is usually used to increase their bioactivity.

On the other hand, polymeric materials are mostly bioinert and need to be functionalized in order to make them bioactive. Their biodegradability rates vary and are usually longer than that of natural biomaterials which might render them unsuitable for some regenerative medicine applications while making them ideal candidates for others. While self-assembling natural biomaterials can be injected to the site of injury with or without stem cells, most polymeric materials need to be placed surgically in the body. Nevertheless, these materials provide excellent efficiency for many tissue engineering and regenerative medicine applications and studies on making them more suitable are still underway.

1.2.2 Biological Nanomaterials

Biological nanomaterials are either derived from biological organisms or have the same structural units as biological systems. Thus, any type of protein, glycoprotein, glycosaminoglycan or lipids that are derived from living cells or dead organisms are considered natural biological materials whereas peptides, carbohydrates or lipids that are synthesized in the laboratory by using their basic units such as amino acids as the beginning material are called synthetic biological materials. Biological materials in general constitute one of the most widely studied nanomaterials for regenerative medicine applications due to their inherent biocompatibility and their ability to induce specific cellular functions. While nanoscale characteristics of the natural biological materials are dictated by their original state most of the time, they can also be modified and used as composites with different types of materials to obtain more control over the material properties and enhance their efficacy. Synthetic biological materials, on the other hand are usually self-assembled due to their rational design.

1.2.2.1 Natural Biological Materials

Natural biomaterials that are used for regenerative medicine applications are usually the materials that have been isolated from the ECM of various organisms, including echinoderms, humans and other mammals, and fish. As they are naturally found molecules, they might carry various signals at the same time complicating their effect on cells for tissue regeneration applications, or they might carry antigenic materials since they are isolated from other organisms. While for some of them, like alginate and chitosan, the original materials are in ample supply and the cost of the materials is mainly due to processing expenses, for other natural biological materials, the original donor material can be less accessible. Although many different types of natural biological materials have been proposed to be used for tissue regeneration, the main ones are collagen, fibronectin, alginate and chitosan.

1.2.2.1.1 Collagen

Collagen is a widely used protein for biomedical applications and is the most widely found fibrous protein in animals. As elongated fibrils, collagen is usually found in both fibrous tissues including ligaments and tendons and other tissues such as bone, cartilage and cornea. Due to its excellent tensile strength, collagens are present in high amounts in these organs. Although different cell types can express collagen, fibroblasts are the main producers. Twenty-eight different types of collagens have been identified and they are found in different amounts in distinct tissues. Their composition and chemical structure (crosslinking) also vary depending on the age of the organism and whether there is a wound or scar tissue. Although over 90% of the collagens that are found in the human body is collagen type I, collagen II is an important structural element in cartilage, collagen IV is found in the basement membranes. Collagens type I, II, III, and V are fibril forming and have triple helix structure with repetitive amino acid sequences that are homologous in many species. In other collagen types, the triple helixes are interrupted with non-helical regions or have distinct structures for specific functions such as anchoring.

As a biomaterial, collagen has excellent tensile strength and weak antigenic properties, and is biocompatible and biodegradable. Through self-assembly, collagen fibers form strong, stable nanofibers that can be further stabilized through crosslinking with different chemical agents. These characteristics have made collagen one of the most widely used biomaterials for regenerative medicine applications both in their natural form and after processing with several nanotechnology methods such as electrospinning and layer-by-layer formation (Fullana & Wnek, 2012; Levingstone et al., 2016; Sell, McClure, Garg, Wolfe, & Bowlin, 2009). Collagen has also been used in combination with other materials such as polymeric materials or biological materials such as chitosan, hyaluronic acid, hydroxyapatite and alginate (Baniasadi & Minary-Jolandan, 2015; Z. G. Chen, Wang, Wei, Mo, & Cui, 2010; Davidenko, Campbell, Thian, Watson, & Cameron, 2010; Fu et al., 2014; Haaparanta et al., 2014; Kane et al., 2015; Kundu, Shim, Jang, Kim, & Cho, 2015; L. Ma et al., 2003; Perez et al., 2014; Raftery et al., 2016; Schnell et al., 2007; R. K. Singh, Seliktar, & Putnam, 2013; Wahl & Czernuszka, 2006; J. Xie et al., 2016; Y. Z. Zhang, Venugopal, Huang, Lim, & Ramakrishna, 2005).

1.2.2.1.2 Fibronectin

Fibronectin is one of the essential components of the ECM and is synthesized by adherent cells. Fibronectin assembly is the beginning step of the self-assembly of other ECM molecules and enables many cellular functions such as cell adhesion and migration. Fibronectin fibers harbor important binding sites for many cell surface receptors and ECM proteins, and has important functions in cell adhesion, proliferation, migration and differentiation. Not only fibronectin itself, but also the signals that are derived from fibronectin are widely used for modulating cellular behavior.

Fibronectin that is used for the production of scaffolds for regenerative medicine studies has been usually obtained from human plasma. Different methods can be used for the assembly of fibronectin fibers including use of reducing or oxidizing agents, force-based assembly and use of peptidic fragments. Fibrin glue, which is made of two components contained in separate vials, one of which has fibrinogen, fibronectin and Factor XIII while the other has thrombin, has been used as a sealant for wounds in the clinics for many years. Thus, the biocompatibility of fibronectin has been well assayed. On the other hand, the complexity of the signals and ligands that are found on fibronectin might be a complicating factor for modulating stem cell fate towards a specific lineage for tissue regeneration. Nevertheless, fibronectin has been widely studied for tissue engineering and regenerative medicine applications. Fibronectin has been used either by itself or in combination with other natural biomaterials or synthetic polymers to induce stem cell differentiation in various studies (Battista et al., 2005; Gjorevski et al., 2016; Prichard, Reichert, & Klitzman, 2007; Rowlands, George, & Cooper-White, 2008; P. Singh & Schwarzbauer, 2012; Tate et al., 2009).

1.2.2.1.3 Alginate

Alginates are anionic polymers that are isolated from algae. The alginate polymers are composed of two types of uronic acid with approximately 500–5000 residues per chain which results in their high molecular weight. The uronic acids that are found in alginate are 1 → 4 linked α-L-guluronic acid (G) or β-D-mannuronic acid (M), and the polymers are usually formed by alternating G and M units. Alginate harbors several characteristics that are looked for in ideal biomaterials such as biocompatibility, low toxicity and comparatively low cost. It also resembles the natural ECM due to its mechanical characteristics and water content; however, it usually needs to be functionalized to gain ECM-like bioactivity. Both functionalized and non-functionalized alginate have been used for various tissue regeneration applications. For example, blends of chitosan and alginate, peptide functionalized alginate, electrospun alginate blends, alginate composite films and other preparation methods have been used in wound healing applications (Balakrishnan, Mohanty, Umashankar, & Jayakrishnan, 2005; Tarun & Gobi, 2012; Hashimoto, Suzuki, Tanihara, Kakimaru, & Suzuki, 2004; Kataria, Gupta, Rath, Mathur, & Dhakate, 2014; Y. H. Lee, Chang, Yang, Chien, & Lai, 2012; Murakami et al., 2010; Rezvanian, Mohd Amin, & Ng, 2016). Alginate has also been widely used for the investigation of the relationship between matrix stiffness and stem cell behavior which provides crucial information for the design and development of more efficient materials for regenerative medicine applications (Banerjee et al., 2009; A. Singh & Elisseeff, 2010; L. Warren et al., 2010). Alginate hydrogels and their composites have also been utilized for bone tissue regeneration, neural regeneration, dentistry, cartilage tissue regeneration and even 3D printing of tissue ECMs (Duan, Hockaday, Kang, & Butcher, 2013; Prang et al., 2006; Srinivasan, Jayasree, Chennazhi, Nair, & Jayakumar, 2012; Venkatesan, Bhatnagar, Manivasagan, Kang, & Kim, 2015). Alginate has also been proposed to be an effective carrier material for cell transplantation applications (Bidarra, Barrias, & Granja, 2014; Perez et al., 2014).

1.2.2.1.4 Chitosan

Chitin is a polycationic polysaccharide that is found in the exoskeletons of crustaceans and insects and it is the second most widely found polysaccharide on earth after cellulose. It is formed by heavily acetylated β-1,4-linked glucosamine units. Chitosan is the alkaline deacylated form of chitin, and has been widely studied as drug delivery agents and for tissue engineering and regenerative medicine applications. Like alginate, chitosan is also biocompatible and has low cost which are both desirable qualities in materials to be used for biomedical applications. Chitosan also has low toxicity and is biodegradable and can be used as both hydrogels and fibers by electrospinning. Hydrogels that are formed by chitosan and its blends have been utilized for osteoregeneration, chondroregeneration, neural regeneration and for enhancing dermal wound healing (Crompton et al., 2007; Dhivya, Saravanan, Sastry, & Selvamurugan, 2015; Miguel, Ribeiro, Brancal, Coutinho, & Correia, 2014; Obara et al., 2005; Park et al., 2009; Ribeiro et al., 2009). Electrospun chitosan nanofibers have been used for neural tissue regeneration in several studies (Du et al., 2014; P. Gupta, Sharan, et al., 2015; W. Wang et al., 2009). In addition, electrospun chitosan nanofibers and chitosan blends have been shown to enhance the rate of wound healing in the skin tissue (Antunes, Moreira, Gaspar, & Correia, 2015; J. P. Chen, Chang, & Chen, 2008; Levengood, Erickson, Chang, & Zhang, 2017; Oryan & Sahvieh, 2017), induce bone regeneration (Oryan & Sahvieh, 2017; Shin et al., 2005; Toskas et al., 2013) and have been used for controlled growth factor release (Z. Xie et al., 2013).

1.2.2.2 Self-Assembling Synthetic Biological Materials

Like other nanomaterials, two types of processes can be utilized for the synthesis of biological materials: top-down and bottom-up approaches. Most of the natural biomaterials that were described above have been fabricated through top-down approach where the materials were isolated from complex biological organisms and processed further. In the bottom-up approach, the most widely applied strategy is to synthesize smaller molecules which later self-assemble into complex biomacromolecules. Synthetic biomaterials that mimic their natural counterparts are such molecules that self-assemble into supramolecular structures and include peptides, peptide-amphiphiles, glycopeptides and sequences of nucleic acids. Self-assembly of such molecules is driven by weak noncovalent interactions such as hydrogen bonds and Van der Waals, hydrophobic and electrostatic interactions. Although these types of intermolecular forces are weak when alone, they form very strong assemblies when there are many of these interactions. This type of self-assembly is similar to that observed in natural molecules after their synthesis inside the organisms, thus the biochemical rules that drive the assembly of natural biomacromolecules are mimicked in synthetic self-assembling nanomaterials. Several examples to self-assembling synthetic biomaterials are peptide nanofibers, nanotubes, nanoribbons and helical ribbons, aptamers, nanostructures that are made by DNA origami, peptide amphiphile nanofibers and nanospheres and glycopeptide nanofibers and nanospheres.

1.2.2.2.1 Peptide Nanofibers

There are 20 natural amino acids that can be combined in a multitude of ways which make up the extraordinary complexity of the biological systems. Self-assembling peptide sequences are tailorable and biocompatible molecules which can be designed to form not only ECM-mimetic scaffolds through their self-assembly into nanofibrous systems, but also supramolecular structures that can mimic growth factors, cytokines, exhibit enzymatic activity, or be utilized as drug delivery vehicles. As smaller examples of these self-assembling peptides, Fmoc dipeptides can form nanofibers through self-assembly, which form networks of fibers that can encapsulate water and form hydrogels (Jayawarna et al., 2009; Orbach et al., 2009; Yan et al., 2013). These hydrogels resemble the natural ECM in terms of their nanofibrous morphology and can be used as scaffolds for cell culture (Castelletto et al., 2011; Gouveia, Jones, Hamley, & Connon, 2014; Yan et al., 2013).

Self-assembling peptides contain short peptide sequences with hydrophobic and hydrophilic domains. In aqueous solutions, the intermolecular interactions between the hydrophobic domains of these peptides and the interactions between the water molecules and the hydrophilic domains drive the self-assembly of these peptides into nanofibers or nanoribbons, depending on the peptide sequence. Some of these self-assembling peptides are already commercially available and have been used for a variety of in vitro and in vivo regenerative medicine applications. These peptides and their composites with other nanomaterials have been used for neural regeneration (T. Y. Cheng et al., 2013; Liedmann, Frech, Morgan, Rolfs, & Frech, 2012; J. Lu et al., 2018; Y. Sun et al., 2016; X. Wu et al., 2017), ligament, cartilage and bone regeneration (K. Chen, Sahoo, et al., 2012; Florine et al., 2013; Humpolíček et al., 2018; Pan et al., 2013; Y. Wu et al., 2016), induction of angiogenesis (Kim, Jung, Kim, & Kim, 2013; X. Liu et al., 2012; X. M. Wang, Qiao, & Horii, 2011; X. Wang, Horii, & Zhang, 2008), and controlled release of drugs including growth factors (Koss, Tsui, & Unsworth, 2016; X. Wang et al., 2008; A. Zhou et al., 2016).

1.2.2.2.2 Peptide Amphiphile Nanofibers

Peptide amphiphile molecules that can self-assemble into nanofibers are composed of an alkyl tail, a β-sheet forming peptide sequence, and a hydrophilic bioactive sequence at the end of the molecule. When mixed with oppositely charged peptide amphiphile molecules, they self-assemble into nanofibers in aqueous solutions through hydrophobic collapse of alkyl tails, β-sheet formation between β-sheet units and electrostatic interactions between oppositely charged amino acids. They can also be induced for self-assembly through pH change or addition of ions into the solution. The hydrophilic bioactive peptide groups form the outside shell of the nanofibers, whereas the alkyl tails form the core. Peptide amphiphile molecules have been used for various purposes from making solar cells to extracellular matrix mimetic scaffolds (Hamsici, Sardan Ekiz, Cinar Ciftci, Tekinay, & Guler, 2017; Khalily, Gulseren, Tekinay, & Guler, 2015; G. A. Silva et al., 2004). Nanofibrous peptide systems mimic the morphological characteristics of the natural extracellular matrix, support cells mechanically and allow diffusion of nutrients and waste products in addition to supporting cell mobility. Glycosaminoglycan mimetic peptide nanofiber systems have previously been used for inducing neurite growth, angiogenesis, chondrogenic differentiation and biomineralization (B. Mammadov, Mammadov, Guler, & Tekinay, 2012; R. Mammadov et al., 2011; Tansik et al., 2016; Ustun, Tombuloglu, Kilinc, Guler, & Tekinay, 2013).

Use of peptide amphiphiles for regenerative medicine applications has several advantages over traditional tissue engineering materials including natural biomacromolecules such as collagen or chitosan, or synthetic polymeric scaffolds such as polyethylene glycol scaffolds.

1.2.2.2.3 Glycopeptide Nanofibers

Carbohydrates in the form of GAGs and glycoproteins are widely observed in the ECM and have crucial functions in regulating cellular behaviors including cell adhesion, proliferation, migration and differentiation. Interaction of ECM carbohydrates with cell surface lectins is also important in health and pathogenesis of tissues. In addition, posttranslational modifications of proteins with carbohydrates through glycosylation are also important for correct functioning of the proteins through modulating intermolecular interactions. All of these reasons have urged scientists to develop glycopeptide nanomaterials that can mimic the natural ECM for regenerative medicine applications. High density presentation of the carbohydrates is morphologically important for maintaining the functions of GAGs and glycoproteins. Thus, self-assembling glycopeptides provide invaluable opportunities for mimicking the characteristics of the natural GAGs and glycoproteins due to their inherent ability for high-density presentation of surface molecules. Peptide monomers that are functionalized with glyco or glyco-similar groups can be used to mimic natural glycoproteins and GAGs. Similarly to their natural counterparts, asparagine or serine/threonine can be used for attaching these functional groups to create N-linked or mucin-like glycosylations.

Supramolecular glycopeptide materials have been investigated for their use in many different scenarios. Glycopeptides have been used as antibiotics (Kahne, Leimkuhler, Lu, & Walsh, 2005), and vaccine materials in order to modulate the immune system (McDonald, Byrne, & Payne, 2015; J. D. Warren, Geng, & Danishefsky, 2006). Self-assembled glycopeptide nanomaterials have also been used for stem cell differentiation, cartilage regeneration and immune system modulation (Caliskan, Sardan Ekiz, Tekinay, & Guler, 2017; Gunay et al., 2017; Restuccia, Tian, Collier, & Hudalla, 2015; Ustun et al., 2013). In addition, self-assembled peptide nanofibers that have GAG-mimetic chemical groups have been utilized for inducing angiogenesis, cardiac regeneration, neural regeneration, stem cell differentiation, wound healing, cartilage and bone regeneration through growth factor binding and presentation (Arslan et al., 2016; Kocabey, Ceylan, Tekinay, & Guler, 2013; B. Mammadov et al., 2016; Rufaihah et al., 2017; Senturk, Mercan, Delibasi, Guler, & Tekinay, 2016; Sever et al., 2016; Tansik et al., 2016; Yaylaci et al., 2016).

1.2.2.2.4 Peptide Nanospheres

Peptide amphiphile systems can also be designed to self-assemble into nanospherical nanostructures. Nanospherical peptide systems are able to mimic non-fibrous components of the extracellular matrix such as growth factors or other cytokines, in addition to being used as drug delivery agents. The nanospherical peptide structures can be self-assembled through rational design of the peptide sequences. The cores of the nanospheres can be composed of either hydrophobic amino acids or lipid-like molecules that come together in aqueous environment through hydrophobic interactions. The outer cores of the nanospheres are usually made of hydrophilic and charged amino acids that interact with each other through electrostatic interactions and hydrogen bonds. Different short peptides that are designed by using this rational design approach can also self-assemble to form non-homogenous nanospheres which can be used to mimic several biological ligands at the same time. In addition, these nanospheres can be used as excellent targeted drug delivery agents with ability to carry both hydrophobic and hydrophilic cargo (S. Li, Zou, et al., 2018; R. Mammadov et al., 2015; Mumcuoglu et al., 2016; Mumcuoglu, Sardan, Tekinay, Guler, & Tekinay, 2015).

1.2.2.3 DNA and RNA as Biomaterials

The importance of DNA as the genetic material has been known since the beginning of the twentieth century. The striking self-replicating ability has been a model for many biomaterial and nanomaterial studies as one of the main pillars of biomimicry. The base-pairing ability on the other hand has enabled many scientists to manufacture materials through a method now known as "DNA origami." Origami is a Japanese word which means folding a plain sheet in different forms in order to produce specific structures. In DNA origami long single strands of DNA are folded into two-dimensional and three-dimensional structures through the use of 200–300 nucleotide long staple strands. The diameter of a single strand of DNA is less than 1 nm, which makes DNA origami a perfect method to produce precisely aligned and organized materials at nanoscale (E. S. Andersen et al., 2009; Castro et al., 2011; Saccà & Niemeyer, 2012). The biocompatibility of the two-dimensional and three-dimensional structures that are formed through DNA origami has been shown in many studies and has been reviewed extensively (Chandrasekaran, Anderson, Kizer, Halvorsen, & Wang, 2016; Z. G. Wang, Song, & Ding, 2013; H. Zhang et al., 2017). The potential of DNA nanoparticles that are made through DNA origami for drug delivery and theranostic purposes has been studied where precisely placed fluorophores or other chemical groups were utilized (Jusuk, Vietz, Raab, Dammeyer, & Tinnefeld, 2015; Selnihhin, Sparvath, Preus, Birkedal, & Andersen, 2018; L. Sun, Gao, et al., 2017). DNA nanoparticles have also been effectively used as drug delivery tools for cancer treatment (Z. G. Wang, Song, & Ding, 2013; H. Zhang et al., 2017; Q. Zhang et al., 2014).

Aptamers, which are short chemically synthesized DNA or RNA strands that fold into three-dimensional structures can selectively bind to any type of ligand, and have been promising tools as part of drug delivery applications for regenerative medicine (N. Chen, Zhang, Soontornworajit, Zhou, & Wang, 2012; Feng, Lyu, Offenhäusser, & Mayer, 2016; Lönne, Zhu, Stahl, & Walter, 2014; Wiraja et al., 2014). The interaction between aptamers and their ligands resemble the antibody–antigen interaction, however, due to their smaller size compared to antibodies, the aptamers can reach target sites more efficiently; hence, they are more preferable for targeted drug delivery applications. In addition, there are fewer immune reactions against aptamers compared to antibodies at respective doses, which again makes them ideal candidates as immunomodulatory regenerative nanomaterials.

1.3 Conclusions and Future Perspectives

Regenerative medicine and tissue engineering are the ultimate treatment methods for tissue and organ failure due to diseases, accidents or aging since many of the current therapeutic methods fail to provide complete or even partial recovery for most of these medical problems. Although much research has been done on the molecular biology of different types of stem cells and the possibility of transplanting them for tissue regeneration or using them for tissue engineering in the laboratory, the cellular therapy is still controversial due to teratogenic activities of some stem cell types and inconclusive clinical trials for others. Nanomaterials can be used to aid in stem cell transplantation without compromising cell viability, protecting the cells from immune cells and providing the right clues to cells for modulating cellular behavior. They can also be used to modulate stem cell function and differentiation for tissue engineering efforts in the laboratory, providing not only mechanical support for cells while allowing diffusion of nutrients, oxygen, and waste, but also the necessary biological signals for cell adhesion, viability, proliferation, and differentiation.

The current nanomaterials have striking abilities to do the abovementioned functions due to adjustable mechanical properties, and well-defined chemical characteristics which is a requirement for clinical validation of biomedical devices and drugs. They also have scaffold-like porous structures that enable not only the migration and proliferation of cells but also the transport of nutrients and waste. On the other hand, the current nanomaterials are limited by our knowledge on the identity and function of precise biological signals that are already present in the natural ECM and our ability to synthesize some of the known signals in an efficient manner.

The complexity of the ECM and its interactions with cells can only be unraveled through molecular studies and complex bioimaging techniques which can also benefit from nanomaterials. Since the natural ECM has a complicated array of natural biomaterials, the use of only some of these materials or signals in three-dimensional cultures of ECM-mimetic scaffolds can enable us to gain more precise information on how individual biological signals affect cells, and use them as part of therapeutic materials. Lab-on-a-chip systems may also provide invaluable tools to study the effects of various signals on cells in a high-throughput manner. In addition, advances in chemical synthesis techniques can help us more efficiently fabricate complex materials such as specific glycopeptides or lipoglycopeptides, expanding the number of building blocks that we can use for mimicking the natural ECM and effectively modulating cell behavior for regenerative medicine purposes.

References

Abagnale, G., Steger, M., Nguyen, V. H., Hersch, N., Sechi, A., Joussen, S., … Wagner, W. (2015). Surface topography enhances differentiation of mesenchymal stem cells towards osteogenic and adipogenic lineages. *Biomaterials, 61*, 316–326. https://doi.org/10.1016/j.biomaterials.2015.05.030

Abbasi, N., Hashemi, S. M., Salehi, M., Jahani, H., Mowla, S. J., Soleimani, M., & Hosseinkhani, H. (2016). Influence of oriented nanofibrous PCL scaffolds on quantitative gene expression during neural differentiation of mouse embryonic stem cells. *Journal of Biomedical Materials Research—Part A, 104*(1), 155–164. https://doi.org/10.1002/jbm.a.35551

Ahadian, S., Davenport Huyer, L., Estili, M., Yee, B., Smith, N., Xu, Z., … Radisic, M. (2017). Moldable elastomeric polyester-carbon nanotube scaffolds for cardiac tissue engineering. *Acta Biomaterialia, 52*, 81–91. https://doi.org/10.1016/j.actbio.2016.12.009

Ahadian, S., Yamada, S., Ramón-Azcón, J., Estili, M., Liang, X., Nakajima, K., … Matsue, T. (2016). Hybrid hydrogel-aligned carbon nanotube scaffolds to enhance cardiac differentiation of embryoid bodies. *Acta Biomaterialia, 31*, 134–143. https://doi.org/10.1016/j.actbio.2015.11.047

Ahn, H. S., Hwang, J. Y., Kim, M. S., Lee, J. Y., Kim, J. W., Kim, H. S., … Hyun, J. K. (2015). Carbon-nanotube-interfaced glass fiber scaffold for regeneration of transected sciatic nerve. *Acta Biomaterialia, 13*, 324–334. https://doi.org/10.1016/j.actbio.2014.11.026

Allur Subramaniyan, S., Sheet, S., Balasubramaniam, S., Berwin Singh, S. V., Rampa, D. R., Shanmugam, S., … Shim, K. S. (2018). Fabrication of nanofiber coated with l-arginine via electrospinning technique: A novel nanomatrix to counter oxidative stress under crosstalk of co-cultured fibroblasts and satellite cells. *Cell Communication & Adhesion, 24*(1), 19–32. https://doi.org/10.1080/15419061.2018.1493107

Al-Nbaheen, M., Vishnubalaji, R., Ali, D., Bouslimi, A., Al-Jassir, F., Megges, M., … Aldahmash, A. (2013). Human stromal (mesenchymal) stem cells from bone marrow, adipose tissue and skin exhibit differences in molecular phenotype and differentiation potential. *Stem Cell Reviews and Reports, 9*(1), 32–43. https://doi.org/10.1007/s12015-012-9365-8

Andersen, A. J., Wibroe, P. P., & Moghimi, S. M. (2012). Perspectives on carbon nanotube-mediated adverse immune effects. *Advanced Drug Delivery Reviews, 64*, 1700–1705. https://doi.org/10.1016/j.addr.2012.05.005

Andersen, E. S., Dong, M., Nielsen, M. M., Jahn, K., Subramani, R., Mamdouh, W., … Kjems, J. (2009). Self-assembly of a nanoscale DNA box with a controllable lid. *Nature, 459*(7243), 73–76. https://doi.org/10.1038/nature07971

Annesini, M. C., Marrelli, L., Piemonte, V., & Turchetti, L. (2017). *Artificial organ engineering*. Cham: Springer. https://doi.org/10.1007/978-1-4471-6443-2

Annur, D., Wang, Z. K., Der Liao, J., & Kuo, C. (2015). Plasma-synthesized silver nanoparticles on electrospun chitosan nanofiber surfaces for antibacterial applications. *Biomacromolecules, 16*(10), 3248–3255. https://doi.org/10.1021/acs.biomac.5b00920

Antonucci, A., Kupis-Rozmysłowicz, J., & Boghossian, A. A. (2017). Noncovalent protein and peptide functionalization of single-walled carbon nanotubes for biodelivery and optical sensing applications. *ACS Applied Materials and Interfaces, 9*, 11321–11331. https://doi.org/10.1021/acsami.7b00810

Antunes, B. P., Moreira, A. F., Gaspar, V. M., & Correia, I. J. (2015). Chitosan/arginine-chitosan polymer blends for assembly of nanofibrous membranes for wound regeneration. *Carbohydrate Polymers, 130*, 104–112. https://doi.org/10.1016/j.carbpol.2015.04.072

Arslan, E., Guler, M. O., & Tekinay, A. B. (2016). Glycosaminoglycan-mimetic signals direct the osteo/chondrogenic differentiation of mesenchymal stem cells in a three-dimensional peptide nanofiber extracellular matrix mimetic environment. *Biomacromolecules, 17*(4), 1280–1291. https://doi.org/10.1021/acs.biomac.5b01637

Aviss, K. J., Gough, J. E., & Downes, S. (2010). Aligned electrospun polymer fibres for skeletal muscle regeneration. *European Cells and Materials, 19*, 193–204. https://doi.org/10.22203/eCM.v019a19

Baei, P., Jalili-Firoozinezhad, S., Rajabi-Zeleti, S., Tafazzoli-Shadpour, M., Baharvand, H., & Aghdami, N. (2016). Electrically conductive gold nanoparticle-chitosan thermosensitive hydrogels for cardiac tissue engineering. *Materials Science and Engineering C, 63*, 131–141. https://doi.org/10.1016/j.msec.2016.02.056

Balakrishnan, B., Mohanty, M., Umashankar, P. R., & Jayakrishnan, A. (2005). Evaluation of an in situ forming hydrogel wound dressing based on oxidized alginate and gelatin. *Biomaterials, 26*(32), 6335–6342. https://doi.org/10.1016/j.biomaterials.2005.04.012

Banerjee, A., Arha, M., Choudhary, S., Ashton, R. S., Bhatia, S. R., Schaffer, D. V., & Kane, R. S. (2009). The influence of hydrogel modulus on the proliferation and differentiation of encapsulated neural stem cells. *Biomaterials, 30*(27), 4695–4699. https://doi.org/10.1016/j.biomaterials.2009.05.050

Baniasadi, M., & Minary-Jolandan, M. (2015). Alginate-collagen fibril composite hydrogel. *Materials, 8*(2), 799–814. https://doi.org/10.3390/ma8020799

Bapat, R. A., Chaubal, T. V., Joshi, C. P., Bapat, P. R., Choudhury, H., Pandey, M., … Kesharwani, P. (2018). An overview of application of silver nanoparticles for biomaterials in dentistry. *Materials Science and Engineering C, 91*, 881–898. https://doi.org/10.1016/j.msec.2018.05.069

Bareket, L., Waiskopf, N., Rand, D., Lubin, G., David-Pur, M., Ben-Dov, J., … Hanein, Y. (2014). Semiconductor nanorod-carbon nanotube biomimetic films for wire-free photostimulation of blind retinas. *Nano Letters, 14*(11), 6685–6692. https://doi.org/10.1021/nl5034304

Bari, A., Bloise, N., Fiorilli, S., Novajra, G., Vallet-Regí, M., Bruni, G., … Vitale-Brovarone, C. (2017). Copper-containing mesoporous bioactive glass nanoparticles as multifunctional agent for bone regeneration. *Acta Biomaterialia, 55*, 493–504. https://doi.org/10.1016/j.actbio.2017.04.012

Battista, S., Guarnieri, D., Borselli, C., Zeppetelli, S., Borzacchiello, A., Mayol, L., … Netti, P. A. (2005). The effect of matrix composition of 3D constructs on embryonic stem cell differentiation. *Biomaterials, 26*(31), 6194–6207. https://doi.org/10.1016/j.biomaterials.2005.04.003

Bellis, S. L. (2011). Advantages of RGD peptides for directing cell association with biomaterials. *Biomaterials, 32*(18), 4205–4210. https://doi.org/10.1016/j.biomaterials.2011.02.029

Ben-David, U., & Benvenisty, N. (2011). The tumorigenicity of human embryonic and induced pluripotent stem cells. *Nature Reviews Cancer, 11*, 268–277. https://doi.org/10.1038/nrc3034

Bhattacharya, K., Andón, F. T., El-Sayed, R., & Fadeel, B. (2013). Mechanisms of carbon nanotube-induced toxicity: Focus on pulmonary inflammation. *Advanced Drug Delivery Reviews, 65*, 2087–2097. https://doi.org/10.1016/j.addr.2013.05.012

Bidarra, S. J., Barrias, C. C., & Granja, P. L. (2014). Injectable alginate hydrogels for cell delivery in tissue engineering. *Acta Biomaterialia, 10*, 1646–1662. https://doi.org/10.1016/j.actbio.2013.12.006

Bilem, I., Chevallier, P., Plawinski, L., Sone, E. D., Durrieu, M. C., & Laroche, G. (2016). RGD and BMP-2 mimetic peptide crosstalk enhances osteogenic commitment of human bone marrow stem cells. *Acta Biomaterialia, 36*, 132–142. https://doi.org/10.1016/j.actbio.2016.03.032

Bishop, C. J., Tzeng, S. Y., & Green, J. J. (2015). Degradable polymer-coated gold nanoparticles for co-delivery of DNA and siRNA. *Acta Biomaterialia, 11*(1), 393–403. https://doi.org/10.1016/j.actbio.2014.09.020

Bölgen, N., Yang, Y., Korkusuz, P., Güzel, E., El Haj, A. J., & Pişkin, E. (2011). 3D ingrowth of bovine articular chondrocytes in biodegradable cryogel scaffolds for cartilage tissue engineering. *Journal of Tissue Engineering and Regenerative Medicine, 5*(10), 770–779. https://doi.org/10.1002/term.375

Bornstein, P., & Sage, H. (2003). Structurally distinct collagen types. *Annual Review of Biochemistry, 49*(1), 957–1003. https://doi.org/10.1146/annurev.bi.49.070180.004521

Bottini, M., Bruckner, S., Nika, K., Bottini, N., Bellucci, S., Magrini, A., … Mustelin, T. (2006). Multi-walled carbon nanotubes induce T lymphocyte apoptosis. *Toxicology Letters, 160*(2), 121–126. https://doi.org/10.1016/j.toxlet.2005.06.020

Brennan, S. A., Ní Fhoghlú, C., Devitt, B. M., O'Mahony, F. J., Brabazon, D., & Walsh, A. (2015). Silver nanoparticles and their orthopaedic applications. *The Bone & Joint Journal, 97-B*(5), 582–589. https://doi.org/10.1302/0301-620x.97b5.33336

Brohem, C. A., Da Silva Cardeal, L. B., Tiago, M., Soengas, M. S., De Moraes Barros, S. B., & Maria-Engler, S. S. (2011). Artificial skin in perspective: Concepts and applications. *Pigment Cell and Melanoma Research, 24*, 35–50. https://doi.org/10.1111/j.1755-148X.2010.00786.x

Brown, S. D., Nativo, P., Smith, J. A., Stirling, D., Edwards, P. R., Venugopal, B., … Wheate, N. J. (2010). Gold nanoparticles for the improved anticancer drug delivery of the active component of oxaliplatin. *Journal of the American Chemical Society, 132*(13), 4678–4684. https://doi.org/10.1021/ja908117a

Cai, L., Dinh, C. B., & Heilshorn, S. C. (2014). One-pot synthesis of elastin-like polypeptide hydrogels with grafted VEGF-mimetic peptides. *Biomaterials Science, 2*(5), 757–765. https://doi.org/10.1039/c3bm60293a

Caliskan, O. S., Sardan Ekiz, M., Tekinay, A. B., & Guler, M. O. (2017). Spatial organization of functional groups on bioactive supramolecular glycopeptide nanofibers for differentiation of mesenchymal stem cells (MSCs) to brown adipogenesis. *Bioconjugate Chemistry, 28*(3), 740–750. https://doi.org/10.1021/acs.bioconjchem.6b00632

Castelletto, V., Moulton, C. M., Cheng, G., Hamley, I. W., Hicks, M. R., Rodger, A., … Alemán, C. (2011). Self-assembly of Fmoc-tetrapeptides based on the RGDS cell adhesion motif. *Soft Matter, 7*(24), 11405–11415. https://doi.org/10.1039/c1sm06550e

Castro, C. E., Kilchherr, F., Kim, D. N., Shiao, E. L., Wauer, T., Wortmann, P., … Dietz, H. (2011). A primer to scaffolded DNA origami. *Nature Methods, 8*(3), 221–229. https://doi.org/10.1038/nmeth.1570

Cen, L., Liu, W., Cui, L., Zhang, W., & Cao, Y. (2008). Collagen tissue engineering: Development of novel biomaterials and applications. *Pediatric Research, 63*, 492–496. https://doi.org/10.1203/PDR.0b013e31816c5bc3

Cha, S. H., Lee, H. J., & Koh, W. G. (2017). Study of myoblast differentiation using multi-dimensional scaffolds consisting of nano and micropatterns. *Biomaterials Research, 21*, 1. https://doi.org/10.1186/s40824-016 0087-x

Chan, T. R., Stahl, P. J., & Yu, S. M. (2011). Matrix-bound VEGF mimetic peptides: Design and endothelial-cell activation in collagen scaffolds. *Advanced Functional Materials, 21*(22), 4252–4262. https://doi.org/10.1002/adfm.201101163

Chandrasekaran, A. R., Anderson, N., Kizer, M., Halvorsen, K., & Wang, X. (2016). Beyond the fold: Emerging biological applications of DNA origami. *Chembiochem, 17*, 1081–1089. https://doi.org/10.1002/cbic.201600038

Chen, F. M., Zhang, M., & Wu, Z. F. (2010). Toward delivery of multiple growth factors in tissue engineering. *Biomaterials, 31*, 6279–6308. https://doi.org/10.1016/j.biomaterials.2010.04.053

Chen, J. P., Chang, G. Y., & Chen, J. K. (2008). Electrospun collagen/chitosan nanofibrous membrane as wound dressing. *Colloids and Surfaces A: Physicochemical and Engineering Aspects, 313–314*, 183–188. https://doi.org/10.1016/j.colsurfa.2007.04.129

Chen, K., Sahoo, S., He, P., Ng, K. S., Toh, S. L., & Goh, J. C. H. (2012). A hybrid silk/RADA-based fibrous scaffold with triple hierarchy for ligament regeneration. *Tissue Engineering Part A, 18*(13–14), 1399–1409. https://doi.org/10.1089/ten.tea.2011.0376

Chen, N., Zhang, Z., Soontornworajit, B., Zhou, J., & Wang, Y. (2012). Cell adhesion on an artificial extracellular matrix using aptamer-functionalized PEG hydrogels. *Biomaterials, 33*(5), 1353–1362. https://doi.org/10.1016/j.biomaterials.2011.10.062

Chen, W., Liu, J., Manuchehrabadi, N., Weir, M. D., Zhu, Z., & Xu, H. H. K. (2013). Umbilical cord and bone marrow mesenchymal stem cell seeding on macroporous calcium phosphate for bone regeneration in rat cranial defects. *Biomaterials, 34*(38), 9917–9925. https://doi.org/10.1016/j.biomaterials.2013.09.002

Chen, Z. G., Wang, P. W., Wei, B., Mo, X. M., & Cui, F. Z. (2010). Electrospun collagen-chitosan nanofiber: A biomimetic extracellular matrix for endothelial cell and smooth muscle cell. *Acta Biomaterialia, 6*(2), 372–382. https://doi.org/10.1016/j.actbio.2009.07.024

Cheng, T. Y., Chen, M. H., Chang, W. H., Huang, M. Y., & Wang, T. W. (2013). Neural stem cells encapsulated in a functionalized self-assembling peptide hydrogel for brain tissue engineering. *Biomaterials, 34*(8), 2005–2016. https://doi.org/10.1016/j.biomaterials.2012.11.043

Cheng, Y., Samia, A. C., Meyers, J. D., Panagopoulos, I., Fei, B., & Burda, C. (2008). Highly efficient drug delivery with gold nanoparticle vectors for in vivo photodynamic therapy of cancer. *Journal of the American Chemical Society, 130*(32), 10643–10647. https://doi.org/10.1021/ja801631c

Chua, K. N., Chai, C., Lee, P. C., Tang, Y. N., Ramakrishna, S., Leong, K. W., & Mao, H. Q. (2006). Surface-aminated electrospun nanofibers enhance adhesion and expansion of human umbilical cord blood hematopoietic stem/progenitor cells. *Biomaterials, 27*(36), 6043–6051. https://doi.org/10.1016/j.biomaterials.2006.06.017

Cinar, G., Ozdemir, A., Hamsici, S., Gunay, G., Dana, A., Tekinay, A. B., & Guler, M. O. (2017). Local delivery of doxorubicin through supramolecular peptide amphiphile nanofiber gels. *Biomaterials Science, 5*(1), 67–76. https://doi.org/10.1039/c6bm00656f

Cohen-Karni, T., Jeong, K. J., Tsui, J. H., Reznor, G., Mustata, M., Wanunu, M., … Kohane, D. S. (2012). Nanocomposite gold-silk nanofibers. *Nano Letters, 12*(10), 5403–5406. https://doi.org/10.1021/nl302810c

Colangelo, A. M., De Gioia, L., Colombo, D., Redaelli, C., Alberghina, L., Martegani, E., … Sarmientos, P. (2008). A new nerve growth factor-mimetic peptide active on neuropathic pain in rats. *Journal of Neuroscience, 28*(11), 2698–2709. https://doi.org/10.1523/JNEUROSCI.5201-07.2008

Corrêa, J. M., Mori, M., Sanches, H. L., Da Cruz, A. D., Poiate, E., & Poiate, I. A. V. P. (2015). Silver nanoparticles in dental biomaterials. *International Journal of Biomaterials, 2015*, 485275. https://doi.org/10.1155/2015/485275

Crompton, K. E., Goud, J. D., Bellamkonda, R. V., Gengenbach, T. R., Finkelstein, D. I., Horne, M. K., & Forsythe, J. S. (2007). Polylysine-functionalised thermoresponsive chitosan hydrogel for neural tissue engineering. *Biomaterials, 28*(3), 441–449. https://doi.org/10.1016/j.biomaterials.2006.08.044

Cui, H., Webber, M. J., & Stupp, S. I. (2010). Self-assembly of peptide amphiphiles: From molecules to nanostructures to biomaterials. *Biopolymers, 94*, 1–18. https://doi.org/10.1002/bip.21328

Curry, T., Kopelman, R., Shilo, M., & Popovtzer, R. (2014). Multifunctional theranostic gold nanoparticles for targeted CT imaging and photothermal therapy. *Contrast Media and Molecular Imaging, 9*(1), 53–61. https://doi.org/10.1002/cmmi.1563

da Silva, E. E., Della Colleta, H. H. M., Ferlauto, A. S., Moreira, R. L., Resende, R. R., Oliveira, S., … Ladeira, L. O. (2009). Nanostructured 3-D collagen/nanotube biocomposites for future bone regeneration scaffolds. *Nano Research, 2*(6), 462–473. https://doi.org/10.1007/s12274-009-9042-7

Dalby, M. J., Gadegaard, N., Tare, R., Andar, A., Riehle, M. O., Herzyk, P., … Oreffo, R. O. C. (2007). The control of human mesenchymal cell differentiation using nanoscale symmetry and disorder. *Nature Materials, 6*(12), 997–1003. https://doi.org/10.1038/nmat2013

Damania, A., Kumar, A., Teotia, A. K., Kimura, H., Kamihira, M., Ijima, H., … Kumar, A. (2018). Decellularized liver matrix-modified cryogel scaffolds as potential hepatocyte carriers in bio-

artificial liver support systems and implantable liver constructs. *ACS Applied Materials and Interfaces, 10*(1), 114–126. https://doi.org/10.1021/acsami.7b13727
Das, A., Kumar, A., Patil, N. B., Viswanathan, C., & Ghosh, D. (2015). Preparation and characterization of silver nanoparticle loaded amorphous hydrogel of carboxymethylcellulose for infected wounds. *Carbohydrate Polymers, 130*, 254–261. https://doi.org/10.1016/j.carbpol.2015.03.082
Davidenko, N., Campbell, J. J., Thian, E. S., Watson, C. J., & Cameron, R. E. (2010). Collagen-hyaluronic acid scaffolds for adipose tissue engineering. *Acta Biomaterialia, 6*(10), 3957–3968. https://doi.org/10.1016/j.actbio.2010.05.005
Dehdilani, N., Shamsasenjan, K., Movassaghpour, A., Akbarzadehlaleh, P., Tabrizi, B. A., Parsa, H., & Sabagi, F. (2016). Improved survival and hematopoietic differentiation of murine embryonic stem cells on electrospun polycaprolactone nanofiber. *Cell Journal, 17*(4), 629–638.
Dhivya, S., Saravanan, S., Sastry, T. P., & Selvamurugan, N. (2015). Nanohydroxyapatite-reinforced chitosan composite hydrogel for bone tissue repair in vitro and in vivo. *Journal of Nanobiotechnology, 13*(1), 40. https://doi.org/10.1186/s12951-015-0099-z
Du, J., Tan, E., Kim, H. J., Zhang, A., Bhattacharya, R., & Yarema, K. J. (2014). Comparative evaluation of chitosan, cellulose acetate, and polyethersulfone nanofiber scaffolds for neural differentiation. *Carbohydrate Polymers, 99*, 483–490. https://doi.org/10.1016/j.carbpol.2013.08.050
Duan, B., Hockaday, L. A., Kang, K. H., & Butcher, J. T. (2013). 3D Bioprinting of heterogeneous aortic valve conduits with alginate/gelatin hydrogels. *Journal of Biomedical Materials Research—Part A, 101A*(5), 1255–1264. https://doi.org/10.1002/jbm.a.34420
Eren Cimenci, C., Uzunalli, G., Uysal, O., Yergoz, F., Karaca Umay, E., Guler, M. O., & Tekinay, A. B. (2017). Laminin mimetic peptide nanofibers regenerate acute muscle defect. *Acta Biomaterialia, 60*, 190–200. https://doi.org/10.1016/j.actbio.2017.07.010
Erol, O., Uyan, I., Hatip, M., Yilmaz, C., Tekinay, A. B., & Guler, M. O. (2018). Recent advances in bioactive 1D and 2D carbon nanomaterials for biomedical applications. *Nanomedicine: Nanotechnology, Biology, and Medicine, 14*(7), 2433–2454. https://doi.org/10.1016/j.nano.2017.03.021
Fabbro, A., Sucapane, A., Toma, F. M., Calura, E., Rizzetto, L., Carrieri, C., ... Ballerini, L. (2013). Adhesion to carbon nanotube conductive scaffolds forces action-potential appearance in immature rat spinal neurons. *PLoS One, 8*(8), e73621. https://doi.org/10.1371/journal.pone.0073621
Fan, L., Feng, C., Zhao, W., Qian, L., Wang, Y., & Li, Y. (2012). Directional neurite outgrowth on superaligned carbon nanotube yarn patterned substrate. *Nano Letters, 12*(7), 3668–3673. https://doi.org/10.1021/nl301428w
Fang, J. H., Lai, Y. H., Chiu, T. L., Chen, Y. Y., Hu, S. H., & Chen, S. Y. (2014). Magnetic core-shell nanocapsules with dual-targeting capabilities and co-delivery of multiple drugs to treat brain gliomas. *Advanced Healthcare Materials, 3*(8), 1250–1260. https://doi.org/10.1002/adhm.201300598
Farrukh, A., Ortega, F., Fan, W., Marichal, N., Paez, J. I., Berninger, B., ... Salierno, M. J. (2017). Bifunctional hydrogels containing the laminin motif IKVAV promote neurogenesis. *Stem Cell Reports, 9*(5), 1432–1440. https://doi.org/10.1016/j.stemcr.2017.09.002
Feng, L., Lyu, Z., Offenhäusser, A., & Mayer, D. (2016). Electrochemically triggered aptamer immobilization via click reaction for vascular endothelial growth factor detection. *Engineering in Life Sciences, 16*(6), 550–559. https://doi.org/10.1002/elsc.201600068
Fenoglio, I., Aldieri, E., Gazzano, E., Cesano, F., Colonna, M., Scarano, D., ... Fubini, B. (2012). Thickness of multiwalled carbon nanotubes affects their lung toxicity. *Chemical Research in Toxicology, 25*(1), 74–82. https://doi.org/10.1021/tx200255h
Ferreira, D. S., Lin, Y. A., Cui, H., Hubbell, J. A., Reis, R. L., & Azevedo, H. S. (2015). Molecularly engineered self-assembling membranes for cell-mediated degradation. *Advanced Healthcare Materials, 4*(4), 602–612. https://doi.org/10.1002/adhm.201400586
Ferreira, M. S. V., & Mousavi, S. H. (2018). Nanofiber technology in the ex vivo expansion of cord blood-derived hematopoietic stem cells. *Nanomedicine: Nanotechnology, Biology, and Medicine, 14*, 1707–1718. https://doi.org/10.1016/j.nano.2018.04.017

Fesharaki, M., Razavi, S., Ghasemi-Mobarakeh, L., Behjati, M., Yarahmadian, R., Kazemi, M., & Hejazi, H. (2018). Differentiation of human scalp adipose-derived mesenchymal stem cells into mature neural cells on electrospun nanofibrous scaffolds for nerve tissue engineering applications. *Cell Journal, 20*(2), 168–176. https://doi.org/10.22074/cellj.2018.4898

Feyzizarnagh, H., Yoon, D. Y., Goltz, M., & Kim, D. S. (2016). Peptide nanostructures in biomedical technology. *Wiley Interdisciplinary Reviews: Nanomedicine and Nanobiotechnology, 8*, 730–743. https://doi.org/10.1002/wnan.1393

Florine, E. M., Miller, R. E., Porter, R. M., Evans, C. H., Kurz, B., & Grodzinsky, A. J. (2013). Effects of dexamethasone on mesenchymal stromal cell chondrogenesis and aggrecanase activity: Comparison of agarose and self-assembling peptide scaffolds. *Cartilage, 4*(1), 63–74. https://doi.org/10.1177/1947603512455196

Francis, M. P. (2010). *Recapitulating osteoblastogenesis with electrospun fibrinogen nanofibers and adipose stem cells and electrospinning adipose tissue-derived basement membrane.* ProQuest Dissertations and Theses. Retrieved from http://search.proquest.com/docview/276104341?accountid=8359

Friedenstein, A. J., Gorskaja, J. F., & Kulagina, N. N. (1976). Fibroblast precursors in normal and irradiated mouse hematopoietic organs. *Experimental Hematology, 4*(5), 267–274. Retrieved from http://www.ncbi.nlm.nih.gov/pubmed/976387

Fu, W., Liu, Z., Feng, B., Hu, R., He, X., Wang, H., … Wang, W. (2014). Electrospun gelatin/PCL and collagen/PLCL scaffolds for vascular tissue engineering. *International Journal of Nanomedicine, 9*(1), 2335–2344. https://doi.org/10.2147/IJN.S61375

Fullana, M. J., & Wnek, G. E. (2012). Electrospun collagen and its applications in regenerative medicine. *Drug Delivery and Translational Research, 2*, 313–322. https://doi.org/10.1007/s13346-012-0087-x

Gerecht, S., Burdick, J. A., Ferreira, L. S., Townsend, S. A., Langer, R., & Vunjak-Novakovic, G. (2007). Hyaluronic acid hydrogel for controlled self-renewal and differentiation of human embryonic stem cells. *Proceedings of the National Academy of Sciences, 104*(27), 11298–11303. https://doi.org/10.1073/pnas.0703723104

Gjorevski, N., Sachs, N., Manfrin, A., Giger, S., Bragina, M. E., Ordóñez-Morán, P., … Lutolf, M. P. (2016). Designer matrices for intestinal stem cell and organoid culture. *Nature, 539*(7630), 560–564. https://doi.org/10.1038/nature20168

González-Sánchez, M. I., Perni, S., Tommasi, G., Morris, N. G., Hawkins, K., López-Cabarcos, E., & Prokopovich, P. (2015). Silver nanoparticle based antibacterial methacrylate hydrogels potential for bone graft applications. *Materials Science and Engineering C, 50*, 332–340. https://doi.org/10.1016/j.msec.2015.02.002

Gopinathan, J., & Noh, I. (2018). Recent trends in bioinks for 3D printing. *Biomaterials Research, 22*, 11. https://doi.org/10.1186/s40824-018-0122-1

Gouveia, R. M., Jones, R. R., Hamley, I. W., & Connon, C. J. (2014). The bioactivity of composite Fmoc-RGDS-collagen gels. *Biomaterials Science, 2*(9), 1222–1229. https://doi.org/10.1039/c4bm00121d

Graf, J., Ogle, R. C., Robey, F. A., Sasaki, M., Martin, G. R., Yamada, Y., & Kleinman, H. K. (1987). A pentapeptide from the laminin B1 chain mediates cell adhesion and binds the 67 000 laminin receptor. *Biochemistry, 26*(22), 6896–6900. https://doi.org/10.1021/bi00396a004

Gulseren, G., Yasa, I. C., Ustahuseyin, O., Tekin, E. D., Tekinay, A. B., & Guler, M. O. (2015). Alkaline phosphatase-mimicking peptide nanofibers for osteogenic differentiation. *Biomacromolecules, 16*(7), 2198–2208. https://doi.org/10.1021/acs.biomac.5b00593

Gunay, G., Sardan Ekiz, M., Ferhati, X., Richichi, B., Nativi, C., Tekinay, A. B., & Guler, M. O. (2017). Antigenic GM3 lactone mimetic molecule integrated mannosylated glycopeptide nanofibers for the activation and maturation of dendritic cells. *ACS Applied Materials and Interfaces, 9*(19), 16035–16042. https://doi.org/10.1021/acsami.7b04094

Guo, Y., Jiang, S., Grena, B. J. B., Kimbrough, I. F., Thompson, E. G., Fink, Y., … Jia, X. (2017). Polymer composite with carbon nanofibers aligned during thermal drawing as a microelectrode for chronic neural interfaces. *ACS Nano, 11*(7), 6574–6585. https://doi.org/10.1021/acsnano.6b07550

Gupta, A., Liberati, T. A., Verhulst, S. J., Main, B. J., Roberts, M. H., Potty, A. G. R., … El-Amin, S. F., III. (2015). Biocompatibility of single-walled carbon nanotube composites for bone regeneration. *Bone & Joint Research, 4*(5), 70–77. https://doi.org/10.1302/2046-3758.45.2000382

Gupta, P., Sharan, S., Roy, P., & Lahiri, D. (2015). Aligned carbon nanotube reinforced polymeric scaffolds with electrical cues for neural tissue regeneration. *Carbon, 95*, 715–724. https://doi.org/10.1016/j.carbon.2015.08.107

Haaparanta, A. M., Järvinen, E., Cengiz, I. F., Ellä, V., Kokkonen, H. T., Kiviranta, I., & Kellomäki, M. (2014). Preparation and characterization of collagen/PLA, chitosan/PLA, and collagen/chitosan/PLA hybrid scaffolds for cartilage tissue engineering. *Journal of Materials Science: Materials in Medicine, 25*(4), 1129–1136. https://doi.org/10.1007/s10856-013-5129-5

Hamsici, S., Cinar, G., Celebioglu, A., Uyar, T., Tekinay, A. B., & Guler, M. O. (2017). Bioactive peptide functionalized aligned cyclodextrin nanofibers for neurite outgrowth. *Journal of Materials Chemistry B, 5*(3), 517–524. https://doi.org/10.1039/c6tb02441f

Hamsici, S., Sardan Ekiz, M., Cinar Ciftci, G., Tekinay, A. B., & Guler, M. O. (2017). Gemcitabine integrated nano-prodrug carrier system. *Bioconjugate Chemistry, 28*(5), 1491–1498. https://doi.org/10.1021/acs.bioconjchem.7b00155

Hashimoto, T., Suzuki, Y., Tanihara, M., Kakimaru, Y., & Suzuki, K. (2004). Development of alginate wound dressings linked with hybrid peptides derived from laminin and elastin. *Biomaterials, 25*(7–8), 1407–1414. https://doi.org/10.1016/j.biomaterials.2003.07.004

Hass, R., Kasper, C., Böhm, S., & Jacobs, R. (2011). Different populations and sources of human mesenchymal stem cells (MSC): A comparison of adult and neonatal tissue-derived MSC. *Cell Communication and Signaling, 9*, 12. https://doi.org/10.1186/1478-811X-9-12

Heo, D. N., Ko, W. K., Bae, M. S., Lee, J. B., Lee, D. W., Byun, W., … Kwon, I. K. (2014). Enhanced bone regeneration with a gold nanoparticle-hydrogel complex. *Journal of Materials Chemistry B, 2*(11), 1584–1593. https://doi.org/10.1039/c3tb21246g

Hinderer, S., Layland, S. L., & Schenke-Layland, K. (2016). ECM and ECM-like materials—Biomaterials for applications in regenerative medicine and cancer therapy. *Advanced Drug Delivery Reviews, 97*, 260–269. https://doi.org/10.1016/j.addr.2015.11.019

Hixon, K. R., Lu, T., & Sell, S. A. (2017). A comprehensive review of cryogels and their roles in tissue engineering applications. *Acta Biomaterialia, 62*, 29–41. https://doi.org/10.1016/j.actbio.2017.08.033

Hofstetter, W. L. (2018). The artificial trachea: A mission of creativity, perspiration, and perseverance. *Journal of Thoracic and Cardiovascular Surgery, 156*(3), 1273–1274. https://doi.org/10.1016/j.jtcvs.2018.05.021

Horii, A., Wang, X., Gelain, F., & Zhang, S. (2007). Biological designer self-assembling peptide nanofiber scaffolds significantly enhance osteoblast proliferation, differentiation and 3-D migration. *PLoS One, 2*(2), e190. https://doi.org/10.1371/journal.pone.0000190

Hu, F., Zhang, X., Liu, H., Xu, P., Doulathunnisa, Teng, G., & Xiao, Z. (2017). Neuronally differentiated adipose-derived stem cells and aligned PHBV nanofiber nerve scaffolds promote sciatic nerve regeneration. *Biochemical and Biophysical Research Communications, 489*(2), 171–178. https://doi.org/10.1016/j.bbrc.2017.05.119

Humpolíček, P., Radaszkiewicz, K. A., Capáková, Z., Pacherník, J., Bober, P., Kašpárková, V., … Stejskal, J. (2018). Polyaniline cryogels: Biocompatibility of novel conducting macroporous material. *Scientific Reports, 8*(1), 135. https://doi.org/10.1038/s41598-017-18290-1

Hwang, N. S., Varghese, S., Zhang, Z., & Elisseeff, J. (2006). Chondrogenic differentiation of human embryonic stem cell-derived cells in arginine-glycine-aspartate-modified hydrogels. *Tissue Engineering, 12*(9), 2695–2706. https://doi.org/10.1089/ten.2006.12.2695

Im, O., Li, J., Wang, M., Zhang, L. G., & Keidar, M. (2012). Biomimetic three-dimensional nanocrystalline hydroxyapatite and magnetically synthesized single-walled carbon nanotube chitosan nanocomposite for bone regeneration. *International Journal of Nanomedicine, 7*, 2087–2099. https://doi.org/10.2147/IJN.S29743

Jang, J. H., Castano, O., & Kim, H. W. (2009). Electrospun materials as potential platforms for bone tissue engineering. *Advanced Drug Delivery Reviews, 61*, 1065–1083. https://doi.org/10.1016/j.addr.2009.07.008

Jayawarna, V., Richardson, S. M., Hirst, A. R., Hodson, N. W., Saiani, A., Gough, J. E., & Ulijn, R. V. (2009). Introducing chemical functionality in Fmoc-peptide gels for cell culture. *Acta Biomaterialia, 5*(3), 934–943. https://doi.org/10.1016/j.actbio.2009.01.006

Jo, Y.-K., Kim, B. H., & Jung, G. (2009). Antifungal activity of silver ions and nanoparticles on phytopathogenic fungi. *Plant Disease, 93*(10), 1037–1043. https://doi.org/10.1094/pdis-93-10-1037

Jung, D., Minami, I., Patel, S., Lee, J., Jiang, B., Yuan, Q., … Nakatsuji, N. (2012). Incorporation of functionalized gold nanoparticles into nanofibers for enhanced attachment and differentiation of mammalian cells. *Journal of Nanobiotechnology, 10*, 23. https://doi.org/10.1186/1477-3155-10-23

Jusuk, I., Vietz, C., Raab, M., Dammeyer, T., & Tinnefeld, P. (2015). Super-resolution imaging conditions for enhanced yellow fluorescent protein (eYFP) demonstrated on DNA origami nanorulers. *Scientific Reports, 5*, 14075. https://doi.org/10.1038/srep14075

Kahne, D., Leimkuhler, C., Lu, W., & Walsh, C. (2005). Glycopeptide and lipoglycopeptide antibiotics. *Chemical Reviews, 105*(2), 425–448. https://doi.org/10.1021/cr030103a

Kane, R. J., Weiss-Bilka, H. E., Meagher, M. J., Liu, Y., Gargac, J. A., Niebur, G. L., … Roeder, R. K. (2015). Hydroxyapatite reinforced collagen scaffolds with improved architecture and mechanical properties. *Acta Biomaterialia, 17*, 16–25. https://doi.org/10.1016/j.actbio.2015.01.031

Kataria, K., Gupta, A., Rath, G., Mathur, R. B., & Dhakate, S. R. (2014). In vivo wound healing performance of drug loaded electrospun composite nanofibers transdermal patch. *International Journal of Pharmaceutics, 469*(1), 102–110. https://doi.org/10.1016/j.ijpharm.2014.04.047

Kenry, Lee, W. C., Loh, K. P., & Lim, C. T. (2018). When stem cells meet graphene: Opportunities and challenges in regenerative medicine. *Biomaterials, 155*, 236–250. https://doi.org/10.1016/j.biomaterials.2017.10.004

Kern, S., Eichler, H., Stoeve, J., Klüter, H., & Bieback, K. (2006). Comparative analysis of mesenchymal stem cells from bone marrow, umbilical cord blood, or adipose tissue. *Stem Cells, 24*(5), 1294–1301. https://doi.org/10.1634/stemcells.2005-0342

Kfoury, Y., & Scadden, D. T. (2015). Mesenchymal cell contributions to the stem cell niche. *Cell Stem Cell, 16*, 239–253. https://doi.org/10.1016/j.stem.2015.02.019

Khalily, M. A., Gulseren, G., Tekinay, A. B., & Guler, M. O. (2015). Biocompatible supramolecular catalytic one-dimensional nanofibers for efficient labeling of live cells. *Bioconjugate Chemistry, 26*(12), 2371–2375. https://doi.org/10.1021/acs.bioconjchem.5b00443

Khan, M., Johnson, J., Xu, Y., Belevych, A., & Gyorke, S. (2015). SAEM annual meeting abstracts. *Academic Emergency Medicine, 22*, S3–S425. https://doi.org/10.1111/acem.12644

Kim, J. H., Jung, Y., Kim, B. S., & Kim, S. H. (2013). Stem cell recruitment and angiogenesis of neuropeptide substance P coupled with self-assembling peptide nanofiber in a mouse hind limb ischemia model. *Biomaterials, 34*(6), 1657–1668. https://doi.org/10.1016/j.biomaterials.2012.11.008

Knight, C. G., Morton, L. F., Peachey, A. R., Tuckwell, D. S., Farndale, R. W., & Barnes, M. J. (2000). The collagen-binding a-domains of integrins $\alpha1/\beta1$ and $\alpha2/\beta1$ recognize the same specific amino acid sequence, GFOGER, in native (triple-helical) collagens. *Journal of Biological Chemistry, 275*(1), 35–40. https://doi.org/10.1074/jbc.275.1.35

Kocabey, S., Ceylan, H., Tekinay, A. B., & Guler, M. O. (2013). Glycosaminoglycan mimetic peptide nanofibers promote mineralization by osteogenic cells. *Acta Biomaterialia, 9*(11), 9075–9085. https://doi.org/10.1016/j.actbio.2013.07.007

Konno, M., Hamabe, A., Hasegawa, S., Ogawa, H., Fukusumi, T., Nishikawa, S., … Ishii, H. (2013). Adipose-derived mesenchymal stem cells and regenerative medicine. *Development Growth and Differentiation, 55*, 309–318. https://doi.org/10.1111/dgd.12049

Koss, K., Tsui, C., & Unsworth, L. D. (2016). Induced neural differentiation of MMP-2 cleaved (RADA)4 drug delivery systems. *Journal of Controlled Release, 243*, 204–213. https://doi.org/10.1016/j.jconrel.2016.09.037

Kuang, S., Kuroda, K., Le Grand, F., & Rudnicki, M. A. (2007). Asymmetric self-renewal and commitment of satellite stem cells in muscle. *Cell, 129*(5), 999–1010. https://doi.org/10.1016/j.cell.2007.03.044

Kundu, J., Shim, J. H., Jang, J., Kim, S. W., & Cho, D. W. (2015). An additive manufacturing-based PCL-alginate-chondrocyte bioprinted scaffold for cartilage tissue engineering. *Journal of Tissue Engineering and Regenerative Medicine, 9*(11), 1286–1297. https://doi.org/10.1002/term.1682

Lam, C. W., James, J. T., McCluskey, R., & Hunter, R. L. (2004). Pulmonary toxicity of single-wall carbon nanotubes in mice 7 and 90 days after intractracheal instillation. *Toxicological Sciences, 77*(1), 126–134. https://doi.org/10.1093/toxsci/kfg243

Lanone, S., Andujar, P., Kermanizadeh, A., & Boczkowski, J. (2013). Determinants of carbon nanotube toxicity. *Advanced Drug Delivery Reviews, 65*, 2063–2069. https://doi.org/10.1016/j.addr.2013.07.019

Lee, J., Kim, K.-J., Sang, W., Guk, J., & Gun, D. (2012). The silver nanoparticle (Nano-Ag): A new model for antifungal agents. In *Silver nanoparticles*. Croatia: Intech. https://doi.org/10.5772/8510

Lee, K. Y., Jeong, L., Kang, Y. O., Lee, S. J., & Park, W. H. (2009). Electrospinning of polysaccharides for regenerative medicine. *Advanced Drug Delivery Reviews, 61*, 1020–1032. https://doi.org/10.1016/j.addr.2009.07.006

Lee, W. C., Lim, C. H. Y. X., Shi, H., Tang, L. A. L., Wang, Y., Lim, C. T., & Loh, K. P. (2011). Origin of enhanced stem cell growth and differentiation on graphene and graphene oxide. *ACS Nano, 5*(9), 7334–7341. https://doi.org/10.1021/nn202190c

Lee, Y. H., Chang, J. J., Yang, M. C., Chien, C. T., & Lai, W. F. (2012). Acceleration of wound healing in diabetic rats by layered hydrogel dressing. *Carbohydrate Polymers, 88*(3), 809–819. https://doi.org/10.1016/j.carbpol.2011.12.045

Lees, D., & Partington, P. (2016). Articular cartilage. *Orthopaedics and Traumatology, 30*(3), 265–272. https://doi.org/10.1016/j.mporth.2016.04.007

Levengood, S. L., Erickson, A. E., Chang, F. C., & Zhang, M. (2017). Chitosan-poly(caprolactone) nanofibers for skin repair. *Journal of Materials Chemistry B, 5*(9), 1822–1833. https://doi.org/10.1039/C6TB03223K

Levingstone, T. J., Thompson, E., Matsiko, A., Schepens, A., Gleeson, J. P., & O'Brien, F. J. (2016). Multi-layered collagen-based scaffolds for osteochondral defect repair in rabbits. *Acta Biomaterialia, 32*, 149–160. https://doi.org/10.1016/j.actbio.2015.12.034

Li, C., Armstrong, J. P., Pence, I. J., Kit-Anan, W., Puetzer, J. L., Correia Carreira, S., … Stevens, M. M. (2018). Glycosylated superparamagnetic nanoparticle gradients for osteochondral tissue engineering. *Biomaterials, 176*, 24–33. https://doi.org/10.1016/j.biomaterials.2018.05.029

Li, J., Song, Y., Liu, X., Zhang, M., He, R., Chang, Y., … Zhang, J. (2013). The effects of $C_{60}(C(COOH)_2)_2$-FITC on proliferation and differentiation of human mesenchymal stem cells *in vitro*. *Journal of Nanoscience and Nanotechnology, 13*(10), 6517–6521. https://doi.org/10.1166/jnn.2013.7166

Li, S., Zou, Q., Li, Y., Yuan, C., Xing, R., & Yan, X. (2018). Smart peptide-based supramolecular photodynamic metallo-nanodrugs designed by multicomponent coordination self-assembly. *Journal of the American Chemical Society, 140*(34), 10794–10802. https://doi.org/10.1021/jacs.8b04912

Liedmann, A., Frech, S., Morgan, P. J., Rolfs, A., & Frech, M. J. (2012). Differentiation of human neural progenitor cells in functionalized hydrogel matrices. *BioResearch Open Access, 1*(1), 16–24. https://doi.org/10.1089/biores.2012.0209

Lim, S. H., & Mao, H. Q. (2009). Electrospun scaffolds for stem cell engineering. *Advanced Drug Delivery Reviews, 61*, 1084–1096. https://doi.org/10.1016/j.addr.2009.07.011

Lin, C., Liu, C., Zhang, L., Huang, Z., Zhao, P., Chen, R., … Liu, B. (2018). Interaction of iPSC-derived neural stem cells on poly(L-lactic acid) nanofibrous scaffolds for possible use in neural tissue engineering. *International Journal of Molecular Medicine, 41*(2), 697–708. https://doi.org/10.3892/ijmm.2017.3299

Lin, C. C., & Fu, S. J. (2016). Osteogenesis of human adipose-derived stem cells on poly(dopamine)-coated electrospun poly(lactic acid) fiber mats. *Materials Science and Engineering C, 58*, 254–263. https://doi.org/10.1016/j.msec.2015.08.009

Lin, X., Takahashi, K., Campion, S. L., Liu, Y., Gustavsen, G. G., Peña, L. A., & Zamora, P. O. (2006). Synthetic peptide F2A4-K-NS mimics fibroblast growth factor-2 in vitro and is angiogenic in vivo. *International Journal of Molecular Medicine, 17*(5), 833–839. https://doi.org/10.3892/ijmm.17.5.833

Liu, D., Yi, C., Zhang, D., Zhang, J., & Yang, M. (2010). Inhibition of proliferation and differentiation of mesenchymal stem cells by carboxylated carbon nanotubes. *ACS Nano, 4*(4), 2185–2195. https://doi.org/10.1021/nn901479w

Liu, J. H., Cao, L., Luo, P. G., Yang, S. T., Lu, F., Wang, H., … Sun, Y. P. (2010). Fullerene-conjugated doxorubicin in cells. *ACS Applied Materials and Interfaces, 2*(5), 1384–1389. https://doi.org/10.1021/am100037y

Liu, S. Q., Tian, Q., Hedrick, J. L., Po Hui, J. H., Rachel Ee, P. L., & Yang, Y. Y. (2010). Biomimetic hydrogels for chondrogenic differentiation of human mesenchymal stem cells to neocartilage. *Biomaterials, 31*(28), 7298–7307. https://doi.org/10.1016/j.biomaterials.2010.06.001

Liu, X., Wang, X., Horii, A., Wang, X., Qiao, L., Zhang, S., & Cui, F. Z. (2012). In vivo studies on angiogenic activity of two designer self-assembling peptide scaffold hydrogels in the chicken embryo chorioallantoic membrane. *Nanoscale, 4*(8), 2720–2727. https://doi.org/10.1039/c2nr00001f

Lönne, M., Zhu, G., Stahl, F., & Walter, J. G. (2014). Aptamer-modified nanoparticles as biosensors. *Advances in Biochemical Engineering/Biotechnology, 140*, 121–154. https://doi.org/10.1007/10_2013_231

Lu, J., Sun, X., Yin, H., Shen, X., Yang, S., Wang, Y., … Wang, X. (2018). A neurotrophic peptide-functionalized self-assembling peptide nanofiber hydrogel enhances rat sciatic nerve regeneration. *Nano Research, 11*(9), 4599–4613. https://doi.org/10.1007/s12274-018-2041-9

Lu, L., Fu, R., Li, C., Yu, C., Li, Z., Guan, H., … Zhao, D. (2013). Silver nanoparticle/chitosan oligosaccharide/poly(vinyl alcohol) nanofibers as wound dressings: a preclinical study. *International Journal of Nanomedicine, 8*, 4131. https://doi.org/10.2147/ijn.s51679

Lv, F. J., Tuan, R. S., Cheung, K. M. C., & Leung, V. Y. L. (2014). Concise review: The surface markers and identity of human mesenchymal stem cells. *Stem Cells, 32*, 1408–1419. https://doi.org/10.1002/stem.1681

Ma, K., Chan, C. K., Liao, S., Hwang, W. Y. K., Feng, Q., & Ramakrishna, S. (2008). Electrospun nanofiber scaffolds for rapid and rich capture of bone marrow-derived hematopoietic stem cells. *Biomaterials, 29*(13), 2096–2103. https://doi.org/10.1016/j.biomaterials.2008.01.024

Ma, L., Gao, C., Mao, Z., Zhou, J., Shen, J., Hu, X., & Han, C. (2003). Collagen/chitosan porous scaffolds with improved biostability for skin tissue engineering. *Biomaterials, 24*(26), 4833–4841.

Madl, C. M., Heilshorn, S. C., & Blau, H. M. (2018). Bioengineering strategies to accelerate stem cell therapeutics. *Nature, 557*, 335–342. https://doi.org/10.1038/s41586-018-0089-z

Magrez, A., Kasas, S., Salicio, V., Pasquier, N., Seo, J. W., Celio, M., … Forró, L. (2006). Cellular toxicity of carbon-based nanomaterials. *Nano Letters, 6*(6), 1121–1125. https://doi.org/10.1021/nl060162e

Mahmoodinia Maymand, M., Soleimanpour-lichaei, H. R., Ardeshirylajimi, A., Soleimani, M., Enderami, S. E., Nojehdehi, S., … Kabir Salmani, M. (2018). Improvement of hepatogenic differentiation of iPS cells on an aligned polyethersulfone compared to random nanofibers. *Artificial Cells, Nanomedicine and Biotechnology, 46*(4), 853–860. https://doi.org/10.1080/21691401.2017.1345929

Malchesky, P. S. (2009). Artificial organs 2008: A year in review. *Artificial Organs, 33*, 273–295. https://doi.org/10.1111/j.1525-1594.2009.00718.x

Mammadov, B., Mammadov, R., Guler, M. O., & Tekinay, A. B. (2012). Cooperative effect of heparan sulfate and laminin mimetic peptide nanofibers on the promotion of neurite outgrowth. *Acta Biomaterialia, 8*(6), 2077–2086. https://doi.org/10.1016/j.actbio.2012.02.006

Mammadov, B., Sever, M., Gecer, M., Zor, F., Ozturk, S., Akgun, H., … Tekinay, A. B. (2016). Sciatic nerve regeneration induced by glycosaminoglycan and laminin mimetic peptide nanofiber gels. *RSC Advances, 6*(112), 110535–110547. https://doi.org/10.1039/C6RA24450E

Mammadov, R., Mammadov, B., Toksoz, S., Aydin, B., Yagci, R., Tekinay, A. B., & Guler, M. O. (2011). Heparin mimetic peptide nanofibers promote angiogenesis. *Biomacromolecules, 12*(10), 3508. https://doi.org/10.1021/bm200957s

Mammadov, R., Cinar, G., Gunduz, N., Goktas, M., Kayhan, H., Tohumeken, S., … Guler, M. O. (2015). Virus-like nanostructures for tuning immune response. *Scientific Reports, 5*, 16728. https://doi.org/10.1038/srep16728

Martínez-Castañón, G. A., Niño-Martínez, N., Martínez-Gutierrez, F., Martínez-Mendoza, J. R., & Ruiz, F. (2008). Synthesis and antibacterial activity of silver nanoparticles with different sizes. *Journal of Nanoparticle Research, 10*(8), 1343–1348. https://doi.org/10.1007/s11051-008-9428-6

Massobrio, P., Massobrio, G., & Martinoia, S. (2016). Interfacing cultured neurons to microtransducers arrays: A review of the neuro-electronic junction models. *Frontiers in Neuroscience, 10*, 282. https://doi.org/10.3389/fnins.2016.00282

McDonald, D. M., Byrne, S. N., & Payne, R. J. (2015). Synthetic self-adjuvanting glycopeptide cancer vaccines. *Frontiers in Chemistry, 3*, 60. https://doi.org/10.3389/fchem.2015.00060

Mehrali, M., Moghaddam, E., Shirazi, S. F. S., Baradaran, S., Mehrali, M., Latibari, S. T., … Osman, N. A. A. (2014). Synthesis, mechanical properties, and in vitro biocompatibility with osteoblasts of calcium silicate-reduced graphene oxide composites. *ACS Applied Materials and Interfaces, 6*(6), 3947–3962. https://doi.org/10.1021/am500845x

Mehta, M., Madl, C. M., Lee, S., Duda, G. N., & Mooney, D. J. (2015). The collagen I mimetic peptide DGEA enhances an osteogenic phenotype in mesenchymal stem cells when presented from cell-encapsulating hydrogels. *Journal of Biomedical Materials Research—Part A, 103*(11), 3516–3525. https://doi.org/10.1002/jbm.a.35497

Mhanna, R., Öztürk, E., Vallmajo-Martin, Q., Millan, C., Müller, M., & Zenobi-Wong, M. (2013). GFOGER-modified MMP-sensitive polyethylene glycol hydrogels induce chondrogenic differentiation of human mesenchymal stem cells. *Tissue Engineering Part A, 20*(7–8), 1165–1174. https://doi.org/10.1089/ten.tea.2013.0519

Miguel, S. P., Ribeiro, M. P., Brancal, H., Coutinho, P., & Correia, I. J. (2014). Thermoresponsive chitosan-agarose hydrogel for skin regeneration. *Carbohydrate Polymers, 111*, 366–373. https://doi.org/10.1016/j.carbpol.2014.04.093

Milovanovic, M., Arsenijevic, A., Milovanovic, J., Kanjevac, T., & Arsenijevic, N. (2017). Nanoparticles in antiviral therapy. In A. Grumezescu (Ed.), *Antimicrobial nanoarchitectonics: From synthesis to applications* (pp. 383–410). Amsterdam: Elsevier. https://doi.org/10.1016/B978-0-323-52733-0.00014-8

Minami, K., Kasuya, Y., Yamazaki, T., Ji, Q., Nakanishi, W., Hill, J. P., … Ariga, K. (2015). Highly ordered 1D fullerene crystals for concurrent control of macroscopic cellular orientation and differentiation toward large-scale tissue engineering. *Advanced Materials, 27*(27), 4020–4026. https://doi.org/10.1002/adma.201501690

Mitchell, A. C., Briquez, P. S., Hubbell, J. A., & Cochran, J. R. (2016). Engineering growth factors for regenerative medicine applications. *Acta Biomaterialia, 30*, 1–12. https://doi.org/10.1016/j.actbio.2015.11.007

Miyahara, Y., Nagaya, N., Kataoka, M., Yanagawa, B., Tanaka, K., Hao, H., … Mori, H. (2006). Monolayered mesenchymal stem cells repair scarred myocardium after myocardial infarction. *Nature Medicine, 12*(4), 459–465. https://doi.org/10.1038/nm1391

Mohtaram, N. K., Ko, J., King, C., Sun, L., Muller, N., Jun, M. B. G., & Willerth, S. M. (2015). Electrospun biomaterial scaffolds with varied topographies for neuronal differentiation of human-induced pluripotent stem cells. *Journal of Biomedical Materials Research—Part A, 103*(8), 2591–2601. https://doi.org/10.1002/jbm.a.35392

Mok, H., & Zhang, M. (2013). Superparamagnetic iron oxide nanoparticle-based delivery systems for biotherapeutics. *Expert Opinion on Drug Delivery, 10*(1), 73–87. https://doi.org/10.1517/17425247.2013.747507

Mok, S. W., Nizak, R., Fu, S. C., Ho, K. W. K., Qin, L., Saris, D. B. F., … Malda, J. (2016). From the printer: Potential of three-dimensional printing for orthopaedic applications. *Journal of Orthopaedic Translation, 6*, 42–49. https://doi.org/10.1016/j.jot.2016.04.003

Montellano, A., Da Ros, T., Bianco, A., & Prato, M. (2011). Fullerene C60 as a multifunctional system for drug and gene delivery. *Nanoscale, 3*(10), 4035–4041. https://doi.org/10.1039/c1nr10783f

Mori, Y., Ono, T., Miyahira, Y., Nguyen, V. Q., Matsui, T., & Ishihara, M. (2013). Antiviral activity of silver nanoparticle/chitosan composites against H1N1 influenza A virus. *Nanoscale Research Letters, 8*(1), 93. https://doi.org/10.1186/1556-276x-8-93

Mukherjee, S., Nandi, S. K., Kundu, B., Chanda, A., Sen, S., & Das, P. K. (2016). Enhanced bone regeneration with carbon nanotube reinforced hydroxyapatite in animal model. *Journal of the Mechanical Behavior of Biomedical Materials, 60*, 243–255. https://doi.org/10.1016/j.jmbbm.2016.02.005

Muller, J., Huaux, F., Moreau, N., Misson, P., Heilier, J. F., Delos, M., … Lison, D. (2005). Respiratory toxicity of multi-wall carbon nanotubes. *Toxicology and Applied Pharmacology, 207*(3), 221–231. https://doi.org/10.1016/j.taap.2005.01.008

Mumcuoglu, D., Sardan Ekiz, M., Gunay, G., Tekinay, T., Tekinay, A. B., & Guler, M. O. (2016). Cellular internalization of therapeutic oligonucleotides by peptide amphiphile nanofibers and nanospheres. *ACS Applied Materials and Interfaces, 8*(18), 11280–11287. https://doi.org/10.1021/acsami.6b01526

Mumcuoglu, D., Sardan, M., Tekinay, T., Guler, M. O., & Tekinay, A. B. (2015). Oligonucleotide delivery with cell surface binding and cell penetrating peptide amphiphile nanospheres. *Molecular Pharmaceutics, 12*(5), 1584–1591. https://doi.org/10.1021/acs.molpharmaceut.5b00007

Murakami, K., Aoki, H., Nakamura, S., Nakamura, S. I., Takikawa, M., Hanzawa, M., … Ishihara, M. (2010). Hydrogel blends of chitin/chitosan, fucoidan and alginate as healing-impaired wound dressings. *Biomaterials, 31*(1), 83–90. https://doi.org/10.1016/j.biomaterials.2009.09.031

Murphy, S. V., & Atala, A. (2014). 3D bioprinting of tissues and organs. *Nature Biotechnology, 32*, 773–785. https://doi.org/10.1038/nbt.2958

Nayak, T. R., Andersen, H., Makam, V. S., Khaw, C., Bae, S., Xu, X., … Özyilmaz, B. (2011). Graphene for controlled and accelerated osteogenic differentiation of human mesenchymal stem cells. *ACS Nano, 5*(6), 4670–4678. https://doi.org/10.1021/nn200500h

Norouzi, M., Boroujeni, S. M., Omidvarkordshouli, N., & Soleimani, M. (2015). Advances in skin regeneration: Application of electrospun scaffolds. *Advanced Healthcare Materials, 4*, 1114–1133. https://doi.org/10.1002/adhm.201500001

Obara, K., Ishihara, M., Fujita, M., Kanatani, Y., Hattori, H., Matsui, T., … Maehara, T. (2005). Acceleration of wound healing in healing-impaired db/db mice with a photocrosslinkable chitosan hydrogel containing fibroblast growth factor-2. *Wound Repair and Regeneration, 13*(4), 390–397. https://doi.org/10.1111/j.1067-1927.2005.130406.x

Odabas, S., Feichtinger, G. A., Korkusuz, P., Inci, I., Bilgic, E., Yar, A. S., … Piskin, E. (2013). Auricular cartilage repair using cryogel scaffolds loaded with BMP-7-expressing primary chondrocytes. *Journal of Tissue Engineering and Regenerative Medicine, 7*(10), 831–840. https://doi.org/10.1002/term.1634

Olson, E. S., Aguilera, T. A., Jiang, T., Ellies, L. G., Nguyen, Q. T., Wong, E. H., … Tsien, R. Y. (2009). In vivo characterization of activatable cell penetrating peptides for targeting protease activity in cancer. *Integrative Biology, 1*(5–6), 382–393. https://doi.org/10.1039/b904890a

Orbach, R., Adler-Abramovich, L., Zigerson, S., Mironi-Harpaz, I., Seliktar, D., & Gazit, E. (2009). Self-assembled Fmoc-peptides as a platform for the formation of nanostructures and hydrogels. *Biomacromolecules, 10*(9), 2646–2651. https://doi.org/10.1021/bm900584m

Oropallo, W., & Piegl, L. A. (2016). Ten challenges in 3D printing. *Engineering with Computers, 32*(1), 135–148. https://doi.org/10.1007/s00366-015-0407-0

Oryan, A., & Sahvieh, S. (2017). Effectiveness of chitosan scaffold in skin, bone and cartilage healing. *International Journal of Biological Macromolecules, 104*, 1003–1011. https://doi.org/10.1016/j.ijbiomac.2017.06.124

Ozdemir, A., Ekiz, M. S., Dilli, A., Guler, M. O., & Tekinay, A. B. (2016). Amphiphilic peptide coated superparamagnetic iron oxide nanoparticles for: In vivo MR tumor imaging. *RSC Advances, 6*(51), 45135–45146. https://doi.org/10.1039/c6ra07380h

Pal, S., Tak, Y. K., & Song, J. M. (2007). Does the antibacterial activity of silver nanoparticles depend on the shape of the nanoparticle? A study of the gram-negative bacterium Escherichia coli. *Applied and Environmental Microbiology, 73*(6), 1712–1720. https://doi.org/10.1128/AEM.02218-06

Pan, H., Hao, S., Zheng, Q., Li, J., Zheng, J., Hu, Z., … Yang, Q. (2013). Bone induction by biomimetic PLGA copolymer loaded with a novel synthetic RADA16-P24 peptide in vivo. *Materials Science and Engineering C, 33*(6), 3336–3345. https://doi.org/10.1016/j.msec.2013.04.019

Pang, M., Shu, T., Chen, R. Q., Liu, C., He, L., Yang, Y., … Rong, L. M. (2016). Neural precursor cells generated from induced pluripotent stem cells with gelatin sponge-electrospun PLGA/PEG nanofibers for spinal cord injury repair. *International Journal of Clinical and Experimental Medicine, 9*(9), 17985–17994.

Panseri, S., Cunha, C., Lowery, J., Del Carro, U., Taraballi, F., Amadio, S., … Gelain, F. (2008). Electrospun micro- and nanofiber tubes for functional nervous regeneration in sciatic nerve transections. *BMC Biotechnology, 8*, 39. https://doi.org/10.1186/1472-6750-8-39

Park, K. M., Lee, S. Y., Joung, Y. K., Na, J. S., Lee, M. C., & Park, K. D. (2009). Thermosensitive chitosan-Pluronic hydrogel as an injectable cell delivery carrier for cartilage regeneration. *Acta Biomaterialia, 5*(6), 1956–1965. https://doi.org/10.1016/j.actbio.2009.01.040

Partha, R., Mitchell, L. R., Lyon, J. L., Joshi, P. P., & Conyers, J. L. (2008). Buckysomes: Fullerene-based nanocarriers for hydrophobic molecule delivery. *ACS Nano, 2*(9), 1950–1958. https://doi.org/10.1021/nn800422k

Patel, A., Mukundan, S., Wang, W., Karumuri, A., Sant, V., Mukhopadhyay, S. M., & Sant, S. (2016). Carbon-based hierarchical scaffolds for myoblast differentiation: Synergy between nano-functionalization and alignment. *Acta Biomaterialia, 32*, 77–88. https://doi.org/10.1016/j.actbio.2016.01.004

Pek, Y. S., Wan, A. C. A., & Ying, J. Y. (2010). The effect of matrix stiffness on mesenchymal stem cell differentiation in a 3D thixotropic gel. *Biomaterials, 31*(3), 385–391. https://doi.org/10.1016/j.biomaterials.2009.09.057

Perez, R. A., Kim, M., Kim, T.-H., Kim, J.-H., Lee, J. H., Park, J.-H., … Kim, H.-W. (2014). Utilizing core–shell fibrous collagen-alginate hydrogel cell delivery system for bone tissue engineering. *Tissue Engineering Part A, 20*(1–2), 103–114. https://doi.org/10.1089/ten.tea.2013.0198

Perugini, M., Varelias, A., Sadlon, T., & D'Andrea, R. J. (2009). Hematopoietic growth factor mimetics: From concept to clinic. *Cytokine and Growth Factor Reviews, 20*, 87–94. https://doi.org/10.1016/j.cytogfr.2009.01.002

Pham, Q. P., Sharma, U., & Mikos, A. G. (2006). Electrospinning of polymeric nanofibers for tissue engineering applications: A review. *Tissue Engineering, 12*(5), 1197–1211. https://doi.org/10.1089/ten.2006.12.ft-65

Philip, D. L., Silantyeva, E. A., Becker, M. L., & Willits, R. K. (2019). RGD-functionalized nanofibers increase early GFAP expression during neural differentiation of mouse embryonic stem cells. *Biomacromolecules, 20*(3), 1443–1454. https://doi.org/10.1021/acs.biomac.9b00018

Pina, S., Oliveira, J. M., & Reis, R. L. (2015). Natural-based nanocomposites for bone tissue engineering and regenerative medicine: A review. *Advanced Materials, 27*, 1143–1169. https://doi.org/10.1002/adma.201403354

Prang, P., Müller, R., Eljaouhari, A., Heckmann, K., Kunz, W., Weber, T., … Weidner, N. (2006). The promotion of oriented axonal regrowth in the injured spinal cord by alginate-based anisotropic capillary hydrogels. *Biomaterials, 27*(19), 3560–3569. https://doi.org/10.1016/j.biomaterials.2006.01.053

Prichard, H. L., Reichert, W. M., & Klitzman, B. (2007). Adult adipose-derived stem cell attachment to biomaterials. *Biomaterials, 28*(6), 936–946. https://doi.org/10.1016/j.biomaterials.2006.09.012

Priya, S. G., Gupta, A., Jain, E., Sarkar, J., Damania, A., Jagdale, P. R., … Kumar, A. (2016). Bilayer cryogel wound dressing and skin regeneration grafts for the treatment of acute skin wounds. *ACS Applied Materials and Interfaces, 8*(24), 15145–15159. https://doi.org/10.1021/acsami.6b04711

Raftery, R. M., Woods, B., Marques, A. L. P., Moreira-Silva, J., Silva, T. H., Cryan, S. A., … O'Brien, F. J. (2016). Multifunctional biomaterials from the sea: Assessing the effects of chitosan incorporation into collagen scaffolds on mechanical and biological functionality. *Acta Biomaterialia, 43*, 160–169. https://doi.org/10.1016/j.actbio.2016.07.009
Rath, G., Hussain, T., Chauhan, G., Garg, T., & Goyal, A. K. (2016). Development and characterization of cefazolin loaded zinc oxide nanoparticles composite gelatin nanofiber mats for postoperative surgical wounds. *Materials Science and Engineering C, 58*, 242–253. https://doi.org/10.1016/j.msec.2015.08.050
Ratheesh, G., Venugopal, J. R., Chinappan, A., Ezhilarasu, H., Sadiq, A., & Ramakrishna, S. (2017). 3D Fabrication of polymeric scaffolds for regenerative therapy. *ACS Biomaterials Science and Engineering, 3*, 1175–1194. https://doi.org/10.1021/acsbiomaterials.6b00370
Raucci, M. G., Alvarez-Perez, M., Giugliano, D., Zeppetelli, S., & Ambrosio, L. (2016). Properties of carbon nanotube-dispersed Sr-hydroxyapatite injectable material for bone defects. *Regenerative Biomaterials, 3*(1), 13–23. https://doi.org/10.1093/rb/rbv026
Ravichandran, R., Venugopal, J. R., Sundarrajan, S., Mukherjee, S., & Ramakrishna, S. (2012). Precipitation of nanohydroxyapatite on PLLA/PBLG/Collagen nanofibrous structures for the differentiation of adipose derived stem cells to osteogenic lineage. *Biomaterials, 33*(3), 846–855. https://doi.org/10.1016/j.biomaterials.2011.10.030
Restuccia, A., Tian, Y. F., Collier, J. H., & Hudalla, G. A. (2015). Self-assembled glycopeptide nanofibers as modulators of galectin-1 bioactivity. *Cellular and Molecular Bioengineering, 8*(3), 471–487. https://doi.org/10.1007/s12195-015-0399-2
Rezvanian, M., Mohd Amin, M. C. I., & Ng, S. F. (2016). Development and physicochemical characterization of alginate composite film loaded with simvastatin as a potential wound dressing. *Carbohydrate Polymers, 137*, 295–304. https://doi.org/10.1016/j.carbpol.2015.10.091
Ribeiro, M. P., Espiga, A., Silva, D., Baptista, P., Henriques, J., Ferreira, C., … Correia, I. J. (2009). Development of a new chitosan hydrogel for wound dressing. *Wound Repair and Regeneration, 17*(6), 817–824. https://doi.org/10.1111/j.1524-475X.2009.00538.x
Romero-Sánchez, L. B., Marí-Beffa, M., Carrillo, P., Medina, M. Á., & Díaz-Cuenca, A. (2018). Copper-containing mesoporous bioactive glass promotes angiogenesis in an in vivo zebrafish model. *Acta Biomaterialia, 68*, 272–285. https://doi.org/10.1016/j.actbio.2017.12.032
Rowlands, A. S., George, P. A., & Cooper-White, J. J. (2008). Directing osteogenic and myogenic differentiation of MSCs: Interplay of stiffness and adhesive ligand presentation. *American Journal of Physiology-Cell Physiology, 295*(4), C1037–C1044. https://doi.org/10.1152/ajpcell.67.2008
Rufaihah, A. J., Yasa, I. C., Ramanujam, V. S., Arularasu, S. C., Kofidis, T., Guler, M. O., & Tekinay, A. B. (2017). Angiogenic peptide nanofibers repair cardiac tissue defect after myocardial infarction. *Acta Biomaterialia, 58*, 102–112. https://doi.org/10.1016/j.actbio.2017.06.009
Ruoslahti, E. (2002). Rgd and other recognition sequences for integrins. *Annual Review of Cell and Developmental Biology, 12*(1), 697–715. https://doi.org/10.1146/annurev.cellbio.12.1.697
Ryoo, S. R., Kim, Y. K., Kim, M. H., & Min, D. H. (2010). Behaviors of NIH-3T3 fibroblasts on graphene/carbon nanotubes: Proliferation, focal adhesion, and gene transfection studies. *ACS Nano, 4*(11), 6587–6598. https://doi.org/10.1021/nn1018279
Sabaghi, F., Shamsasenjan, K., Movasaghpour, A. A., Amirizadeh, N., Nikougoftar, M., & Bagheri, N. (2016). Evaluation of human cord blood CD34+ hematopoietic stem cell differentiation to megakaryocyte on aminated PES nanofiber scaffold compare to 2-D culture system. *Artificial Cells, Nanomedicine and Biotechnology, 44*(4), 1062–1068. https://doi.org/10.3109/21691401.2015.1011800
Saccà, B., & Niemeyer, C. M. (2012). DNA origami: The art of folding DNA. *Angewandte Chemie—International Edition, 51*, 58–66. https://doi.org/10.1002/anie.201105846
Salinas, C. N., & Anseth, K. S. (2008). The enhancement of chondrogenic differentiation of human mesenchymal stem cells by enzymatically regulated RGD functionalities. *Biomaterials, 29*(15), 2370–2377. https://doi.org/10.1016/j.biomaterials.2008.01.035
Schnell, E., Klinkhammer, K., Balzer, S., Brook, G., Klee, D., Dalton, P., & Mey, J. (2007). Guidance of glial cell migration and axonal growth on electrospun nanofibers of poly-ε-

caprolactone and a collagen/poly-ε-caprolactone blend. *Biomaterials, 28*(19), 3012–3025. https://doi.org/10.1016/j.biomaterials.2007.03.009

Sell, S. A., McClure, M. J., Garg, K., Wolfe, P. S., & Bowlin, G. L. (2009). Electrospinning of collagen/biopolymers for regenerative medicine and cardiovascular tissue engineering. *Advanced Drug Delivery Reviews, 61*, 1007–1019. https://doi.org/10.1016/j.addr.2009.07.012

Selnihhin, D., Sparvath, S. M., Preus, S., Birkedal, V., & Andersen, E. S. (2018). Multifluorophore DNA origami beacon as a biosensing platform. *ACS Nano, 12*(6), 5699–5708. https://doi.org/10.1021/acsnano.8b01510

Senturk, B., Mercan, S., Delibasi, T., Guler, M. O., & Tekinay, A. B. (2016). Angiogenic peptide nanofibers improve wound healing in STZ-induced diabetic rats. *ACS Biomaterials Science and Engineering, 2*(7), 1180–1189. https://doi.org/10.1021/acsbiomaterials.6b00238

Sever, M., Mammadov, B., Guler, M. O., & Tekinay, A. B. (2014). Tenascin-C mimetic peptide nanofibers direct stem cell differentiation to osteogenic lineage. *Biomacromolecules, 15*(12), 4480–4487. https://doi.org/10.1021/bm501271x

Sever, M., Turkyilmaz, M., Sevinc, C., Cakir, A., Ocalan, B., Cansev, M., … Tekinay, A. B. (2016). Regenerative effects of peptide nanofibers in an experimental model of Parkinson's disease. *Acta Biomaterialia, 46*, 79–90. https://doi.org/10.1016/j.actbio.2016.09.011

Shekaran, A., García, J. R., Clark, A. Y., Kavanaugh, T. E., Lin, A. S., Guldberg, R. E., & García, A. J. (2014). Bone regeneration using an alpha 2 beta 1 integrin-specific hydrogel as a BMP-2 delivery vehicle. *Biomaterials, 35*(21), 5453–5461. https://doi.org/10.1016/j.biomaterials.2014.03.055

Shi, J., Yu, X., Wang, L., Liu, Y., Gao, J., Zhang, J., … Zhang, Z. (2013). PEGylated fullerene/iron oxide nanocomposites for photodynamic therapy, targeted drug delivery and MR imaging. *Biomaterials, 34*(37), 9666–9677. https://doi.org/10.1016/j.biomaterials.2013.08.049

Shi, J., Zhang, H., Wang, L., Li, L., Wang, H., Wang, Z., … Zhang, Z. (2013). PEI-derivatized fullerene drug delivery using folate as a homing device targeting to tumor. *Biomaterials, 34*(1), 251–261. https://doi.org/10.1016/j.biomaterials.2012.09.039

Shih, Y. R. V., Tseng, K. F., Lai, H. Y., Lin, C. H., & Lee, O. K. (2011). Matrix stiffness regulation of integrin-mediated mechanotransduction during osteogenic differentiation of human mesenchymal stem cells. *Journal of Bone and Mineral Research, 26*(4), 730–738. https://doi.org/10.1002/jbmr.278

Shin, S.-Y., Park, H.-N., Kim, K.-H., Lee, M.-H., Choi, Y. S., Park, Y.-J., … Chung, C.-P. (2005). Biological evaluation of chitosan nanofiber membrane for guided bone regeneration. *Journal of Periodontology, 76*(10), 1778–1784. https://doi.org/10.1902/jop.2005.76.10.1778

Shoval, A. (2009). Carbon nanotube electrodes for effective interfacing with retinal tissue. *Frontiers in Neuroengineering, 2*, 4. https://doi.org/10.3389/neuro.16.004.2009

Shrestha, S., Shrestha, B. K., Lee, J., Joong, O. K., Kim, B. S., Park, C. H., & Kim, C. S. (2019). A conducting neural interface of polyurethane/silk-functionalized multiwall carbon nanotubes with enhanced mechanical strength for neuroregeneration. *Materials Science and Engineering C, 102*, 511–523. https://doi.org/10.1016/j.msec.2019.04.053

Shvedova, A., Castranova, V., Kisin, E., Schwegler-Berry, D., Murray, A., Gandelsman, V., & Baron, P. (2003). Exposure to carbon nanotube material: Assessment of nanotube cytotoxicity using human keratinocyte cells. *Journal of Toxicology and Environmental Health—Part A, 66*(20), 1909–1926. https://doi.org/10.1080/713853956

Silva, G. A., Czeisler, C., Niece, K. L., Beniash, E., Harrington, D. A., Kessler, J. A., & Stupp, S. I. (2004). Selective differentiation of neural progenitor cells by high-epitope density nanofibers. *Science, 303*(5662), 1352–1355. https://doi.org/10.1126/science.1093783

Singh, A., & Elisseeff, J. (2010). Biomaterials for stem cell differentiation. *Journal of Materials Chemistry, 20*(40), 8832–8847. https://doi.org/10.1039/c0jm01613f

Singh, D., Zo, S. M., Kumar, A., & Han, S. S. (2013). Engineering three-dimensional macroporous hydroxyethyl methacrylate-alginate-gelatin cryogel for growth and proliferation of lung epithelial cells. *Journal of Biomaterials Science, Polymer Edition, 24*(11), 1343–1359. https://doi.org/10.1080/09205063.2012.759505

Singh, N., Chen, J., Koziol, K. K., Hallam, K. R., Janas, D., Patil, A. J., … Rahatekar, S. S. (2016). Chitin and carbon nanotube composites as biocompatible scaffolds for neuron growth. *Nanoscale, 8*(15), 8288–8299. https://doi.org/10.1039/c5nr06595j
Singh, P., & Schwarzbauer, J. E. (2012). Fibronectin and stem cell differentiation—Lessons from chondrogenesis. *Journal of Cell Science, 125*(16), 3703–3712. https://doi.org/10.1242/jcs.095786
Singh, R. K., Seliktar, D., & Putnam, A. J. (2013). Capillary morphogenesis in PEG-collagen hydrogels. *Biomaterials, 34*(37), 9331–9340. https://doi.org/10.1016/j.biomaterials.2013.08.016
Smith, L. A., Liu, X., Hu, J., & Ma, P. X. (2010). The enhancement of human embryonic stem cell osteogenic differentiation with nano-fibrous scaffolding. *Biomaterials, 31*(21), 5526–5535. https://doi.org/10.1016/j.biomaterials.2010.03.065
Song, J., Wang, F., Yang, X., Ning, B., Harp, M. G., Culp, S. H., … Chen, X. (2016). Gold nanoparticle coated carbon nanotube ring with enhanced raman scattering and photothermal conversion property for theranostic applications. *Journal of the American Chemical Society, 138*(22), 7005–7015. https://doi.org/10.1021/jacs.5b13475
Srinivasan, S., Jayasree, R., Chennazhi, K. P., Nair, S. V., & Jayakumar, R. (2012). Biocompatible alginate/nano bioactive glass ceramic composite scaffolds for periodontal tissue regeneration. *Carbohydrate Polymers, 87*(1), 274–283. https://doi.org/10.1016/j.carbpol.2011.07.058
Sudheesh Kumar, P. T., Lakshmanan, V. K., Anilkumar, T. V., Ramya, C., Reshmi, P., Unnikrishnan, A. G., … Jayakumar, R. (2012). Flexible and microporous chitosan hydrogel/nano ZnO composite bandages for wound dressing: In vitro and in vivo evaluation. *ACS Applied Materials and Interfaces, 4*(5), 2618–2629. https://doi.org/10.1021/am300292v
Sudheesh Kumar, P. T., Lakshmanan, V. K., Raj, M., Biswas, R., Hiroshi, T., Nair, S. V., & Jayakumar, R. (2013). Evaluation of wound healing potential of β-chitin hydrogel/nano zinc oxide composite bandage. *Pharmaceutical Research, 30*(2), 523–537. https://doi.org/10.1007/s11095-012-0898-y
Sun, L., Gao, Y., Xu, Y., Chao, J., Liu, H., Wang, L., … Fan, C. (2017). Real-time imaging of single-molecule enzyme cascade using a DNA origami raft. *Journal of the American Chemical Society, 139*(48), 17525–17532. https://doi.org/10.1021/jacs.7b09323
Sun, W., Incitti, T., Migliaresi, C., Quattrone, A., Casarosa, S., & Motta, A. (2017). Viability and neuronal differentiation of neural stem cells encapsulated in silk fibroin hydrogel functionalized with an IKVAV peptide. *Journal of Tissue Engineering and Regenerative Medicine, 11*(5), 1532–1541. https://doi.org/10.1002/term.2053
Sun, Y., Li, W., Wu, X., Zhang, N., Zhang, Y., Ouyang, S., … Wu, W. (2016). Functional self-assembling peptide nanofiber hydrogels designed for nerve degeneration. *ACS Applied Materials and Interfaces, 8*(3), 2348–2359. https://doi.org/10.1021/acsami.5b11473
Tahlawi, A., Klontzas, M. E., Allenby, M. C., Morais, J. C. F., Panoskaltsis, N., & Mantalaris, A. (2019). RGD-functionalized polyurethane scaffolds promote umbilical cord blood mesenchymal stem cell expansion and osteogenic differentiation. *Journal of Tissue Engineering and Regenerative Medicine, 13*(2), 232–243. https://doi.org/10.1002/term.2784
Tansik, G., Kilic, E., Beter, M., Demiralp, B., Kiziltas Sendur, G., Can, N., … Tekinay, A. B. (2016). A glycosaminoglycan mimetic peptide nanofiber gel as an osteoinductive scaffold. *Biomaterials Science, 4*(9), 1328–1339. https://doi.org/10.1039/c6bm00179c
Tarun, K., & Gobi, N. (2012). Calcium alginate/PVA blended nano fibre matrix for wound dressing. *Indian Journal of Fibre and Textile Research, 37*(2), 127–132.
Tate, C. C., Shear, D. A., Tate, M. C., Archer, D. R., Stein, D. G., & LaPlaca, M. C. (2009). Laminin and fibronectin scaffolds enhance neural stem cell transplantation into the injured brain. *Journal of Tissue Engineering and Regenerative Medicine, 3*(3), 208–217. https://doi.org/10.1002/term.154
Tekinay, A. B., Guler, M. O., Mumcuoglu, D., & Ustun, S. (2013). Peptide nanofibers for controlled growth factor release. *Therapeutic Delivery, 4*(6), 651–654. https://doi.org/10.4155/tde.13.35
Thomas, R., Soumya, K. R., Mathew, J., & Radhakrishnan, E. K. (2015). Electrospun polycaprolactone membrane incorporated with biosynthesized silver nanoparticles as effective wound

dressing material. *Applied Biochemistry and Biotechnology, 176*(8), 2213–2224. https://doi.org/10.1007/s12010-015-1709-9

Topal, A. E., Tansik, G., Ozkan, A. D., Guler, M. O., Dana, A., & Tekinay, A. B. (2017). Nanomechanical characterization of osteogenic differentiation of mesenchymal stem cells on bioactive peptide nanofiber hydrogels. *Advanced Materials Interfaces, 4*(20), 1700090. https://doi.org/10.1002/admi.201700090

Toskas, G., Cherif, C., Hund, R. D., Laourine, E., Mahltig, B., Fahmi, A., … Hanke, T. (2013). Chitosan(PEO)/silica hybrid nanofibers as a potential biomaterial for bone regeneration. *Carbohydrate Polymers, 94*(2), 713–722. https://doi.org/10.1016/j.carbpol.2013.01.068

Uccelli, A., Moretta, L., & Pistoia, V. (2008). Mesenchymal stem cells in health and disease. *Nature Reviews Immunology, 8*, 726–736. https://doi.org/10.1038/nri2395

Ustun, S., Tombuloglu, A., Kilinc, M., Guler, M. O., & Tekinay, A. B. (2013). Growth and differentiation of prechondrogenic cells on bioactive self-assembled peptide nanofibers. *Biomacromolecules, 14*(1), 17–26. https://doi.org/10.1021/bm301538k

Varaprasad, K., Mohan, Y. M., Vimala, K., & Mohana Raju, K. (2011). Synthesis and characterization of hydrogel-silver nanoparticle-curcumin composites for wound dressing and antibacterial application. *Journal of Applied Polymer Science, 121*(2), 784–796. https://doi.org/10.1002/app.33508

Venkatesan, J., Bhatnagar, I., Manivasagan, P., Kang, K. H., & Kim, S. K. (2015). Alginate composites for bone tissue engineering: A review. *International Journal of Biological Macromolecules, 72*, 269–281. https://doi.org/10.1016/j.ijbiomac.2014.07.008

Viguet-Carrin, S., Garnero, P., & Delmas, P. D. (2006). The role of collagen in bone strength. *Osteoporosis International, 17*, 319–336. https://doi.org/10.1007/s00198-005-2035-9

Wagner, W., Wein, F., Seckinger, A., Frankhauser, M., Wirkner, U., Krause, U., … Ho, A. D. (2005). Comparative characteristics of mesenchymal stem cells from human bone marrow, adipose tissue, and umbilical cord blood. *Experimental Hematology, 33*(11), 1402–1416. https://doi.org/10.1016/j.exphem.2005.07.003

Wahl, D. A., & Czernuszka, J. T. (2006). Collagen-hydroxyapatite composites for hard tissue repair. *European Cells and Materials, 11*, 43–56. https://doi.org/10.22203/eCM.v011a06

Wang, W., Itoh, S., Konno, K., Kikkawa, T., Ichinose, S., Sakai, K., … Watabe, K. (2009). Effects of Schwann cell alignment along the oriented electrospun chitosan nanofibers on nerve regeneration. *Journal of Biomedical Materials Research—Part A, 91*(4), 994–1005. https://doi.org/10.1002/jbm.a.32329

Wang, X. M., Qiao, L., & Horii, A. (2011). Screening of functionalized self-assembling peptide nanofiber scaffolds with angiogenic activity for endothelial cell growth. *Progress in Natural Science: Materials International, 21*(2), 111–116. https://doi.org/10.1016/S1002-0071(12)60043-4

Wang, X., Horii, A., & Zhang, S. (2008). Designer functionalized self-assembling peptide nanofiber scaffolds for growth, migration, and tubulogenesis of human umbilical vein endothelial cells. *Soft Matter, 4*(12), 2388–2395. https://doi.org/10.1039/b807155a

Wang, X., Yan, C., Ye, K., He, Y., Li, Z., & Ding, J. (2013). Effect of RGD nanospacing on differentiation of stem cells. *Biomaterials, 34*(12), 2865–2874. https://doi.org/10.1016/j.biomaterials.2013.01.021

Wang, Y., Lee, W. C., Manga, K. K., Ang, P. K., Lu, J., Liu, Y. P., … Loh, K. P. (2012). Fluorinated graphene for promoting neuro-induction of stem cells. *Advanced Materials, 24*(31), 4285–4290. https://doi.org/10.1002/adma.201200846

Wang, Z. G., Song, C., & Ding, B. (2013). Functional DNA nanostructures for photonic and biomedical applications. *Small, 9*, 2210–2222. https://doi.org/10.1002/smll.201300141

Warheit, D. B., Laurence, B. R., Reed, K. L., Roach, D. H., Reynolds, G. A. M., & Webb, T. R. (2004). Comparative pulmonary toxicity assessment of single-wall carbon nanotubes in rats. *Toxicological Sciences, 77*(1), 117–125. https://doi.org/10.1093/toxsci/kfg228

Warren, J. D., Geng, X., & Danishefsky, S. J. (2006). Synthetic glycopeptide-based vaccines. In *Glycopeptides and glycoproteins* (pp. 109–141). Berlin: Springer. https://doi.org/10.1007/128_031

Warren, L., Manos, P. D., Ahfeldt, T., Loh, Y. H., Li, H., Lau, F., … Rossi, D. J. (2010). Highly efficient reprogramming to pluripotency and directed differentiation of human cells with synthetic modified mRNA. *Cell Stem Cell, 7*(5), 618–630. https://doi.org/10.1016/j.stem.2010.08.012

Wei, M., Li, S., Yang, Z., Zheng, W., & Le, W. (2017). Gold nanoparticles enhance the differentiation of embryonic stem cells into dopaminergic neurons via mTOR/p70S6K pathway. *Nanomedicine, 12*(11), 1305–1317. https://doi.org/10.2217/nnm-2017-0001

Wen, J. H., Vincent, L. G., Fuhrmann, A., Choi, Y. S., Hribar, K. C., Taylor-Weiner, H., … Engler, A. J. (2014). Interplay of matrix stiffness and protein tethering in stem cell differentiation. *Nature Materials, 13*(10), 979–987. https://doi.org/10.1038/nmat4051

Wiraja, C., Yeo, D., Lio, D., Labanieh, L., Lu, M., Zhao, W., & Xu, C. (2014). Aptamer technology for tracking cells' status & function. *Molecular and Cellular Therapies, 2*(1), 33. https://doi.org/10.1186/2052-8426-2-33

Wu, X., He, L., Li, W., Li, H., Wong, W.-M., Ramakrishna, S., & Wu, W. (2017). Functional self-assembling peptide nanofiber hydrogel for peripheral nerve regeneration. *Regenerative Biomaterials, 4*(1), 21–30. https://doi.org/10.1093/rb/rbw034

Wu, Y., Jia, Z., Liu, L., Zhao, Y., Li, H., Wang, C., … Ruan, D. (2016). Functional self-assembled peptide nanofibers for bone marrow mesenchymal stem cell encapsulation and regeneration in nucleus pulposus. *Artificial Organs, 40*(6), E112–E119. https://doi.org/10.1111/aor.12694

Xiao, L., Aoshima, H., Saitoh, Y., & Miwa, N. (2010). The effect of squalane-dissolved fullerene-C60 on adipogenesis-accompanied oxidative stress and macrophage activation in a preadipocyte-monocyte co-culture system. *Biomaterials, 31*(23), 5976–5985. https://doi.org/10.1016/j.biomaterials.2010.04.032

Xie, J., Peng, C., Zhao, Q., Wang, X., Yuan, H., Yang, L., … Zhang, Y. (2016). Osteogenic differentiation and bone regeneration of iPSC-MSCs supported by a biomimetic nanofibrous scaffold. *Acta Biomaterialia, 29*, 365–379. https://doi.org/10.1016/j.actbio.2015.10.007

Xie, J., MacEwan, M. R., Schwartz, A. G., & Xia, Y. (2010). Electrospun nanofibers for neural tissue engineering. *Nanoscale, 2*(1), 35–44. https://doi.org/10.1039/b9nr00243j

Xie, Z., Paras, C. B., Weng, H., Punnakitikashem, P., Su, L. C., Vu, K., … Nguyen, K. T. (2013). Dual growth factor releasing multi-functional nanofibers for wound healing. *Acta Biomaterialia, 9*(12), 9351–9359. https://doi.org/10.1016/j.actbio.2013.07.030

Xu, S., Wang, H., Cheng, M., Hu, J., Wang, C., & Han, C. C. (2013). Preparation and optimization of silver nanoparticles embedded electrospun membrane for implant associated infections prevention. *ACS Applied Materials and Interfaces, 5*(21), 11014–11021. https://doi.org/10.1021/am403250t

Xu, X., Clément, P., Eklöf-Österberg, J., Kelley-Loughnane, N., Moth-Poulsen, K., Chávez, J. L., & Palma, M. (2018). Reconfigurable carbon nanotube multiplexed sensing devices. *Nano Letters, 18*(7), 4130–4135. https://doi.org/10.1021/acs.nanolett.8b00856

Yan, L., Correia, C., Pereira, D., Sousa, R., Oliveira, J., Oliveira, A., & Reis, R. (2013). Injectable silk fibroin hydrogels with ionic strength and pH response for tissue engineering and regenerative medicine applications. *Chemistry of Materials, Journal of Tissue Engineering and Regenerative Medicine, 7*(S1), 14. https://doi.org/10.1002/term.1822

Yang, D., Li, T., Xu, M., Gao, F., Yang, J., Yang, Z., & Le, W. (2014). Graphene oxide promotes the differentiation of mouse embryonic stem cells to dopamine neurons. *Nanomedicine, 9*(16), 2445–2455. https://doi.org/10.2217/nnm.13.197

Yata, T., Takahashi, Y., Tan, M., Nakatsuji, H., Ohtsuki, S., Murakami, T., … Nishikawa, M. (2017). DNA nanotechnology-based composite-type gold nanoparticle-immunostimulatory DNA hydrogel for tumor photothermal immunotherapy. *Biomaterials, 146*, 136–145. https://doi.org/10.1016/j.biomaterials.2017.09.014

Yaylaci, S. U., Sen, M., Bulut, O., Arslan, E., Guler, M. O., & Tekinay, A. B. (2016). Chondrogenic differentiation of mesenchymal stem cells on glycosaminoglycan-mimetic peptide nanofibers. *ACS Biomaterials Science and Engineering, 2*(5), 871–878. https://doi.org/10.1021/acsbiomaterials.6b00099

Yi, C., Liu, D., Fong, C. C., Zhang, J., & Yang, M. (2010). Gold nanoparticles promote osteogenic differentiation of mesenchymal stem cells through p38 MAPK pathway. *ACS Nano, 4*(11), 6439–6448. https://doi.org/10.1021/nn101373r

Yoon, J. H., Kim, J., Lee, H., Kim, S. Y., Jang, H. H., Ryu, S. H., … Lee, T. G. (2012). Laminin peptide YIGSR induces collagen synthesis in Hs27 human dermal fibroblasts. *Biochemical and Biophysical Research Communications, 428*(3), 416–421. https://doi.org/10.1016/j.bbrc.2012.10.070

Zafar, M., Najeeb, S., Khurshid, Z., Vazirzadeh, M., Zohaib, S., Najeeb, B., & Sefat, F. (2016). Potential of electrospun nanofibers for biomedical and dental applications. *Materials, 9*, 73. https://doi.org/10.3390/ma9020073

Zakrzewski, J. L., Tuckett, A. Z., & van den Brink, M. R. M. (2009). Stem cell and nanofiber-based engineering of a T-cell producing three-dimensional tissue construct. *Pediatric Blood and Cancer, 52*(6), 692. https://doi.org/10.1002/pbc.21972. Retrieved from http://www.embase.com/search/results?subaction=viewrecord&from=export&id=L70021941. http://www3.interscience.wiley.com/cgi-bin/fulltext/122265268/PDFSTART. http://sfx.library.uu.nl/utrecht?sid=EMBASE&issn=1545500

Zhang, H., Qu, X., Chen, H., Kong, H., Ding, R., Chen, D., … Weitz, D. A. (2017). Fabrication of calcium phosphate-based nanocomposites incorporating DNA origami, gold nanorods, and anticancer drugs for biomedical applications. *Advanced Healthcare Materials, 6*(20). https://doi.org/10.1002/adhm.201700664

Zhang, Q., Jiang, Q., Li, N., Dai, L., Liu, Q., Song, L., … Du, Y. (2014). DNA origami as an in vivo drug delivery vehicle for cancer therapy. *ACS Nano, 8*(7), 6633–6643. https://doi.org/10.1021/nn502058j

Zhang, Y. Z., Venugopal, J., Huang, Z. M., Lim, C. T., & Ramakrishna, S. (2005). Characterization of the surface biocompatibility of the electrospun PCL-Collagen nanofibers using fibroblasts. *Biomacromolecules, 6*(5), 2583–2589. https://doi.org/10.1021/bm050314k

Zhang, Z. Q., & Song, S. C. (2016). Thermosensitive/superparamagnetic iron oxide nanoparticle-loaded nanocapsule hydrogels for multiple cancer hyperthermia. *Biomaterials, 106*, 13–23. https://doi.org/10.1016/j.biomaterials.2016.08.015

Zhou, A., Chen, S., He, B., Zhao, W., Chen, X., & Jiang, D. (2016). Controlled release of TGF-beta 1 from RADA self-assembling peptide hydrogel scaffolds. *Drug Design, Development and Therapy, 10*, 3043–3051. https://doi.org/10.2147/DDDT.S109545

Zhou, Y., Yang, D., Chen, X., Xu, Q., Lu, F., & Nie, J. (2008). Electrospun water-soluble carboxyethyl chitosan/poly(vinyl alcohol) nanofibrous membrane as potential wound dressing for skin regeneration. *Biomacromolecules, 9*(1), 349–354. https://doi.org/10.1021/bm7009015

Zhu, L., Wang, K., Ma, T., Huang, L., Xia, B., Zhu, S., … Luo, Z. (2017). Noncovalent bonding of RGD and YIGSR to an electrospun poly(ε-caprolactone) conduit through peptide self-assembly to synergistically promote sciatic nerve regeneration in rats. *Advanced Healthcare Materials, 6*(8), 1600860. https://doi.org/10.1002/adhm.201600860

Zouani, O. F., Chollet, C., Guillotin, B., & Durrieu, M. C. (2010). Differentiation of pre-osteoblast cells on poly(ethylene terephthalate) grafted with RGD and/or BMPs mimetic peptides. *Biomaterials, 31*(32), 8245–8253. https://doi.org/10.1016/j.biomaterials.2010.07.042

Chapter 2
Design and Development of Electrospun Nanofibers in Regenerative Medicine

Brabu Balusamy, Anitha Senthamizhan, and Tamer Uyar

Abstract The regenerative medicine field has promising solutions to overcome existing clinical challenges in the repair or regrowth of injured tissues. To date, an enormous progress has been made in developing numerous strategies for enhanced regeneration. The nanofibers fabricated by electrospinning offer excellent characteristics mimicking the extracellular matrix that support cell adhesion, migration, and differentiation, which are responsible for the regeneration of tissues. Furthermore, due to their ease of production, cost-effectiveness, and ability to have various compositions and different morphologies, the electrospun nanofibers have been extensively explored for their possibilities in the regeneration of various tissues. In the present chapter, we summarize the examples of electrospun nanofibers fabricated for the regeneration of dermal, neural, and orthopedic tissues.

2.1 Introduction

Regenerative medicine encompasses numerous strategies involving principles of engineering and life sciences and aims to restore the form and function of injured and diseased tissues and organs using different materials and cells in order to overcome the obstacles of transplantation therapy. Over the last decade, several efforts have been made in the regeneration of various tissues including dermal, neural, bone, dental, liver, heart, and so on which has revolutionized clinical medicine (Atala, 2012; Colombo, Sampogna, Cocozza, Guraya, & Forgione, 2016; Cui, Nowicki, Fisher, & Zhang, 2017; Dzobo et al., 2018; Miao et al., 2017; Vacanti, 2010; Vijayavenkataraman, Yan, Lu, Wang, & Fuh, 2018). Bio-nano research

---

B. Balusamy · A. Senthamizhan
Fondazione Istituto Italiano di Tecnologia, Genoa, Italy

T. Uyar (✉)
Department of Fiber Science & Apparel Design, College of Human Ecology, Cornell University, Ithaca, NY, USA
e-mail: tu46@cornell.edu

A. B. Tekinay (ed.), *Nanomaterials for Regenerative Medicine*, Stem Cell Biology and Regenerative Medicine, https://doi.org/10.1007/978-3-030-31202-2_2

enables the application of exciting concepts for efficient regeneration of several tissues using different nanomaterials (Arora et al., 2012; Gu, Wu, Chen, & Xiao, 2013; E. S. Kim, Ahn, Dvir, & Kim, 2014; Kumar, Griffin, & Butler, 2017; Verma, Domb, & Kumar, 2011). Among various nanomaterials, nanofibers exhibit excellent features including higher surface area, interconnected porous structure in the range of submicron to nanoscale, suitable mechanical properties, and easy functionalization that are more beneficial in all applications from environmental protection to biomedical purposes including regenerative medicine (Babitha et al., 2017; B. Ding, Wang, & Yu, 2019; Liu, Thomopoulos, & Xia, 2012; Senthamizhan, Balusamy, & Uyar, 2016; Shahriar et al., 2019; Shan et al., 2015; Sridhar et al., 2015; Uyar & Kny, 2017). Although several approaches exist for the preparation of nanofibers, electrospinning is a widely acknowledged and viable method for the fabrication of nanofibers owing to its simplicity, cost-effectiveness, ability to be effectively scaled up for mass production and versatility for fabrication of nanofibers with various structures from a wide range of materials. As the term "Electrospinning" is a composite word of electrostatic and spinning, the process applies principles of the electrostatic force to spin the fibers. Basically, the main components of the electrospinning setup include a high-voltage power supply, a syringe pump, a needle spinneret and a grounded conductive collector. During the electrospinning process, the solution is extruded from the spinneret and produces a pendant drop due to the surface tension caused by the application of high voltage. Further, the electrostatic repulsion on the surface charge deforms the air–liquid interface to a Taylor cone which facilitates the ejection of the charged liquid jet that primarily extends into a straight line followed by extensive whipping owing to the bending instabilities. The stretched jet forms structures in finer diameters that solidifies and is collected as solid fibers in the grounded collector.

Several parameters including applied voltage, flow rate, collecting distance, and field strength generally influence the electrospinning process (Li & Wang, 2013; Lin & Fang, 2017; Mitchell, 2014). Over the last two decades, intensive efforts have been made on modification of the electrospinning setup for fabricating different types of nanofibers including a variety of structures and alignment forms for enhancing the performance in specific applications (Aravindan et al., 2015; Y. Ding, Hou, Zhao, Zhu, & Fong, 2016; Kny, Ghosal, & Thomas, 2018; C. Wang, Wang, et al., 2019; Wendorff, Agarwal, & Greiner, 2012; Wu, Wang, Zhao, & Jiang, 2013; Xue, Wu, Dai, & Xia, 2019). Figure 2.1 illustrates the schematic representation of the electrospinning setup with variable needles, collecting substrates, and syringes and examples of nanofibers with different morphologies obtained through electrospinning (Aravindan et al., 2015; Y. Ding et al., 2016).

The cellular environment of the human body consists of dynamic networks of fibrillar proteins, proteoglycans and glycosaminoglycans (GAGs) which are collectively named as the extracellular matrix (ECM) that plays key roles in tissue regeneration and maintenance. The components of the ECM together with cell adhesion receptors make a complex system that offers an ideal environment to regulate diverse cellular functions including cell adhesion, migration, differentiation, proliferation, and maintaining the cell homeostasis in every tissue and organ

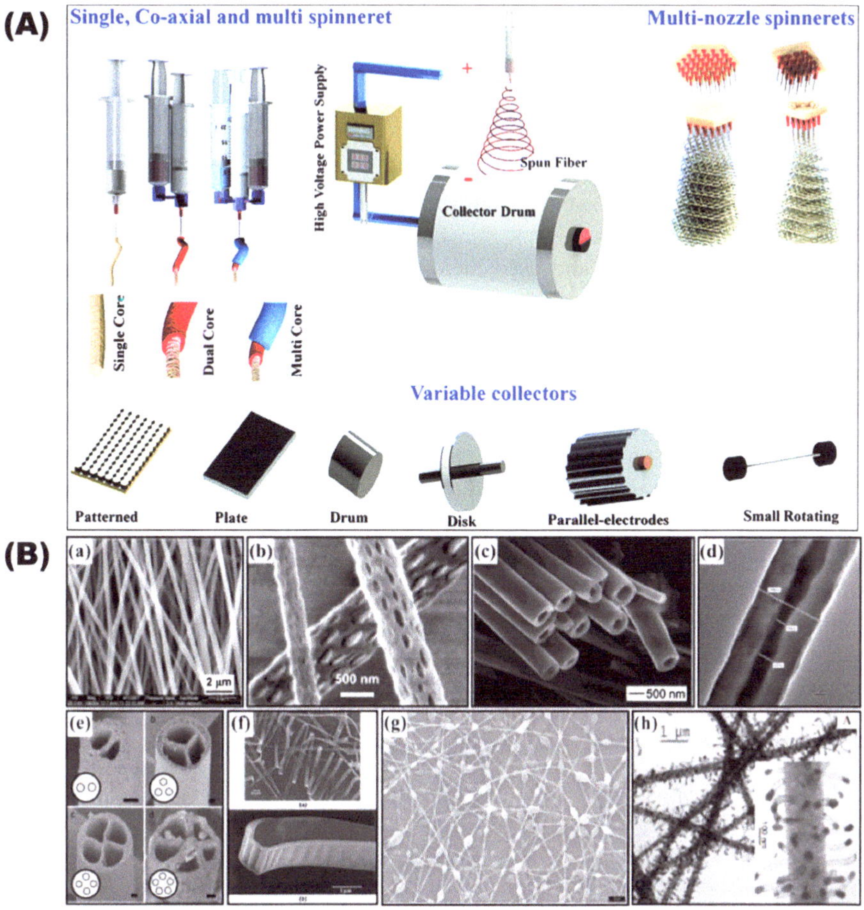

Fig. 2.1 (**A**) Schematic representation of the electrospinning setup with variable needles, collecting substrates, and syringes. (Reproduced with permission from (Aravindan et al., 2015) © 2015 The Royal Society of Chemistry). (**B**) Microscopic images of electrospun nanofibers with different morphologies: (a) cylindrical (b) porous (c) hollow (d) core–shell (e) multichannel nanotubes (f) nano-ribbons (g) bead-on-string and (h) carbon nanotube growth on electrospun carbon nanofibers. (Reproduced with permission from (Y. Ding et al., 2016) © 2016 Elsevier)

(Fernandes, Moroni, Van Blitterswijk, & De Boer, 2009; Hynes, 2009; Y. Kim, Ko, Kwon, & Shin, 2016; Kular, Basu, & Sharma, 2014; Kusindarta & Wihadmadyatami, 2018; Mackiewicz et al., 2016). The excellent characteristics of the electrospun nanofibers offer a great deal of potential in tissue regeneration applications by mimicking the ECM and facilitate the possibilities to control anchoring of cells, and their migration, proliferation and cell behavior. As the electrospun nanofibers mimic the size and structure of the native ECM, accountable attempts have been made using a wide array of materials with different compositions, orientations and dimensions to imitate the roles of native ECM in tissue regeneration (Baker, Handorf,

Ionescu, Li, & Mauck, 2009; Grafahrend, Heffels, Möller, Klee, & Groll, 2010; McLane, Schaub, Gilbert, & Ligon, 2013; Sell et al., 2007; Wu & Hong, 2016). Furthermore, the orientation of electrospun nanofibers is a crucial factor in regenerative medicine by influencing the cell adhesion, and directing the cell migration and proliferation (Fee, Surianarayanan, Downs, Zhou, & Berry, 2016; Gui, Hu, & Han, 2019; Huang, Hu, & Wei, 2016; Jahani et al., 2012; K. Wang et al., 2018). For instance, Nedjari et al. prepared electrospun hybrid nanofibers consisting of poly(L-lactide ε-caprolactone) and fibrinogen (PLCL/FBG) with random, aligned and 3D honeycomb architectures and used them to support osteogenic differentiation of human adipose derived mesenchymal stem cells (ADMSCs). The outcome of the study showed that the nanofibers with honeycomb architecture had synergistic effects of on the behavior of ADMSCs entering the path of osteogenic differentiation. Thus, this electrospun scaffold is expected to support the bone regeneration process through mimicking the conditions in the osteogenic stem cells niche (Nedjari, Awaja, & Altankov, 2017). Owing to the exceptional characteristics, and modification and functionalization options of electrospun nanofibers, their potential in regeneration of numerous tissues has been intensively explored. This chapter presents various electrospinning approaches demonstrated for dermal, neural, and orthopedic regeneration applications.

2.2 Designing Electrospun Nanofibers for Regenerative Medicine

2.2.1 Skin Regeneration

Skin is largest organ of the body that covers about 15% of the adult body weight and acts as a barrier between internal and external environment, protects the body against unfavorable factors and plays key roles in thermoregulation. In addition, the skin is considered as a sensory organ which also helps to maintain homeostasis. In anatomical perspective, skin consists of three layers namely epidermis, dermis, and hypodermis, which possess distinguished functionalities (Reinke & Sorg, 2012; Romanovsky, 2014; Vig et al., 2017; Yannas, Tzeranis, & So, 2017a, 2017b). The epidermis is the external layer and the dermis is the inner layer between the epidermis and hypodermis. The skin has the highest possibility of injury among all tissues owing to numerous factors. In general, the injured epidermis can be self-regenerated due to the presence of stem cells but this is not the same in the case of deep injuries as it requires clinical attention (Reinke & Sorg, 2012; Romanovsky, 2014; Vig et al., 2017; Yannas et al., 2017a, 2017b). Therefore, in the recent years, skin regenerative medicine has attracted significant attention in various aspects.

The electrospun nanofibrous scaffolds have been extensively explored for their potential in skin regeneration applications by using various materials and approaches (Norouzi, Boroujeni, Omidvarkordshouli, & Soleimani, 2015). One of the early studies demonstrated the preparation of ECM mimicking polycaprolactone (PCL) and collagen nanofiber matrices that induced enhanced proliferation of human dermal fibroblasts, which shows that it can be used as a dermal substitute for skin regeneration (Venugopal & Ramakrishna, 2005). Chitin nanofibrous (Chi-N) matrices prepared by electrospinning were tested for their biodegradability and their effect on behavior of normal human keratinocytes in comparison to commercial chitin microfibers (Chi-M, Beschitin W®). The electrospun chitosan nanofibrous matrices were found to degrade in 15 and 28 days in vitro and in vivo, respectively. The Chi-N demonstrated relatively high cell attachment and spreading of cells than the Chi-M, and the Chi-N treated with type I collagen significantly promoted the cellular responses showing its potential in enhancing skin regeneration (Noh et al., 2006). Zhou et al. developed an electrospun water-soluble carboxyethyl chitosan/poly (vinyl alcohol) (CECS/PVA) nanofibrous membrane for skin regeneration applications (Zhou et al., 2008), where the membrane was fabricated through blending various ratios of CECS/PVA solution and then electrospun to obtain nanofibrous mats and further cross-linked using glutaraldehyde. Afterwards, the potential of CECS/PVA electrospun fiber mats as scaffolding materials in skin regeneration was evaluated in vitro using mouse fibroblasts (L929) as a reference cell line and the nanofibrous mats were found to promote adhesion and proliferation of mouse fibroblasts (L929) cell line (Zhou et al., 2008).

Similarly, ultrafine fiber membranes of polyhydroxy butyrate (PHB) and organic-soluble chitosan (O-CS) was prepared by electrospinning a mixed PHB and O-CS solution. The cellular studies indicated that this nanofibrous membrane was beneficial in promoting the cell attachment and proliferation of mouse fibroblast cells (L929) which shows their potential for skin regeneration (G. Ma et al., 2010). In another approach, plasma treated electrospun silk-fibroin (SF) nanofibers were used for dermal regeneration applications. The SF nanofibers were prepared by electrospinning a regenerated SF solution and were further treated with plasma in the presence of oxygen or methane gas to modify their surface characteristics. After a detailed investigation of their surface characteristics, the effects of the nanofibers on both normal human epidermal keratinocytes (NHEK) and fibroblasts (NHEF) were studied. The biological tests of O_2 plasma-treated SF nanofibers showed a higher level of cell attachment on these nanofibers owing to their increase in hydrophilicity as a result of the O_2 plasma treatment which can be beneficial for skin regeneration application (Jeong et al., 2009). Yang et al. adopted emulsion electrospinning technique in order to obtain core–sheath nanofibers for encapsulating basic fibroblast growth factors (bFGF) in the core of the nanofiber and the shell was prepared using poly (ethylene glycol)-poly(DL-lactide) (PELA) (Y. Yang et al., 2011). The core–sheath bFGF/PELA fibers were found to gradually release the growth factor in about 4 weeks. The in vitro studies on mouse embryonic fibroblasts showed that bFGF-loaded fibrous mats enhanced the cell adhesion and proliferation, and secretion of extracellular matrix elements (ECM). Similarly, in vivo studies in rats with diabetic skin ulcers revealed a higher wound recovery rate with complete re-epithelialization and regeneration of skin appendages. The gradual release of bFGF from fibrous mats enhanced collagen deposition and ECM remodeling, and the arrangement and composition of the collagen fibers were similar to normal tissues. Likewise, chitosan nanofibrillar scaffolds prepared by electrospinning has been demonstrated to enhance regeneration of both epidermis and dermis compared to the evaporated films and freeze-dried sponges, when implanted in mice (Tchemtchoua et al., 2011).

In the recent years, similar efforts were made by using a variety of natural or synthetic materials and growth factors in regeneration of skin tissues using dermal and keratinocyte cells. For instance, a novel gellan–PVA nanofibrous scaffold has been prepared for skin tissue regeneration by Vashisth et al. by blending gellan with polyvinyl alcohol (PVA) (Vashisth et al., 2016). Furthermore, to enhance the stability, the nanofibers were cross-linked with heat at 150 °C for 30 min and then maintained at room temperature. For the purpose of comparison, gellan–PVA hydrogels and films were also prepared. Following detailed characterization, cytocompatibility studies using human dermal fibroblast (3T3L1) cells were performed. Phase contrast (Fig. 2.2A) and FESEM micrographs (Fig. 2.2B) of fibroblasts cultured on gellan–PVA nanofibers, hydrogels and films for a week revealed the adhesion of cells on prepared scaffolding materials. Figure 2.2C reveals the fluorescence microscopic images of fibroblast cells cultured on different gellan-based formulations. The overall results indicated that the nanofiber-based scaffold exhibited excellent cell adhesion, viability, and proliferation. Importantly, the nanofibrous scaffold was

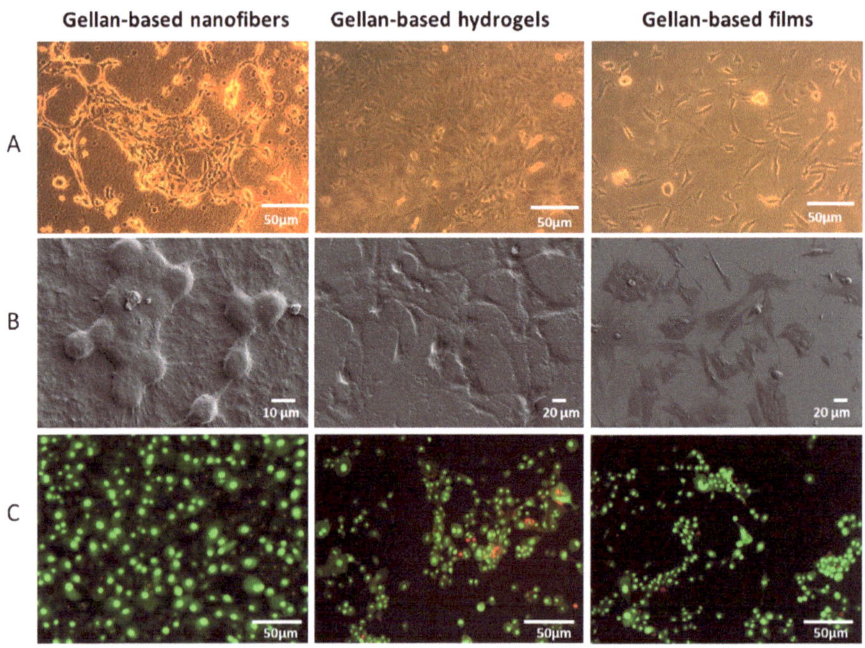

Fig. 2.2 (**A**) Phase contrast, (**B**) FESEM, (**C**) fluorescent micrographs of HDF cells after 1 week of cultivation on different gellan-based formulations. (Reproduced with permission from (Vashisth et al., 2016) © 2016 Elsevier)

biodegradable and could be potentially used as a temporary substrate/or biomedical graft to induce skin tissue regeneration.

Gandhimathi et al. (2014) prepared poly(L-lactic acid)-co-poly-(ε-caprolactone) (PLACL)/silk fibroin (SF)/vitamin E (VE)/curcumin (Cur) nanofibrous scaffolds and evaluated their potential in skin regeneration applications using human dermal fibroblasts cultures (Gandhimathi, Venugopal, Bhaarathy, Ramakrishna & Kumar 2014). Human dermal fibroblasts were cultured on the PLACL/SF/VE/Cur nanofibrous scaffolds and were evaluated for proliferation, cell morphology, F-actin expression, 5-chloromethylfluorescein diacetate (CMFDA) staining, and secretion of collagen in comparison to PLACL. The results of PLACL/SF/VE/Cur nanofibrous scaffolds in every parameter suggested that this scaffold has a promising potential to be used for skin tissue regeneration. Another approach which used cord blood-derived unrestricted somatic stem cells (USSCs) demonstrated that gelatin-modified poly(3-hydroxybutyrate-co-3-hydroxyvalerate) (PHBV) scaffolds can promote skin regeneration (Biazar & Keshel, 2014). For this study, the PHBV was dissolved in 2,2,2-trifluoroethanol (TFE) at a concentration of 2% w/v and electrospun into a nanofibrous mat, dried in the vacuum to remove the solvents and then cross-linked using gelatin. The USSCs were isolated from fresh umbilical cord blood, cultured and analyzed for their cell proliferation and viability. The USSCs were loaded on the nanofibrous scaffold and grafted on the damaged skin of the rats. The unmodified nanofibrous scaffolds and gelatin cross linked nanofibrous scaffolds without stem cells were used for comparison. The results of cytokeratin staining, histological assessment and epidermis thickness assessment were investigated for understanding the regeneration potential. On the post-operative day, gelatin cross linked scaffolds loaded with USSCs enhanced formation of an intact epithelium together with the development of new hair follicles and sebaceous glands and the collagen deposition, which were reminiscent of the structures of the natural skin. The results of immunostaining with cytokeratin and DAPI and also measuring the thickness of the epidermis layer showed that implanting the cell-laden scaffolds lead to the formation of a thicker epidermis layer due to the differentiation of stem cells into keratinocytes and fibroblasts. The overall findings confirm an excellent skin regeneration induced by gelatin cross-linked scaffold loaded with stem cells.

A highly porous nano-/microfibrous cotton-wool-like 3D scaffold with core–shell architecture was created by emulsion electrospinning, and its effects on skin tissue regeneration have been demonstrated by Pal et al. (2017). The 3D structure was produced by using a PCL-chitosan emulsion. Under the electric field, the emulsion containing encapsulated charged chitosan droplets enhance charge of the spinning solution and residual charge in the core of the deposited fiber, thereby creating core–shell, cotton-like fluffy structure with an average pore size of 62 μm, fiber diameter of ~1.62 μm, contact angle of 72° and 80% water uptake capacity. The emulsions were prepared by mixing of PCL and chitosan at ratios of 1:1 (P1C1), 2:1 (P2C1), 3:1 (P3C1), and 1:2 (P1C2) for 5 min and electrospun. Similarly, electrospinning was also carried out for 12 h by using stirred P2C1 emulsion and 10 wt% PCL. Figure 2.3 indicates the preparation of the emulsion solution and morphology of the scaffolds. The in vitro and in vivo studies demonstrated that the gradual

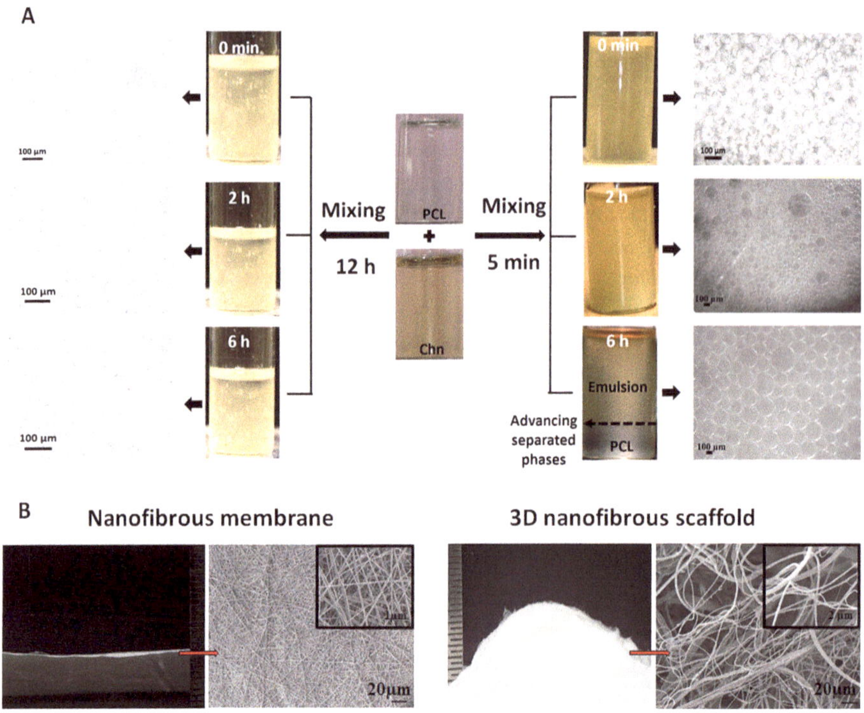

Fig. 2.3 (**A**) Microscopic observation of PCL, chitosan (Chn), and emulsion mix as in composition P2C1 at different time points after being kept under static conditions for 5 min and 12 h. (**B**) Structural evaluation of the sample prepared using P2C1 emulsion through side view optical imaging of as-spun sample and their scanning electron microscopy (SEM) images: cotton-wool-like 3D nano-/microfibrous scaffold prepared by 5 min of stirring and the compact nanofibrous membrane prepared by 12 h of stirring. (Reproduced with permission from (Pal et al., 2017) © 2017 American Chemical Society)

release of chitosan from the scaffold was beneficial in their performances. The 3D nano-/microfibrous scaffold was found to have efficient cellular infiltration and was capable of healing full-thickness excision wounds created in a rat model with accelerated healing within 3 weeks, which proved the efficiency of the scaffold as skin substitute.

Xu et al. developed a battery-operated e-spinning apparatus (BOEA) based on miniaturization and integration that liberates the researchers from conventional power supply and can be operated with a single hand owing to their small volume and light weight as can be seen from Fig. 2.4 (Xu et al., 2015). The apparatus was tested for its stable performance and real-time control capability by electrospinning different polymers including polyvinylpyrrolidone (PVP), PCL, PS, poly(lactic acid) (PLA) and poly(vinylidene fluoride) (PVDF). The study results indicated that the BOEA has a potential to be utilized mainly in biomedical applications including skin damage. Similarly, another technique reported the development of new prototype

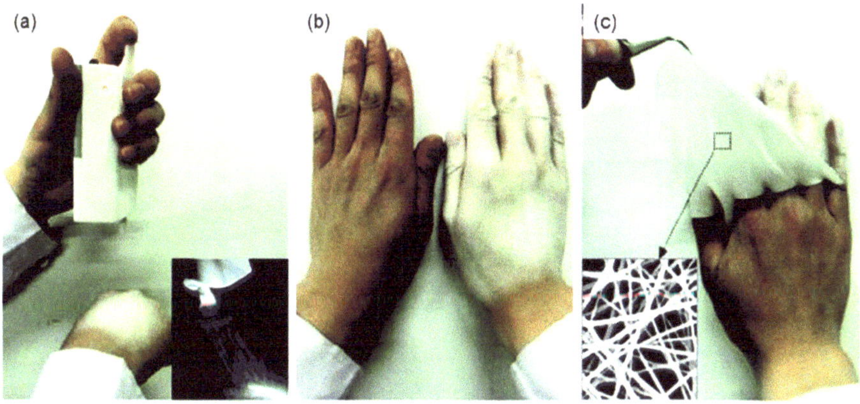

Fig. 2.4 Optical images showing the process of PLA fibers directly electrospun onto the skin using the BOEA in 2 min. (**a**) BOEA was operated by one hand and the inset shows the spinning process of the BOEA in a dark environment. (**b**) A PLA fibrous membrane was fabricated on another hand within 2 min. (**c**) The electrospun fibrous membrane has good flexibility and compactness. The inset is the SEM image of the electrospun fibers. (Reproduced with permission from (Xu et al., 2015) © 2015 The Royal Society of Chemistry)

with epidermal-like layers containing pseudo-rete ridge structures for studying the effect of topographical cues on cell behavior (Asencio et al., 2018). Briefly, stereolithography was used to produce photocurable pre-polymer polyethylene glycol diacrylate (PEGDA) templates using a blue laser beam focused into a DMD (digital multimirror device). Then, the template was fixed on a carbon tape using SEM on an electrospinning mandrel. The poly(3-hydrroxybutyrate-co-3-hydroxyvalerate) (PHBV) polymeric solution was electrospun into templates and the scaffold was used for studying the keratinocyte cell behavior. Overall fabrication process is illustrated in Fig. 2.5. The outcome of the cell behavior study indicated that the rete ridge-like electrospun membranes demonstrated excellent cell attachment, proliferation and migration. The metabolic assessment revealed that the keratinocytes were more viable in microfabricated scaffolds than the cells placed on the plain scaffolds thus indicating their potency in skin regeneration (Asencio et al., 2018).

Another effort was made by Sheikh et al. by using a facile and efficient strategy for the production of 3D silk fibroin nanofibers through cold-plate electrospinning that counteract limitations of the conventional and salt-leaching electrospinning (Sheikh et al., 2015). The developed 3D nanofibrous scaffolds possess higher porosity with controllable thickness which is more beneficial in overall yield and cell infiltration. This cold plate electrospinning technique can be directly applied for the fabrication of biomaterials with exact shapes and morphologies; in particular, such nanofibers are highly desirable for creating structures that resemble the dermis, nose, or ear or for repairing other facial defects. In the continuous progress of developing several strategies for fabricating electrospun nanofibers for regenerative medicine applications, cross-linking of nanofibers and plasma treatment also demonstrated a

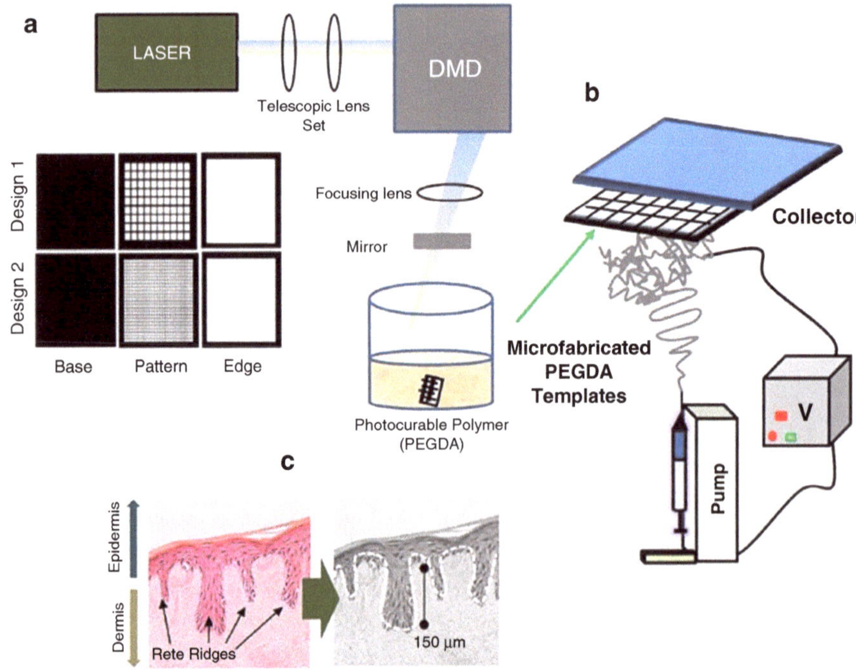

Fig. 2.5 Schematic of the manufacturing of the constructs. Panel (**a**) shows a schematic of the in-house developed microstereolithography setup in which a blue laser is focused into a digital multimirror device via the use of a telescopic lens set; the beam is later directed to a focusing lens followed by a mirror; a bath containing a photocurable polymer (PEGDA) is placed on a xyz stage. Panel (**a**) also shows a schematic of the individual projected layers for two types of microfeature. Panel (**b**) shows a schematic of the electrospinning process performed using the PEGDA templates; these templates are attached to a metallic base in order to create electrospinning collectors in which to spin a PHBV solution. Panel (**c**) shows a histology image of the native Rete Ridges in the skin; this specific image corresponds to a sample of tissue engineered skin produced in our laboratory and exemplifies the type of native topography that was aimed to be emulated in this work. (Reproduced from (Asencio et al., 2018))

significant improvement. As an example, Dias et al. demonstrated that electrospun gelatin nanofiber meshes can be cross-linked in situ using 4-butanediol diglycidyl ether (BDDGE) at different concentrations (2, 4, and 6 wt%) and incubation time-points (24, 48 and 72 h) at 37 °C. The physicochemical and biological studies revealed the well-defined morphology and mechanical properties of the cross-linked meshes, in addition to their induction of biocompatibility, cell adhesion, proliferation and new extracellular matrix synthesis, indicating their potential to be used for skin regeneration (Dias et al., 2017). Román-Doval et al. reported enhancing electrospun scaffolds of PVP with polypyrrole/iodine for tissue engineering of skin regeneration by coating via plasma process (Román-Doval et al., 2019). The polypyrrole/iodine coating of polyvinylpyrrolidone fibers enhanced viability, adhesion, and healing of HaCaT cells.

2.2.2 Neural Regeneration

The functional and anatomical recovery of the nervous system following injury to normalize the human life remains an important clinical challenge. Generally lower organisms have potential of neural regeneration, while the higher evolutionary organisms like humans comparatively lack regeneration capacity. Great strides in neural regeneration of peripheral and central nervous system have been made over the decades using various approaches and materials, but only limited treatments are available (López-Cebral, Silva-Corrcia, Reis, Silva, & Oliveira, 2017; So & Xu, 2015; Stoll, 2014). Consequently, electrospun nanofibers have also been explored in the neural regeneration research (Xie, MacEwan, Schwartz, & Xia, 2010). Apart from the general studies, the effect of alignment of nanofibrous scaffolds on the neural regeneration was also investigated extensively. For instance, Xie et al. reported neurite outgrowth on the nanofibrous scaffold with different orders, structures and surface characteristics by culturing primary DRGs (Xie, MacEwan, Li, Sakiyama-Elbert, & Xia, 2009). The outcome of this study suggests that the neurites were evenly distributed on the scaffolds consisting of random fibers and grew preferentially along the fibers on uniaxially aligned samples as depicted in Fig. 2.6. In the case of stacked nanofibers with double layer meshes, the neurites were guided to grow into complex patterns.

In another early demonstration, Yang et al. investigated the efficacy of aligned poly(L-lactic acid) (PLLA) nano/micro fibrous scaffolds for neural tissue engineering, and their performance was compared with random PLLA scaffolds (F. Yang, Murugan, Wang, & Ramakrishna, 2005). The PLLA fibrous scaffolds were named as aligned nanofibers (ANF), aligned micro fibers (AMF), random nanofibers (RNF) and random micro fibers (RMF), based on their alignment and fiber diameter. Further, the suitability of the scaffolds in neural tissue engineering was evaluated using neural stem cells (NSCs) and the cell morphology, differentiation and neurite outgrowth were studied by various microscopic techniques. The phase contrast light microscopy (PCLM) images confirmed that culturing for 1 day on different scaffolds resulted attachment of the NSCs on all the scaffolds and the cells changed their original round shape to elongated and spindle-like shapes indicating extensive neurite-like outgrowth. Furthermore, the direction of NSC elongation and neurite outgrowth was exactly parallel to the direction of fibers. Although no significant changes were observed on the cell attachment between the fibers with different diameters, the rate of NSC differentiation was higher for PLLA nanofibers than that of micro fibers, suggesting that the aligned nanofibrous PLLA scaffold could be used as a potential cell carrier in neural tissue engineering.

In the recent years, various efforts have been made towards fabricating aligned electrospun nanofibers and their application in neural regeneration. For example, a simple method for fabrication of electrospun fibers with controlled degrees of alignment having potential for nerve regeneration applications has been reported (Vimal, Ahamad, & Katti, 2016). In this study, an in-house developed electrospinning setup was used to fabricate electrospun fibers with controlled degrees of alignment

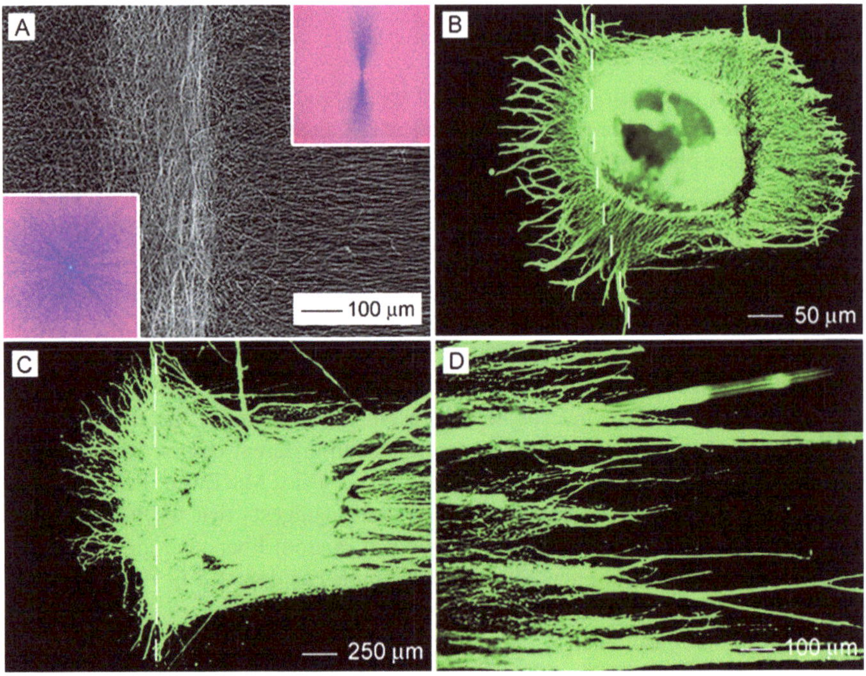

Fig. 2.6 (**A**) SEM image of disorder-to-aligned fiber mat. The Fourier fast transfer (FFT) patterns in the inset indicate that the fibers were aligned on one side and randomly oriented on the other side. (**B**) Typical morphology of DRGs cultured at the border between random and aligned PCL nanofibers (bare). (**C**) Typical morphology of DRGs cultured at the border between random and aligned PCL fibers coated with laminin. The dashed line indicates the borderline between aligned (right side) and randomly oriented (left side) fibers. (**D**) An enlarged view of (**C**). (Reproduced with permission from (Xie et al., 2009) © 2009 American Chemical Society)

including non-aligned (NA), moderately aligned (MA, 75%) to highly aligned (HA, 95%) submicron fibers, while keeping other physical properties unchanged. The modified electrospinning setup used for fabrication and the aligned fibers are depicted in Fig. 2.7i. For this study, human astrocytoma epithelial like cells (U373 cell line) were chosen due to their ability to exhibit sensitivity for topographical cues on a substrate and were seeded on the fibers exhibiting different alignments. The SEM images clearly indicated that the fibers were deposited with different degrees of alignment—from non-aligned (NA), moderately aligned (MA), to highly aligned fibers (HA) as can be seen from Fig. 2.7ii. Furthermore, cell viability and morphology were analyzed using SEM and fluorescence microscopy. The cell viability analysis indicated that the cells were able to proliferate irrespective of fiber alignment. The morphological and structural changes observed in cells cultured on fibrous surfaces using fluorescence microscopy indicated that the cells grown on NA fibers acquired elongated morphology at initial incubation for 12 h, however, this effect was lost when incubated for 24 h and also did not show any preferred orientation. On the contrary, cells grown on MA, and HA fibers aligned themselves

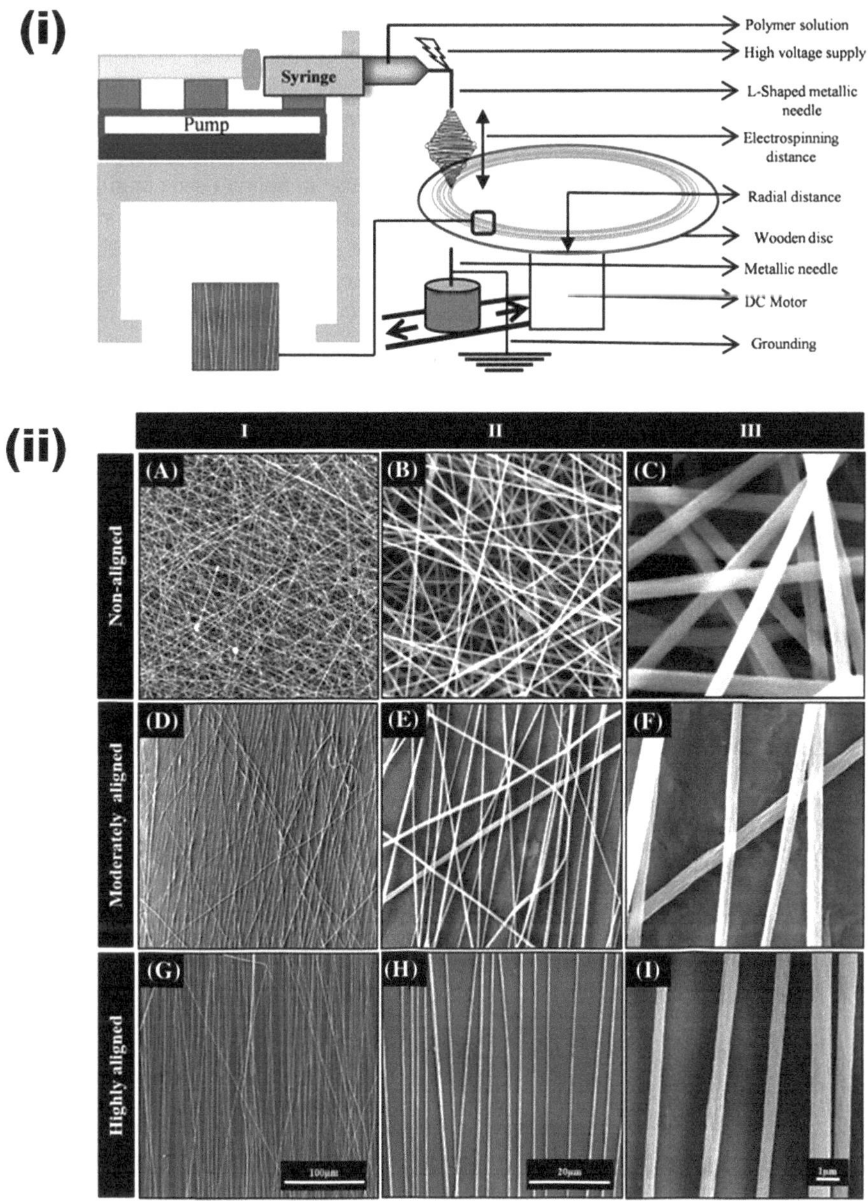

Fig. 2.7 (**i**) Schematic representation of the electrospinning setup used for the fabrication of PS submicron fibers with controlled alignment. (**ii**) Scanning electron micrographs of electrospun PS submicron fibers showing morphology at varying magnifications. Where A, D, and G are at 1000× (scale bar 100 μm); B, E, and H are at 5000× (scale bar 20 μm); and C, F, and I are at 30,000× (scale bar 1 μm). A–C are non-aligned (NA), D–F are moderately aligned (MA), and G–I are highly aligned (HA) submicron fibers. (Reproduced with permission from (Vimal et al., 2016) © 2016 Elsevier)

parallel to the direction of the alignment of fibers, which clearly demonstrated that the degree of alignment showed direct effect on the cell growth. Interestingly, the cells grown on HA fibers appeared to have greater elongation than that of cells grown on MA fibers at both 12 and 24 h of incubation. In summary, this study demonstrated a modified electrospinning setup to fabricate differentially aligned fibrous scaffolds with HA fibers showing their potential use in neural tissue engineering.

Similarly, Xue et al. also demonstrated that uniaxial alignment of the nanofibers not only promoted the differentiation of BMSCs into Schwann cells but also dictated the morphology and alignment of the differentiated cells (Xue et al., 2017). Briefly, four different types of scaffolds comprising random fibers with an average diameter of 488 ± 23 nm (RF500), aligned fibers with an average diameter of 521 ± 15 nm (AF500), aligned fibers with an average diameter of 986 ± 31 nm (AF1000), and aligned fibers with an average diameter of 1001 ± 24 nm, whose surface was coated with laminin (AF1000L) were prepared for possible application in the neural regeneration. The immunofluorescence images indicated that following 19 days of culture, the transdifferentiated BMSCs were observed to have a random distribution in terms of orientation when cultured on TCP and random fibers as shown in Fig. 2.8. In contrast, the cells differentiated on the aligned fibers showed a uniaxial alignment parallel to the fibers, indicating the presence of contact guidance provided by the underlying fibers. After deeper observation, AF1000L was observed to contain the largest number of cells among the different scaffolds. These results

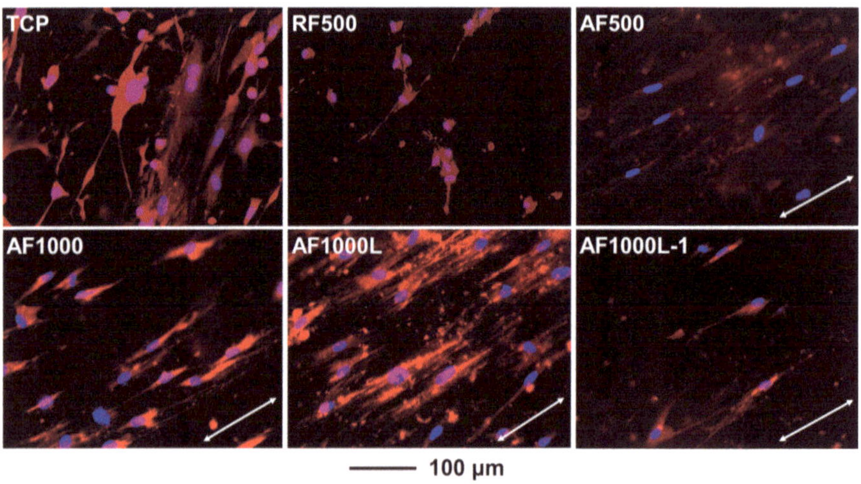

Fig. 2.8 Representative immunofluorescence micrographs of the cells transdifferentiated from BMSCs on different types of scaffolds after culturing for 19 days according to the in vitro chemical induction procedures. The BMSCs were seeded separately at a cell density of 250 cells/mm^2 on TCP, RF500, AF500, AF1000, and AF1000L, and at a density of 50 cells/mm^2 on AF1000L-1. S100 expressed from the differentiated cells was stained with anti-S100 (red) and the cell nuclei were stained with DAPI (blue). The double-headed arrows indicate the alignment directions of the underlying electrospun fibers. (Reproduced with permission from (Xue et al., 2017) © 2017 American Chemical Society)

show that the BMSCs could be successfully transdifferentiated into Schwann cells on both TCP and electrospun fibers, with the AF1000L scaffold giving the highest differentiation induction efficiency. Further to prove the function of the derived Schwann cells, PC12 cells and DRG were co-cultured separately for observing the induction capability of the cell differentiation. The overall results indicated that the neurite outgrowth on the aligned nanofibrous scaffold was significantly higher than the neurites grown on TCP under similar conditions.

As many studies demonstrated, nanofibers can guide the neurites to extend along the direction of alignment, mimicking the native hierarchy of the nerve tissue. The neurite outgrowth on electrospun nanofibers with uniaxial alignment was evaluated by considering the effects of fiber density, surface coating, and supporting substrate (Xie, Liu, Macewan, Bridgman, & Xia, 2014). This study was an effort to determine the effects of the contact cues provided by the nanofibers that can be far more complicated than just guiding the neurites to extend along them. The DRGs were used as a model system to systematically investigate the interactions between neurites and uniaxially aligned nanofibers. The investigation demonstrated for the first time that the neurites could not only project along the nanofibers, but also can be directed to grow along a direction perpendicular to the aligned nanofibers, depending on the following parameters: (1) the density of nanofibers, (2) the proteins deposited on the surfaces of the nanofibers, and (3) surface properties of the substrate on which the nanofibers were placed. As an initial experimental step, the fiber density on the neurite outgrowth was investigated by culturing DRGs on free-standing scaffolds of uniaxially aligned nanofibers that were prepared by collecting for 4 min and 15 min, respectively. When DRGs were cultured on the free-standing scaffolds with a low fiber density (collected for 4 min), the neurites tended to grow in parallel to the direction of fiber alignment (Fig. 2.9A). The trend remained the same on the low density fibers coated with laminin (Fig. 2.9B), while the DRGs tended to grow perpendicular to the fiber alignment in the case of high fiber density (collected for 15 min) as illustrated in Fig. 2.9C. Surprisingly, when cultured on scaffolds with a high fiber density (collected for 15 min) and laminin coating, the resultant neurite fields presented a pattern of parallel growth (Fig. 2.9D), similar to what was observed for the scaffolds with a low fiber density (regardless of laminin-coating).

Furthermore, the effects of support on the neurite growth were evaluated by immobilizing a set of nanofibers with different fiber densities on a bare glass coverslip. It was observed that the neurites extending from DRGs grew in parallel to the direction of fiber alignment when the fiber densities were relatively low (3 min), whereas the neurites tended to form bundles and grow perpendicular to the direction of fiber alignment when the fiber collection time was equal to or longer than 8 min. As a next attempt, the effects of coating were observed by coating the nanofibers with poly(L-lysine) (PLL), laminin and PEG. On the basis of the experimental results, two models were proposed to account for the outgrowth of DRG neurites on uniaxially aligned nanofibers. In the first model, uniaxially aligned nanofibers coated with laminin could interact strongly with the neurites through transmembrane receptors. The neurites could adhere to the fibers very well and leading to the formation of filopodia and lamellipodia that elongated along the long axis of the

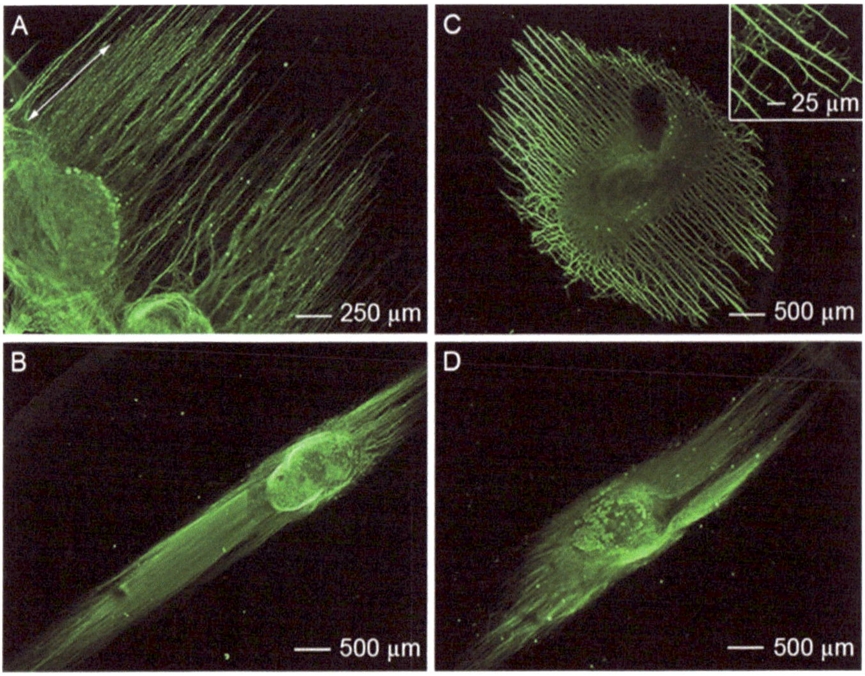

Fig. 2.9 Fluorescence micrographs showing the typical neurite fields of DRGs cultured on free-standing scaffolds of uniaxially aligned nanofibers that were prepared by collecting for (**A** and **B**) 4 min and (**C** and **D**) 15 min, respectively. The surfaces of the nanofibers in (**B**) and (**D**) were coated with laminin prior to DRG culture. The arrow in (**A**) indicates the direction of fiber alignment and it applies to all other samples. The inset in (**C**) shows a blow-up of the splitting tips. All the samples were stained with anti-neurofilament 200. (Reproduced from (Xie et al., 2014) © 2014 American Chemical Society)

uniaxially aligned fibers. By depositing laminin on the surfaces of the nanofibers with a relatively high density, the projection of the neurites would switch from perpendicular to parallel outgrowth. In the second model, the interaction between the neurites was much stronger than the interaction between neurites and fibers. The neurites tended to form bundles or fascicules due to their poor adhesion to the pristine fibers. As the fibers were deposited on PEG-coated coverslips, the neurites switched to perpendicular outgrowth. Due to the repelling effect of the PEG, the neurites kept exploring the microenvironment around them, protruding, and retracting until they found the right direction to pursue.

As next stage of advancements in fabricating electrospun aligned nanofibrous scaffold for neural regeneration, several methodologies were recently developed. To evident for the purpose of peripheral nerve regeneration, aligned polycaprolactone (PCL) scaffolds were prepared by combining use of electrospinning and micropatterning template and their potential in inducing the Schwann cells growth was evaluated (Zhang et al., 2018). The obtained fibers were analyzed for their morphological behavior, wettability, stability, chemical structure and crystalline phases.

The outcome of the study indicated that the micropatterned PCL scaffolds demonstrated a porous micro/nano complex structure with enhanced hydrophobicity and mechanical properties and possess good stability, thus can effectively regulate the attachment and orientation of Schwann cells at the early stages after cell culture. This study is considered as a first report of combining use of electrospinning and micromolding methods for preparing artificial nerve implants, and the technology is anticipated to have potential applications in peripheral nerve regeneration.

In another study, Zhu et al. prepared highly aligned nanocomposite scaffolds for neural tissue regeneration by combining electrospinning and electrospraying techniques (Zhu, Masood, O'Brien, & Zhang, 2015). Briefly, a novel tissue engineered scaffold, which possesses highly aligned PCL microfibrous framework and adjustable bioactive factor embedded poly(DL-lactide-co-glycolide) (PLGA) core–shell nanospheres was fabricated by combining electrospinning and electrospraying techniques. Figure 2.10a schematically shows several important parameters for an ideal tissue engineered neural construct and electrospinning setup used for the fabrication of aligned (Fig. 2.10b) and random microfibers (Fig. 2.10c) and the schematic illustration of the coaxial electrospraying technique for producing PLGA nanospheres into PCL electrospun scaffolds (Fig. 2.10d). Briefly, the electrospun PCL fibrous scaffolds with aligned and random orientations were fabricated using electrospinning, whereas core–shell PLGA nanospheres encapsulating BSA were produced by coaxial electrospraying process. The nanospheres were collected on an electrically grounded PCL mat placed at 15 cm vertical distance to the needle tip. The obtained scaffold was subjected to surface topography, hydrophilicity, tensile strength, release profile, PC-12 cells proliferation, immunocytochemistry of PC-12 cells and astrocytes analysis. The overall outcome of the study indicated that the highly aligned scaffold increased the average length of the neurites and directed neurite extension along the fibers in both PC-12 and astrocyte cell lines, which indicates that the scaffold is promising for guiding neural tissue growth and regeneration.

In another approach, fabrication of 3D tubular nerve guides was reported using simulation of a two pole (2P) electrospinning system (Panahi-Joo et al., 2016). The scaffold structure was optimized through Taguchi statistical method and then morphology, crystallinity, tensile strength and protein adsorption capacity of these 2P highly aligned fibers were studied and further compared with semi-aligned and random fibers fabricated through conventional mandrel electrospinning. In vitro studies were conducted using PC12 and PC6 cells by measuring cell attachment, proliferation and migration. The results indicated that the 2P fibers can be potent stimulators for elongation of neurons and promote the growth of PC12 cells uniformly over the scaffold. In summary, the distribution, growth, elongation and migration of the cells were enhanced on uniform aligned fibers and higher adsorption was obtained in a protein adsorption assay. The results of this study indicated that the conduits fabricated using 2P electrospinning are highly suitable to be used for peripheral and even spinal nerve regeneration application.

3D aligned electrospun fibers has promising applications in a wide range of biomedical areas, since aligned nanofibers (3D AFs) offer a larger surface area for cell adhesion and growth than the 2D nanofibers. Therefore, Jin et al. reported a novel

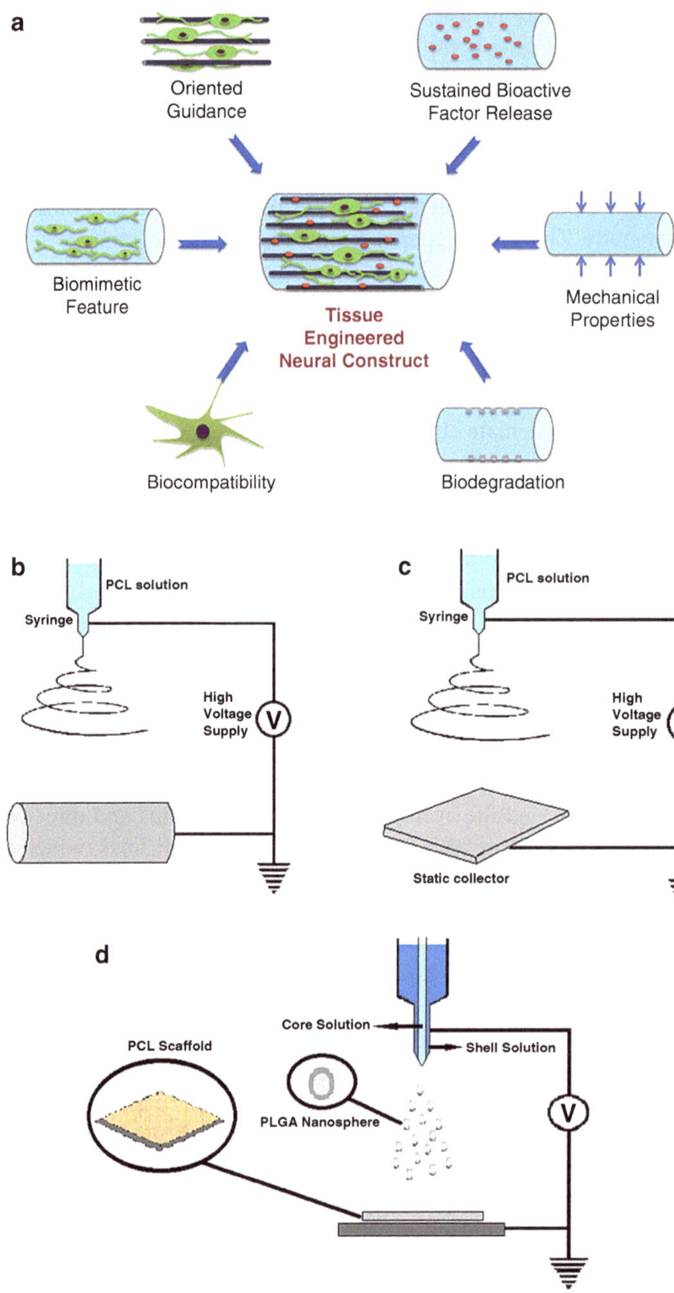

Fig. 2.10 Schematic illustration of (**a**) the key components of an ideal neural scaffold, and the electrospinning setup for (**b**) aligned and (**c**) random microfiber fabrication. (**d**) Schematic diagram of the coaxial electrospraying technique for producing PLGA nanospheres into PCL electrospun scaffolds. (Reproduced with permission from (Zhu et al., 2015) © 2015 Elsevier)

yet facile preparation process of 3D AFs by an improved electrospinning technique that can be used for neural regeneration (Jin et al., 2017). In a typical experimental step, the electrospun nanofibers are ejected toward a plate collector or a rotating mandrel that is covered with aluminum foil to obtain 2D structures with random or aligned arrangement. In the novel fabrication method that was proposed to achieve 3D aligned nanofibers with fluffy structures, the nanofibers were electrospun onto a special collector with a constantly rotating mandrel (500 rpm), half of which was immersed in an ethanol solution. After a certain period of time, the fluffy 3D aligned nanofibers were successfully fabricated through a three-step treatment procedure. In order to further evaluate the effects of obtained fibers on the neural regeneration, neurons were seeded into the 3D AFs and 2D AFs, which were used as a control. The confocal images obtained following 7 days of culture suggested that the neurons could attach tightly on the individual fibers with highly aligned orientations. Whereas, in the case of the neurons cultured on 2D AFs, they grew along multiple fibers instead of one individual fiber and the orientation was not well-defined. This might be ascribed from the tightly packed fibers on the 2D plane. These overall observations indicated that the neurons seeded on the fibers of the 3D AFs could interact with individual fibers, grow along an individual fiber, the neurites could grasp and pull fibers together because the 3D fibrous scaffold could mimic the in vivo cellular environment that leads for enhanced cell behavior.

Similarly, a controlled design of aligned and random nanofibers for 3D bi-functionalized nerve conduits were fabricated via a novel electrospinning setup (J. I. Kim, Hwang, Aguilar, Park, & Kim, 2016). Using this approach, a mat with both aligned and randomly oriented nanofibers made from a co-polymer of poly(lactic-co-glycolic acid) (PLGA) and polyurethane (PU) has been prepared in one step as schematically depicted in Fig. 2.11. In this method, copper wires were set up horizontally on the collector and cellophane tapes were set over the fixed copper wires in the vertical and horizontal axis (Fig. 2.11a, b). The direction of the nanofiber alignment was influenced by the way the cellophane tape was attached on to the copper wires. A horizontal cellophane tape attachment yields a vertical nanofiber alignment, while a vertical cellophane tape yields a horizontal nanofiber alignment. After optimization trials, the ideal rotation speed for the collector was found to be 1000 rpm. The nerve guide conduits (NGCs) were manufactured by rolling up the highly aligned nanofibrous mats fabricated in an angled U-shape, as seen in Fig. 2.11d–f. The electrospinning set up was capable of producing a single mat with multiple angled U-shaped nanofibrous mats with a diameter of 1–7 mm in one session and also could produce more than 50 nerve guide conduits at once, which indicates the mass production capability of the modified electrospinning device. The obtained scaffolds were subjected FFT analysis, SEM, transparency, porosity, mechanical, wettability electric field analysis.

Each one of the angled U-shaped patterns was used as the final layout for the neural guidance conduit. In order to create an neural guidance conduit with an aligned nanofiber interior, the mat was rolled from part A to part B, as seen on Fig. 2.12a and the actual protocol of neural guidance conduit formation is shown in Fig. 2.12b, c. This protocol of neural guidance conduit fabrication ensures that the

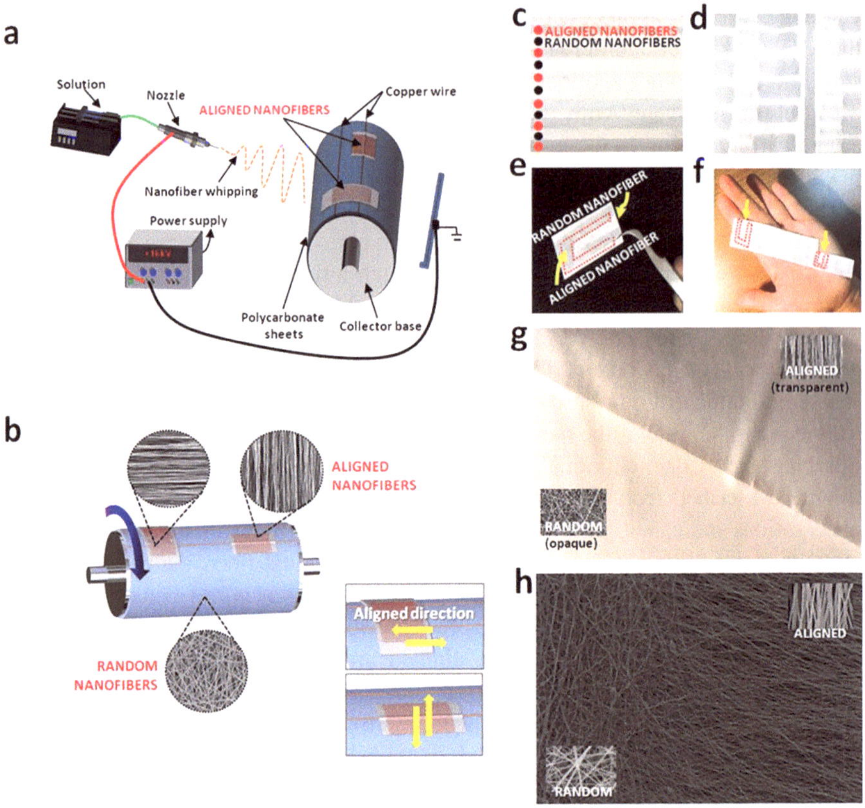

Fig. 2.11 (**a**) Schematic illustration of fabrication for the controlled design of aligned and randomly oriented nanofibers. (**b**) Schematic illustration of the direction of the nanofiber alignment influenced by the cellophane tape attachment. (**c**) Macrograph of the mat with randomly oriented nanofibers with aligned nanofibers at regular intervals. (**d**) Single mat with multiple angled U shape aligned nanofibrous mat. (**e**) Single angled U shape aligned nanofibrous mat. (**f**) Different sized angled U shape aligned nanofibrous mat. (**g**) Macrograph of the border of aligned and random nanofibrous mat. (**h**) The SEM image of the border of aligned and random nanofibrous mat. (Reproduced from (J. I. Kim et al., 2016) © 2016 Nature Publishing Group)

interior part with aligned nanofibers support the nerve cells proliferation in line with the orientation of the nanofibers and the exterior part with random nanofiber orientation increases the mechanical properties of the neural guidance conduit and further reduce the risk of fibrous tissue infiltration inside the conduit. The effects of the scaffold on adhesion and proliferation of cells were evaluated through culturing PC12 cells and S42 cells on PLGA and PU nanofibrous mats with aligned and random orientations. Each scaffold with cells were cultured for 1, 3 and 5 days and in vitro biocompatibility evaluations were performed by using CCK-8 assay, SEM and confocal laser scanning microscopy. The overall results indicated a superiority of the aligned nanofibrous scaffold to randomly oriented nanofibrous scaffolds as a substrate for tissue engineering that offer selective permeability, hydrophilicity and directional steering of nerve growth.

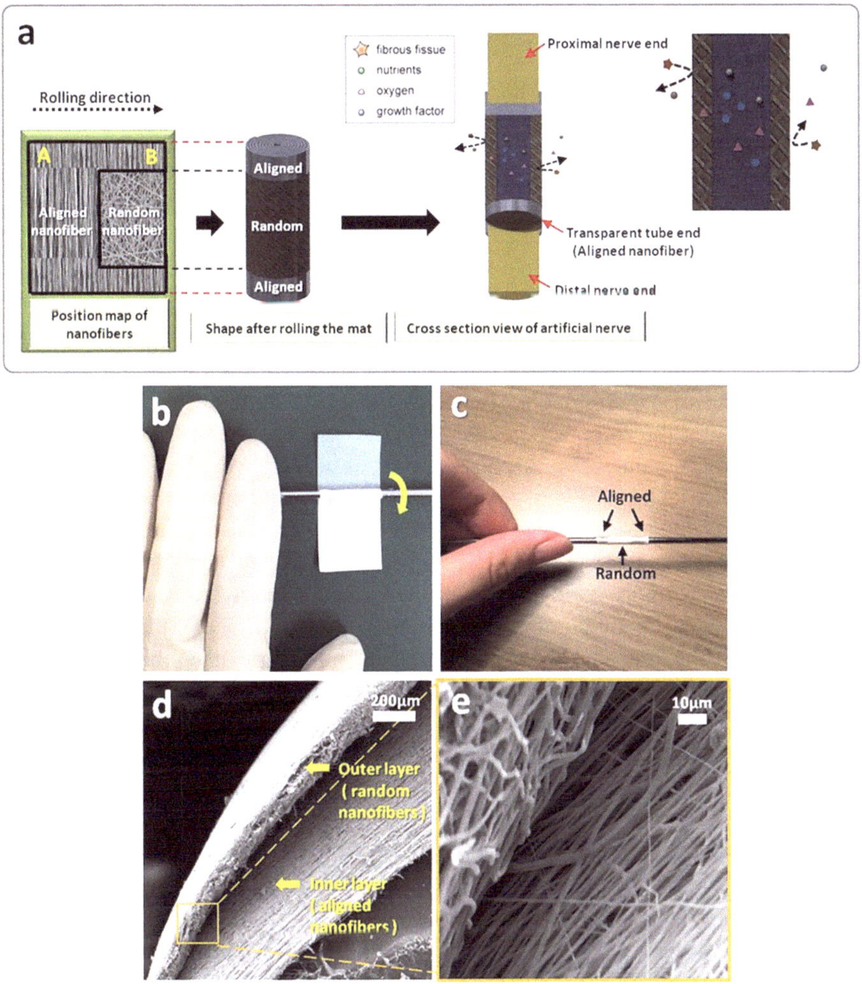

Fig. 2.12 (**a**) Schematic 3D illustration of the nerve guidance conduit (the position map of the nanofiber, shape after rolling the mat and cross section view of the conduit is shown in this figure). (**b**, **c**) SEM images of the cross section view of the nerve tube. (**d**) A digital photo of rolling the angled U shape mat on a rod. (**e**) A digital photo of the nerve guide conduit after rolling. (Reproduced from (J. I. Kim et al., 2016) © 2016 Nature Publishing Group)

2.2.3 *Orthopedic Tissue Regeneration*

Orthopedic regenerative medicine holds promising features for healing injured or diseased musculoskeletal tissues. Although several surgical procedures are available for the treatment of musculoskeletal injuries, such practices suffer from several limitations (Corsi, Schwarz, Mooney, & Huard, 2007; Zamborsky, Kilian, Csobonyeiova, & Danisovic, 2018). Various musculoskeletal tissues including bone, cartilage, tendon, muscle, ligament, tendon-to-bone insertion and meniscus possess fibrous organization. Indisputably, the electrospun nanofibers embraces promising applications in orthopedic regeneration as the electrospinning can produce nanofiber assemblies mimicking the structural organization of musculoskeletal tissues (Liu et al., 2012; B. Ma, Xie, Jiang, Shuler, & Bartlett, 2013; Narayanan et al., 2016). To date, electrospun nanofibers have been extensively explored for their tremendous potential in regeneration of musculoskeletal tissues including bone, tendon and cartilage.

2.2.3.1 Bone Regeneration

The designing of nanofibrous scaffolds for bone regeneration requires complete understanding of the composition and structural organization of the bone tissue as the bone consists of an inorganic phase that is constituted by hydroxyapatite and an organic phase which is made of mainly glycoproteins and proteoglycans. The collagens connect the nearby fibers and hydroxyapatite crystals that arrange them as layers within each fiber. Thus, the designed electrospun nanofibrous scaffolds should possess a highly porous nature, growth factors and hydroxyapatite/collagen or mimicking agents. Taking this requirements in account, one of the early studies demonstrated the preparation of biocomposite polymeric nanofibers containing nanohydroxyapatite by electrospinning and evaluated their potency in bone regeneration (Prabhakaran, Venugopal, & Ramakrishna, 2009). Poly-L-lactide/hydroxyapatite (PLLA/HA) and PLLA/collagen/hydroxyapatite scaffold were fabricated by electrospinning and tested biologically. The outcome of the biocompatibility studies with osteoblasts revealed that PLLA/collagen/hydroxyapatite nanofibers exhibit a synergistic effect of the collagen and hydroxyapatite in nanofibers and provide cell recognition sites for excellent cell proliferation and osteoconduction which is necessary for mineralization and bone formation.

Similarly, cellulose based nanofibers coated with hydroxyapatite were fabricated for tissue regeneration applications. The cellulose acetate nanofibers were deacetylated in order to produce cellulose nanofibers that further converted to sodium carboxymethyl cellulose and then coated with hydroxyapatite. The cell proliferation of osteoblastic MC3T3-E1 cells on hydroxyapatite coated carboxy methyl cellulose nanofibers confirmed its suitability for bone regeneration (Yamaguchi et al., 2016). In addition to the simple coating approaches, there have been numerous strategies adopted in fabrication of electrospun nanofibers aiming bone regeneration. For instance, basic fibroblast growth factor (bFGF)-encapsulated PCL nano/microfibrous scaffolds were fabricated by combining emulsion electrospinning and melt-electrospinning, which was designated as hybrid electrospinning, for improving cell viability and tissue regeneration (Park et al., 2015). The in vitro and in vivo evaluations of the bFGF-encapsulated nano/microfibrous scaffolds revealed that the scaffolds were capable of generating new bone at the defect site.

Recent studies demonstrated more approaches in fabricating electrospun nanofibers for bone regeneration. Wang et al. fabricated dual functional electrospun core–shell nanofibers for anti-infection bone regeneration membranes (Y. Wang, Jiang, et al., 2019). The purpose of this approach was to not only enhance the bone induction but also prevent bacterial growth. On a brief note, metronidazole (MNA) and nanohydroxyapatite (nHA) loaded core–shell nanofibers were prepared via coaxial electrospinning. The shell of the nanofiber was composed of PCL and nHA, whilst the core contained gelatin and MNA. The osteogenic measurements of the fabricated core–shell nanofibers showed that these nanofibers enhanced bone formation and the anti-bacterial experiments confirmed preventing colonization of anaerobic bacteria. Overall, this approach offers enhanced osteogenesis and anti-infection protection for optimal clinical application using guided bone regeneration membranes.

Another report demonstrated the preparation of mineralized nanofiber segments coupled with calcium-binding bone morphogenetic protein 2 (BMP-2) peptides for alveolar bone regeneration (Boda et al., 2019). The electrospun nanofibers of PLGA-collagen-gelatin were mineralized in simulated body fluid and further cryo-cut into segments and then loaded with various amounts of heptaglutamate E7-domain-conjugated BMP-2 peptide. Mineralized short fiber grafts with and without E7-BMP-2 peptides were implanted. A sustained release profile of E7-BMP-2 from the mineralized nanofiber segments, greater new bone volume and bone mineral density and histopathology findings indicated that the BMP2 peptide-loaded nanofibers can be an effective treatment option for alveolar bone loss and defects. Figure 2.13 illustrates the schematic for application of mineralized nanofiber segments loaded with the E7-BMP-2 peptide for healing critical-sized in defects in rat maxillae.

In continuation of developing different electrospinning based approaches for bone regeneration, ultralight 3D hybrid nanofiber aerogels composed of electrospun PLGA-collagen-gelatin and Sr–Cu co-doped bioactive glass fibers with incorporation of heptaglutamate E7 domain specific BMP-2 peptides were developed and evaluated for their potential in cranial bone defect healing (Weng et al., 2018). Wang et al. developed mesoporous silicate nanoparticle (MSN)-based electrospun PCL/gelatin nanofibrous scaffold for dual delivery of alendronate (ALN) and silicate aiming for modulating bone remodeling where the ALN inhibited the bone-resorbing process through preventing guanosine triphosphate-related protein expression and silicate promoted the bone-forming process by improving vascularization and bone calcification (Y. Wang, Cui, et al., 2019).

2.2.3.2 Tendon Regeneration

Tendons are the complex structures with dense connective tissues that attach muscle to bone and are capable of transferring muscle force to skeleton. The microstructural compositions of tendons are highly organized by the ECM (80%) and cellular collagen type I (MacLean, Khan, Malik, Snow, & Anand, 2012; Paredes & Andarawis-Puri, 2016; G. Yang, Rothrauff, & Tuan, 2013). Designing scaffolds for tendon regeneration requires good mechanical strength, shape and good

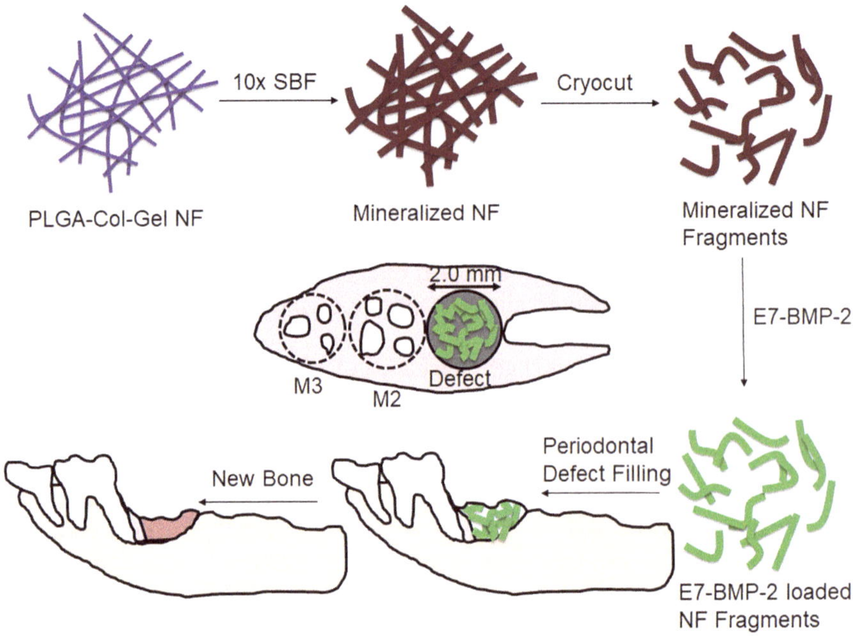

Fig. 2.13 Schematic illustrating the application of mineralized nanofiber segments immobilized with calcium coupling of the E7-BMP-2 peptide for periodontal bone regeneration in maxillary defects (2 mm diameter × 2 mm depth) created following the extraction of the first molar tooth (M1). (Reproduced with permission from (Boda et al., 2019) © 2019 Elsevier)

degradation behavior. As the tendon tissues possess aligned fibrous ECM architecture, the electrospun nanofibers have great stride in tendon regeneration applications. Manning et al. developed a scaffold capable of delivering growth factors and cells in a surgically manageable form for tendon repair (Manning et al., 2013). Briefly, platelet-derived growth factor BB (PDGF-BB) and adipose-derived mesenchymal stem cells (ASCs) were first incorporated into a heparin/fibrin-based delivery system (HBDS) and then the hydrogel was layered on a PLGA electrospun nanofiber. Figure 2.14 shows the HBDS/nanofiber scaffold consisted of 11 alternating layers of aligned electrospun PLGA nanofiber mats and HBDS (i.e., six layers of PLGA and five layers of fibrin). The in vitro studies demonstrated the sustained release of growth factors and in vivo studies in large animals confirmed the developed tendon model was clinically relevant and the cells remained viable. Therefore, this novel layered scaffold was proposed to offer potential tendon healing abilities due to the delivery of both cells and growth factors simultaneously.

Lipner et al. reported improved tendon-to-bone healing by promotion of aligned collagen deposition and increased bone formation using a biomimetic scaffold seeded with pluripotent cells (Lipner et al., 2015). In another study, a tendon construct was fabricated using electrospun aligned PLLA nanofibers to mimic the aligned collagen fiber bundles and layering PLLA fibers with chitosan-collagen hydrogels to mimic the glycosaminoglycans for tendon regeneration (Deepthi, Nivedhitha Sundaram, Deepti

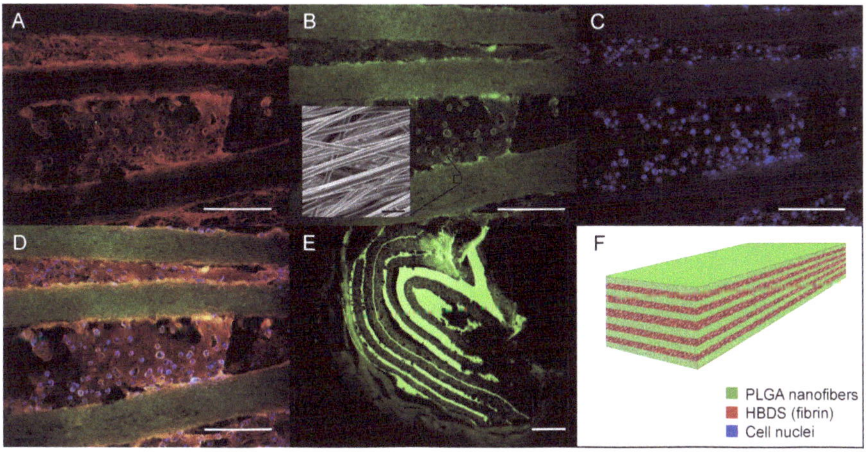

Fig. 2.14 A representative HBDS/nanofiber scaffold with 11 alternating layers of aligned electrospun PLGA nanofiber mats separated by HBDS containing 1 × 10^6 ASCs is shown. (**A–D**) Micrograph showing the HBDS/nanofiber scaffold in vitro; the PLGA was labelled with FITC (green), the HBDS was labeled with Alexa Fluor 546 (red) and the ASC nuclei were labeled with Hoechst 33258 (blue) (scale bar = 200 μm). (**B**, inset) SEM image of the scaffold showing PLGA nanofiber alignment. (**E**) Micrograph showing the HBDS/nanofiber scaffold in vivo 9 days after implantation in a tendon repair. Eleven alternating layers of PLGA and HBDS can be seen (i.e., six layers of PLGA and five layers of fibrin); the PLGA was labelled with FITC (green) (scale bar = 100 μm). (**F**) A schematic of the layered scaffold is shown. (Reproduced with permission from (Manning et al., 2013) © 2013 Elsevier)

Kadavan, & Jayakumar, 2016). Based on the biological studies, the alginate gel coated chitosan-collagen/PLLA scaffolds induced excellent cell proliferation thus indicating the scaffold might be ideal for flexor tendon regeneration.

In another study, a layered electrospun and woven surgical scaffold to enhance endogenous tendon repair was reported by Hakimi et al. (2015). In this study, a bonding technique that enables the processing of electrospun sheets into multilayered, robust, implantable fabrics was developed resulting in a prototype scaffold where an electrospun sheet was reinforced with a woven layer as depicted in Fig. 2.15. The resulting scaffold holds a maximum suture pull out strength of 167 N, which closely matched with human rotator cuff tendons, and the desired nanofiber-mediated bioactivity in vitro and in vivo. Overall, the approach combining a woven medical fabric with an aligned electrospun layer created a promising surgically implantable scaffold with potential to enhance the endogenous repair of rotator cuff tendon tears.

2.2.3.3 Cartilage Regeneration

Cartilage is a critical tissue that does not possess nerves, blood vessels and lymphatics. The cartilage tissues can be easily damaged with many factors including aging, trauma and congenital issues. Therefore, to date significant efforts have been made

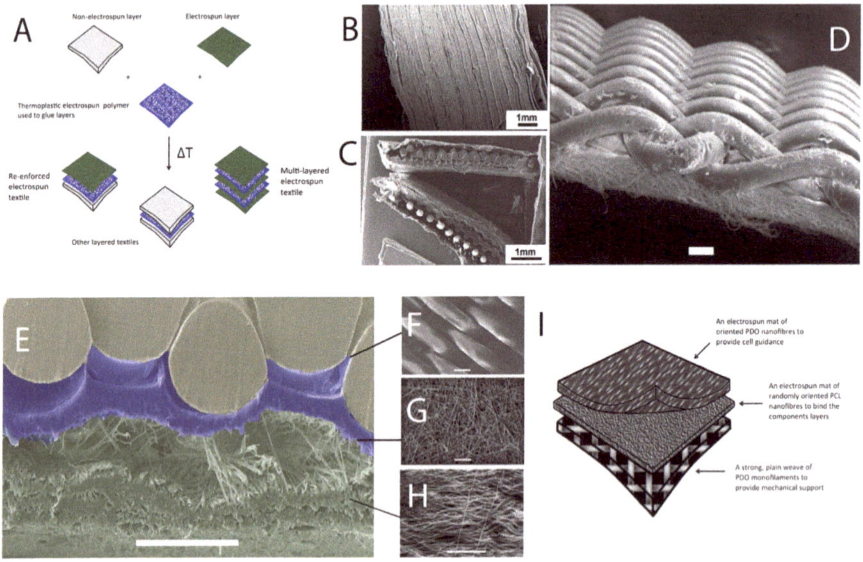

Fig. 2.15 Creating multilayered scaffolds from electrospun and non-electrospun mats and the architecture of the prototype layered electrospun/woven scaffold. (**A**) The technique used to stack and bond electrospun and non-electrospun layers using a thermoplastic mat. Resulting constructs can be adjusted to have the appropriate thickness, texture, and tensile properties depending on the application. (**B**) and (**C**) are SEM images of different prototypes that can be assembled using this method: (**B**) Multilayered electrospun sheets; and (**C**) a woven polydioxanone textile sandwiched between two electrospun mats. (**D**) A schematic diagram of the prototype scaffold designed and tested in this study. (**E**) A cross-sectional view of the layered scaffold. (**F**) The woven layer. (**G**) The random electrospun PCL thermoplastic adhesive layer. (**H**) The aligned PDO electrospun mat, which is the tendon-facing layer of the scaffold. (**I**) A schematic description of the design rational behind the layered scaffold. Unless specified scale bars are 100 μm. (Reproduced with permission from (Hakimi et al., 2015) © 2015 Elsevier)

in the area of cartilage regeneration using several strategies (Nam, Rim, Lee, & Ju, 2018; Pak et al., 2018; Tiku & Sabaawy, 2015). Owing to the exceptional characteristics of electrospun nanofibers, accountable attempts using these nanofibers were also made on cartilage regeneration. As an example, Thorvaldsson et al. used electrospinning technique to coat microfibers with nanofibers. The nanofiber coated microfibers were expected to have benefits of nanostructured surface morphology for cellular growth and tailored porosity for infiltration. The biological studies conducted by seeding human chondrocytes on these scaffolds indicated their potential in cartilage regeneration (Thorvaldsson, Stenhamre, Gatenholm, & Walkenström, 2008). Shafiee et al. used poly(vinyl alcohol)/PCL (PVA/PCL) nanofiber scaffolds seeded with rabbit bone marrow-mesenchymal stem cells (BM-MSC) for cartilage tissue regeneration (Shafiee et al., 2011). The study results indicated that that PVA/PCL scaffolds supported the proliferation and chondrogenic differentiation of MSCs and the in vivo studies showed good healing of defects. The electrospun nanofibers are often immobilized with chondroitin sulfate and hyaluronic acid for cartilage regeneration (Lee et al., 2014; Piai et al., 2017). In another approach, a hydrogel–fiber composite scaffold was used to encapsulate chondrocytes and used in cartilage tissue regeneration (Mohabatpour, Karkhaneh, & Sharifi, 2016).

2.3 Conclusion and Future Outlook

As the nanofibers produced by electrospinning offers excellent characteristics for cellular environment, significant efforts have been exploited in the development of nanofibers and nanofibrous mats for various tissue regeneration applications. In this chapter, we have summarized the overview of different approaches for functionalizing, various compositions, diverse electrospinning technologies and fiber alignment methods improving the regeneration of dermal, neural, and orthopedic tissues. The exceptional characteristics of the electrospun nanofibers have brought numerous advancements in the field of regenerative medicine. Although electrospun nanofibers were extensively explored for their potential in regenerative medicine, several challenges exist to transform electrospun based scaffolds into clinical settings. To date, to the best of our knowledge, we are not aware of any electrospun nanofiber based products that have been approved by FDA to be used in clinical practice for regenerative medicine applications. Apparently, several in vivo studies have been conducted using numerous animal models for different tissue regeneration applications; however, their performance may vary in clinical trial phases thus resulting failure of transforming the scaffolds to clinical practice. Therefore, a concentrated effort is needed in designing of electrospun nanofibrous scaffold architecture and their composition to mimic the spatial distribution of native tissues to enhance the regeneration of different tissues and meet the remaining clinical challenges. Furthermore, scaling up of such developed scaffolds with similar performance and high reproducibility need to be optimized. Overall, the future efforts should be directed in order to transform the electrospun nanofibrous scaffolds from laboratory bench to clinical practice.

References

Aravindan, V., Sundaramurthy, J., Suresh Kumar, P., Lee, Y. S., Ramakrishna, S., & Madhavi, S. (2015). Electrospun nanofibers: A prospective electro-active material for constructing high performance Li-ion batteries. *Chemical Communications, 51*, 2225–2234. https://doi.org/10.1039/c4cc07824a

Arora, P., Sindhu, A., Dilbaghi, N., Chaudhury, A., Rajakumar, G., & Rahuman, A. A. (2012). Nano-regenerative medicine towards clinical outcome of stem cell and tissue engineering in humans. *Journal of Cellular and Molecular Medicine, 16*(9), 1991–2000. https://doi.org/10.1111/j.1582-4934.2012.01534.x

Asencio, I. O., Mittar, S., Sherborne, C., Raza, A., Claeyssens, F., & MacNeil, S. (2018). A methodology for the production of microfabricated electrospun membranes for the creation of new skin regeneration models. *Journal of Tissue Engineering, 9*, 1–8. https://doi.org/10.1177/2041731418799851

Atala, A. (2012). Regenerative medicine strategies. *Journal of Pediatric Surgery, 47*, 17–28. https://doi.org/10.1016/j.jpedsurg.2011.10.013

Babitha, S., Rachita, L., Karthikeyan, K., Shoba, E., Janani, I., Poornima, B., & Purna Sai, K. (2017). Electrospun protein nanofibers in healthcare: A review. *International Journal of Pharmaceutics, 523*, 52–90. https://doi.org/10.1016/j.ijpharm.2017.03.013

Baker, B. M., Handorf, A. M., Ionescu, L. C., Li, W. J., & Mauck, R. L. (2009). New directions in nanofibrous scaffolds for soft tissue engineering and regeneration. *Expert Review of Medical Devices, 6*, 515–532. https://doi.org/10.1586/erd.09.39

Biazar, E., & Keshel, S. H. (2014). Unrestricted somatic stem cells loaded in nanofibrous scaffolds as potential candidate for skin regeneration. *International Journal of Polymeric Materials and Polymeric Biomaterials, 63*(14), 741–752. https://doi.org/10.1080/00914037.2013.879447

Boda, S. K., Almoshari, Y., Wang, H., Wang, X., Reinhardt, R. A., Duan, B., … Xie, J. (2019). Mineralized nanofiber segments coupled with calcium-binding BMP-2 peptides for alveolar bone regeneration. *Acta Biomaterialia, 85*, 282–293. https://doi.org/10.1016/j.actbio.2018.12.051

Colombo, F., Sampogna, G., Cocozza, G., Guraya, S., & Forgione, A. (2016). Regenerative medicine: Clinical applications and future perspectives. *Journal of Microscopy and Ultrastructure, 5*(1), 1. https://doi.org/10.1016/j.jmau.2016.05.002

Corsi, K. A., Schwarz, E. M., Mooney, D. J., & Huard, J. (2007). Regenerative medicine in orthopaedic surgery. *Journal of Orthopaedic Research, 25*, 1261–1268. https://doi.org/10.1002/jor.20432

Cui, H., Nowicki, M., Fisher, J. P., & Zhang, L. G. (2017). 3D bioprinting for organ regeneration. *Advanced Healthcare Materials, 6*, 1601118. https://doi.org/10.1002/adhm.201601118

Deepthi, S., Nivedhitha Sundaram, M., Deepti Kadavan, J., & Jayakumar, R. (2016). Layered chitosan-collagen hydrogel/aligned PLLA nanofiber construct for flexor tendon regeneration. *Carbohydrate Polymers, 153*, 492–500. https://doi.org/10.1016/j.carbpol.2016.07.124

Dias, J. R., Baptista-Silva, S., de Oliveira, C. M. T., Sousa, A., Oliveira, A. L., Bártolo, P. J., & Granja, P. L. (2017). In situ crosslinked electrospun gelatin nanofibers for skin regeneration. *European Polymer Journal, 95*, 161–173. https://doi.org/10.1016/j.eurpolymj.2017.08.015

Ding, B., Wang, X., & Yu, J. (2019). *Electrospinning: Nanofabrication and applications*. https://doi.org/10.1016/c2016-0-01374-8

Ding, Y., Hou, H., Zhao, Y., Zhu, Z., & Fong, H. (2016). Electrospun polyimide nanofibers and their applications. *Progress in Polymer Science, 61*, 67–103. https://doi.org/10.1016/j.progpolymsci.2016.06.006

Dzobo, K., Thomford, N. E., Senthebane, D. A., Shipanga, H., Rowe, A., Dandara, C., … Motaung, K. S. C. M. (2018). Advances in regenerative medicine and tissue engineering: Innovation and transformation of medicine. *Stem Cells International, 2018*, 2495848. https://doi.org/10.1155/2018/2495848

Fee, T., Surianarayanan, S., Downs, C., Zhou, Y., & Berry, J. (2016). Nanofiber alignment regulates NIH3T3 cell orientation and cytoskeletal gene expression on electrospun PCL+gelatin nanofibers. *PLoS One, 11*(5), e0154806. https://doi.org/10.1371/journal.pone.0154806

Fernandes, H., Moroni, L., Van Blitterswijk, C., & De Boer, J. (2009). Extracellular matrix and tissue engineering applications. *Journal of Materials Chemistry, 19*(31), 5474–5484. https://doi.org/10.1039/b822177d

Gandhimathi, C., Venugopal, J. R., Bhaarathy, V., & Ramakrishna, Kumar. (2014). Biocomposite nanofibrous strategies for the controlled release of biomolecules for skin tissue regeneration. *International Journal of Nanomedicine, 9*, 4709. https://doi.org/10.2147/ijn.s65335

Grafahrend, D., Heffels, K. H., Möller, M., Klee, D., & Groll, J. (2010). Electrospun, biofunctionalized fibers as tailored in vitro substrates for keratinocyte cell culture. *Macromolecular Bioscience, 10*(9), 1022–1027. https://doi.org/10.1002/mabi.201000068

Gu, W., Wu, C., Chen, J., & Xiao, Y. (2013). Nanotechnology in the targeted drug delivery for bone diseases and bone regeneration. *International Journal of Nanomedicine, 8*, 2305–2317. https://doi.org/10.2147/IJN.S44393

Gui, X., Hu, J., & Han, Y. (2019). Random and aligned electrospun gelatin nanofiber mats for human mesenchymal stem cells. *Materials Research Innovations, 23*(4), 208–215. https://doi.org/10.1080/14328917.2018.1428073

Hakimi, O., Mouthuy, P. A., Zargar, N., Lostis, E., Morrey, M., & Carr, A. (2015). A layered electrospun and woven surgical scaffold to enhance endogenous tendon repair. *Acta Biomaterialia, 26*, 124–135. https://doi.org/10.1016/j.actbio.2015.08.007

Huang, C. Y., Hu, K. H., & Wei, Z. H. (2016). Comparison of cell behavior on PVA/PVA-gelatin electrospun nanofibers with random and aligned configuration. *Scientific Reports, 6*, 37960. https://doi.org/10.1038/srep37960

Hynes, R. O. (2009). The extracellular matrix: Not just pretty fibrils. *Science, 326*, 1216–1219. https://doi.org/10.1126/science.1176009

Jahani, H., Kaviani, S., Hassanpour-Ezatti, M., Soleimani, M., Kaviani, Z., & Zonoubi, Z. (2012). The effect of aligned and random electrospun fibrous scaffolds on rat mesenchymal stem cell proliferation. *Cell Journal, 14*(1), 31–38.

Jeong, L., Yeo, I. S., Kim, H. N., Yoon, Y. I., Jang, D. H., Jung, S. Y., … Park, W. H. (2009). Plasma-treated silk fibroin nanofibers for skin regeneration. *International Journal of Biological Macromolecules, 44*(3), 222–228. https://doi.org/10.1016/j.ijbiomac.2008.12.008

Jin, L., Xu, Q., Li, C., Huang, J., Zhang, Y., Wu, D., & Wang, Z. (2017). Engineering 3D aligned nanofibers for regulation of cell growth behavior. *Macromolecular Materials and Engineering, 302*(4), 1600448. https://doi.org/10.1002/mame.201600448

Kim, E. S., Ahn, E. H., Dvir, T., & Kim, D. H. (2014). Emerging nanotechnology approaches in tissue engineering and regenerative medicine. *International Journal of Nanomedicine, 9*, 1–5. https://doi.org/10.2147/IJN.S61212

Kim, J. I., Hwang, T. I., Aguilar, L. E., Park, C. H., & Kim, C. S. (2016). A controlled design of aligned and random nanofibers for 3D bi-functionalized nerve conduits fabricated via a novel electrospinning set-up. *Scientific Reports, 6*, 23761. https://doi.org/10.1038/srep23761

Kim, Y., Ko, H., Kwon, I. K., & Shin, K. (2016). Extracellular matrix revisited: Roles in tissue engineering. *International Neurourology Journal, 20*, S23–S29. https://doi.org/10.5213/inj.1632600.318

Kny, E., Ghosal, K., & Thomas, S. (2018). *Electrospinning: From basic research to commercialization* (Soft matter series) (Vol. 7). London: The Royal Society of Chemistry. https://doi.org/10.1039/9781788010580-FP007

Kular, J. K., Basu, S., & Sharma, R. I. (2014). The extracellular matrix: Structure, composition, age-related differences, tools for analysis and applications for tissue engineering. *Journal of Tissue Engineering, 5*, 2041731414557112. https://doi.org/10.1177/2041731414557112

Kumar, R., Griffin, M., & Butler, P. E. (2017). A review of current regenerative medicine strategies that utilize nanotechnology to treat cartilage damage. *The Open Orthopaedics Journal, 10*(1), 862–876. https://doi.org/10.2174/1874325001610010862

Kusindarta, D. L., & Wihadmadyatami, H. (2018). The role of extracellular matrix in tissue regeneration. In H. H. E.-S. Kaoud (Ed.), *Tissue regeneration*. Croatia: Intech. https://doi.org/10.5772/intechopen.75728

Lee, P., Tran, K., Chang, W., Shelke, N. B., Kumbar, S. G., & Yu, X. (2014). Influence of chondroitin sulfate and hyaluronic acid presence in nanofibers and its alignment on the bone marrow stromal cells: Cartilage regeneration. *Journal of Biomedical Nanotechnology, 10*(8), 1469–1479.

Li, Z., & Wang, C. (2013). *One dimensional nanostructures: Electrospinning technique and unique nanofibers*. Berlin: Springer. https://doi.org/10.1007/978-3-642-36427-3

Lin, T., & Fang, J. (2017). *Fundamentals of electrospinning & electrospun nanofibers*. Pennsylvania, PA: DEStech Publications Inc.

Lipner, J., Shen, H., Cavinatto, L., Liu, W., Havlioglu, N., Xia, Y., … Thomopoulos, S. (2015). In vivo evaluation of adipose-derived stromal cells delivered with a nanofiber scaffold for tendon-to-bone repair. *Tissue Engineering Part A, 21*(21–22), 2766–2774. https://doi.org/10.1089/ten.tea.2015.0101

Liu, W., Thomopoulos, S., & Xia, Y. (2012). Electrospun nanofibers for regenerative medicine. *Advanced Healthcare Materials, 1*(1), 10–25. https://doi.org/10.1002/adhm.201100021
López-Cebral, R., Silva-Correia, J., Reis, R. L., Silva, T. H., & Oliveira, J. M. (2017). Peripheral nerve injury: Current challenges, conventional treatment approaches, and new trends in biomaterials-based regenerative strategies. *ACS Biomaterials Science and Engineering, 3*, 3098–3122. https://doi.org/10.1021/acsbiomaterials.7b00655
Ma, B., Xie, J., Jiang, J., Shuler, F. D., & Bartlett, D. E. (2013). Rational design of nanofiber scaffolds for orthopedic tissue repair and regeneration. *Nanomedicine, 8*, 1459–1481. https://doi.org/10.2217/nnm.13.132
Ma, G., Yang, D., Wang, K., Han, J., Ding, S., Song, G., & Nie, J. (2010). Organic-soluble chitosan/polyhydroxybutyrate ultrafine fibers as skin regeneration prepared by electrospinning. *Journal of Applied Polymer Science, 118*(6), 3619–3624. https://doi.org/10.1002/app.32671
Mackiewicz, Z., Konttinen, Y., Kaivosoja, E., Stegajev, V., Wagner, H., Levón, J., & Tiainen, V. (2016). Extracellular matrix and tissue regeneration. In G. Steinhoff (Ed.), *Regenerative medicine—From protocol to patient*. Cham: Springer.
MacLean, S., Khan, W. S., Malik, A. A., Snow, M., & Anand, S. (2012). Tendon regeneration and repair with stem cells. *Stem Cells International, 2012*, 316281. https://doi.org/10.1155/2012/316281
Manning, C. N., Schwartz, A. G., Liu, W., Xie, J., Havlioglu, N., Sakiyama-Elbert, S. E., … Thomopoulos, S. (2013). Controlled delivery of mesenchymal stem cells and growth factors using a nanofiber scaffold for tendon repair. *Acta Biomaterialia, 9*(6), 6905–6914. https://doi.org/10.1016/j.actbio.2013.02.008
McLane, J. S., Schaub, N. J., Gilbert, R. J., & Ligon, L. A. (2013). Electrospun nanofiber scaffolds for investigating cell–matrix adhesion. In A. Coutts (Ed.), *Adhesion protein protocols. Methods in molecular biology (Methods and protocols)*. Totowa, NJ: Humana Press.
Miao, S., Castro, N., Nowicki, M., Xia, L., Cui, H., Zhou, X., … Zhang, L. G. (2017). 4D printing of polymeric materials for tissue and organ regeneration. *Materials Today, 20*, 577–591. https://doi.org/10.1016/j.mattod.2017.06.005
Mitchell, G. (2014). *Electrospinning: Principles, practice and possibilities*. The Royal Society of Chemistry: Cambridge, UK.
Mohabatpour, F., Karkhaneh, A., & Sharifi, A. M. (2016). A hydrogel/fiber composite scaffold for chondrocyte encapsulation in cartilage tissue regeneration. *RSC Advances, 6*(86), 83135–83145. https://doi.org/10.1039/c6ra15592h
Nam, Y., Rim, Y. A., Lee, J., & Ju, J. H. (2018). Current therapeutic strategies for stem cell-based cartilage regeneration. *Stem Cells International, 2018*, 8490489. https://doi.org/10.1155/2018/8490489
Narayanan, N., Jiang, C., Uzunalli, G., Thankappan, S. K., Laurencin, C. T., & Deng, M. (2016). Polymeric electrospinning for musculoskeletal regenerative engineering. *Regenerative Engineering and Translational Medicine, 2*, 69–84. https://doi.org/10.1007/s40883-016-0013-8
Nedjari, S., Awaja, F., & Altankov, G. (2017). Three dimensional honeycomb patterned fibrinogen based nanofibers induce substantial osteogenic response of mesenchymal stem cells. *Scientific Reports, 7*(1), 15947. https://doi.org/10.1038/s41598-017-15956-8
Noh, H. K., Lee, S. W., Kim, J. M., Oh, J. E., Kim, K. H., Chung, C. P., … Min, B. M. (2006). Electrospinning of chitin nanofibers: Degradation behavior and cellular response to normal human keratinocytes and fibroblasts. *Biomaterials, 27*(21), 3934–3944. https://doi.org/10.1016/j.biomaterials.2006.03.016
Norouzi, M., Boroujeni, S. M., Omidvarkordshouli, N., & Soleimani, M. (2015). Advances in skin regeneration: Application of electrospun scaffolds. *Advanced Healthcare Materials, 4*, 1114–1133. https://doi.org/10.1002/adhm.201500001
Pak, J., Lee, J. H., Pak, N., Pak, Y., Park, K. S., Jeon, J. H., … Lee, S. H. (2018). Cartilage regeneration in humans with adipose tissue-derived stem cells and adipose stromal vascular fraction cells: Updated status. *International Journal of Molecular Sciences, 19*, E2146. https://doi.org/10.3390/ijms19072146

Pal, P., Srivas, P. K., Dadhich, P., Das, B., Maulik, D., & Dhara, S. (2017). Nano-/microfibrous cotton-wool-like 3D scaffold with core-shell architecture by emulsion electrospinning for skin tissue regeneration. *ACS Biomaterials Science and Engineering, 3*(12), 3563–3575. https://doi.org/10.1021/acsbiomaterials.7b00681

Panahi-Joo, Y., Karkhaneh, A., Nourinia, A., Abd-Emami, B., Negahdari, B., Renaud, P., & Bonakdar, S. (2016). Design and fabrication of a nanofibrous polycaprolactone tubular nerve guide for peripheral nerve tissue engineering using a two-pole electrospinning system. *Biomedical Materials (Bristol), 11*(2), 025017. https://doi.org/10.1088/1748-6041/11/2/025017

Paredes, J. J., & Andarawis-Puri, N. (2016). Therapeutics for tendon regeneration: A multidisciplinary review of tendon research for improved healing. *Annals of the New York Academy of Sciences, 1383*(1), 125–138. https://doi.org/10.1111/nyas.13228

Park, K. E., Kim, B. S., Kim, M. H., You, H. K., Lee, J., & Park, W. H. (2015). Basic fibroblast growth factor-encapsulated PCL nano/microfibrous composite scaffolds for bone regeneration. *Polymer, 76*, 8–16. https://doi.org/10.1016/j.polymer.2015.08.024

Piai, J. F., da Silva, M. A., Martins, A., Torres, A. B., Faria, S., Reis, R. L., … Neves, N. M. (2017). Chondroitin sulfate immobilization at the surface of electrospun nanofiber meshes for cartilage tissue regeneration approaches. *Applied Surface Science, 403*, 112–125. https://doi.org/10.1016/j.apsusc.2016.12.135

Prabhakaran, M. P., Venugopal, J., & Ramakrishna, S. (2009). Electrospun nanostructured scaffolds for bone tissue engineering. *Acta Biomaterialia, 5*(8), 2884–2893. https://doi.org/10.1016/j.actbio.2009.05.007

Reinke, J. M., & Sorg, H. (2012). Wound repair and regeneration. *European Surgical Research, 49*(1), 35–43. https://doi.org/10.1159/000339613

Román-Doval, R., Tellez-Cruz, M. M., Rojas-Chávez, H., Cruz-Martínez, H., Carrasco-Torres, G., & Vásquez-Garzón, V. R. (2019). Enhancing electrospun scaffolds of PVP with polypyrrole/iodine for tissue engineering of skin regeneration by coating via a plasma process. *Journal of Materials Science, 54*(4), 3342–3353. https://doi.org/10.1007/s10853-018-3024-7

Romanovsky, A. A. (2014). Skin temperature: Its role in thermoregulation. *Acta Physiologica, 210*, 498–507. https://doi.org/10.1111/apha.12231

Sell, S., Barnes, C., Smith, M., McClure, M., Madurantakam, P., Grant, J., … Bowlin, G. (2007). Extracellular matrix regenerated: Tissue engineering via electrospun biomimetic nanofibers. *Polymer International, 56*, 1349–1360. https://doi.org/10.1002/pi.2344

Senthamizhan, A., Balusamy, B., & Uyar, T. (2016). Glucose sensors based on electrospun nanofibers: A review fiber-based platforms for bioanalytics. *Analytical and Bioanalytical Chemistry, 408*(5), 1285–1306. https://doi.org/10.1007/s00216-015-9152-x

Shafiee, A., Soleimani, M., Chamheidari, G. A., Seyedjafari, E., Dodel, M., Atashi, A., & Gheisari, Y. (2011). Electrospun nanofiber-based regeneration of cartilage enhanced by mesenchymal stem cells. *Journal of Biomedical Materials Research—Part A, 99A*(3), 467–478. https://doi.org/10.1002/jbm.a.33206

Shahriar, S., Mondal, J., Hasan, M., Revuri, V., Lee, D., & Lee, Y.-K. (2019). Electrospinning nanofibers for therapeutics delivery. *Nanomaterials, 9*(4), 532. https://doi.org/10.3390/nano9040532

Shan, X., Liu, C., Li, F., Ouyang, C., Gao, Q., & Zheng, K. (2015). Nanoparticles vs. nanofibers: A comparison of two drug delivery systems on assessing drug release performance in vitro. *Designed Monomers and Polymers, 18*(7), 678–689. https://doi.org/10.1080/15685551.2015.1070500

Sheikh, F. A., Ju, H. W., Lee, J. M., Moon, B. M., Park, H. J., Lee, O. J., … Park, C. H. (2015). 3D electrospun silk fibroin nanofibers for fabrication of artificial skin. *Nanomedicine: Nanotechnology, Biology, and Medicine, 11*(3), 681–691. https://doi.org/10.1016/j.nano.2014.11.007

So, K. F., & Xu, X. M. (2015). *Neural regeneration*. Cambridge, MA: Academic Press. https://doi.org/10.1016/C2014-0-00519-9

Sridhar, R., Lakshminarayanan, R., Madhaiyan, K., Barathi, V. A., Limh, K. H. C., & Ramakrishna, S. (2015). Electrosprayed nanoparticles and electrospun nanofibers based on natural materials: Applications in tissue regeneration, drug delivery and pharmaceuticals. *Chemical Society Reviews, 44*, 790–814. https://doi.org/10.1039/c4cs00226a

Stoll, E. A. (2014). Advances toward regenerative medicine in the central nervous system: Challenges in making stem cell therapy a viable clinical strategy. *Molecular and Cellular Therapies, 2*(1), 12. https://doi.org/10.1186/2052-8426-2-12

Tchemtchoua, V. T., Atanasova, G., Aqil, A., Filée, P., Garbacki, N., Vanhooteghem, O., … Colige, A. (2011). Development of a chitosan nanofibrillar scaffold for skin repair and regeneration. *Biomacromolecules, 12*(9), 3194–3204. https://doi.org/10.1021/bm200680q

Thorvaldsson, A., Stenhamre, H., Gatenholm, P., & Walkenström, P. (2008). Electrospinning of highly porous scaffolds for cartilage regeneration. *Biomacromolecules, 9*(3), 1044–1049. https://doi.org/10.1021/bm701225a

Tiku, M. L., & Sabaawy, H. E. (2015). Cartilage regeneration for treatment of osteoarthritis: A paradigm for nonsurgical intervention. *Therapeutic Advances in Musculoskeletal Disease, 7*, 76–87. https://doi.org/10.1177/1759720X15576866

Uyar, T., & Kny, E. (2017). *Electrospun materials for tissue engineering and biomedical applications* (1st ed.). https://doi.org/10.1016/c2015-0-05802-6

Vacanti, J. (2010). Tissue engineering and regenerative medicine: From first principles to state of the art. *Journal of Pediatric Surgery, 45*, 291–294. https://doi.org/10.1016/j.jpedsurg.2009.10.063

Vashisth, P., Nikhil, K., Roy, P., Pruthi, P. A., Singh, R. P., & Pruthi, V. (2016). A novel gellan-PVA nanofibrous scaffold for skin tissue regeneration: Fabrication and characterization. *Carbohydrate Polymers, 136*, 851–859. https://doi.org/10.1016/j.carbpol.2015.09.113

Venugopal, J., & Ramakrishna, S. (2005). Biocompatible nanofiber matrices for the engineering of a dermal substitute for skin regeneration. *Tissue Engineering, 11*(5–6), 847–854. https://doi.org/10.1089/ten.2005.11.847

Verma, S., Domb, A. J., & Kumar, N. (2011). Nanomaterials for regenerative medicine. *Nanomedicine, 6*, 157–181. https://doi.org/10.2217/nnm.10.146

Vig, K., Chaudhari, A., Tripathi, S., Dixit, S., Sahu, R., Pillai, S., … Singh, S. R. (2017). Advances in skin regeneration using tissue engineering. *International Journal of Molecular Sciences, 18*, 789. https://doi.org/10.3390/ijms18040789

Vijayavenkataraman, S., Yan, W. C., Lu, W. F., Wang, C. H., & Fuh, J. Y. H. (2018). 3D bioprinting of tissues and organs for regenerative medicine. *Advanced Drug Delivery Reviews, 132*, 296–332. https://doi.org/10.1016/j.addr.2018.07.004

Vimal, S. K., Ahamad, N., & Katti, D. S. (2016). A simple method for fabrication of electrospun fibers with controlled degree of alignment having potential for nerve regeneration applications. *Materials Science and Engineering C, 63*, 616–627. https://doi.org/10.1016/j.msec.2016.03.008

Wang, C., Wang, J., Zeng, L., Qiao, Z., Liu, X., Liu, H., … Ding, J. (2019). Fabrication of electrospun polymer nanofibers with diverse morphologies. *Molecules, 24*(5), 834. https://doi.org/10.3390/molecules24050834

Wang, K., Liu, L., Xie, J., Shen, L., Tao, J., & Zhu, J. (2018). Facile strategy to generate aligned polymer nanofibers: Effects on cell adhesion. *ACS Applied Materials and Interfaces, 10*(2), 1566–1574. https://doi.org/10.1021/acsami.7b16057

Wang, Y., Cui, W., Zhao, X., Wen, S., Sun, Y., Han, J., & Zhang, H. (2019). Bone remodeling-inspired dual delivery electrospun nanofibers for promoting bone regeneration. *Nanoscale, 11*(1), 60–71. https://doi.org/10.1039/c8nr07329e

Wang, Y., Jiang, Y., Zhang, Y., Wen, S., Wang, Y., & Zhang, H. (2019). Dual functional electrospun core-shell nanofibers for anti-infective guided bone regeneration membranes. *Materials Science and Engineering C, 98*, 134–139. https://doi.org/10.1016/j.msec.2018.12.115

Wendorff, J., Agarwal, S., & Greiner, A. (2012). *Electrospinning: Materials, processing, and applications*. Weinheim, Germany: Wiley-VCH Verlag GmbH & Co. KGaA.

Weng, L., Boda, S. K., Wang, H., Teusink, M. J., Shuler, F. D., & Xie, J. (2018). Novel 3D hybrid nanofiber aerogels coupled with BMP-2 peptides for cranial bone regeneration. *Advanced Healthcare Materials, 7*(10), e1701415. https://doi.org/10.1002/adhm.201701415

Wu, J., Wang, N., Zhao, Y., & Jiang, L. (2013). Electrospinning of multilevel structured functional micro-/nanofibers and their applications. *Journal of Materials Chemistry A, 1*(25), 7290–7305. https://doi.org/10.1039/c3ta10451f

Wu, J., & Hong, Y. (2016). Enhancing cell infiltration of electrospun fibrous scaffolds in tissue regeneration. *Bioactive Materials, 1*, 56–64. https://doi.org/10.1016/j.bioactmat.2016.07.001

Xie, J., Liu, W., Macewan, M. R., Bridgman, P. C., & Xia, Y. (2014). Neurite outgrowth on electrospun nanofibers with uniaxial alignment: The effects of fiber density, surface coating, and supporting substrate. *ACS Nano, 8*(2), 1878–1885. https://doi.org/10.1021/nn406363j

Xie, J., MacEwan, M. R., Li, X., Sakiyama-Elbert, S. E., & Xia, Y. (2009). Neurite outgrowth on nanofiber scaffolds with different orders, structures, and surface properties. *ACS Nano, 3*(5), 1151–1159. https://doi.org/10.1021/nn900070z

Xie, J., MacEwan, M. R., Schwartz, A. G., & Xia, Y. (2010). Electrospun nanofibers for neural tissue engineering. *Nanoscale, 2*(1), 35–44. https://doi.org/10.1039/b9nr00243j

Xu, S. C., Qin, C. C., Yu, M., Dong, R. H., Yan, X., Zhao, H., … Long, Y. Z. (2015). A battery-operated portable handheld electrospinning apparatus. *Nanoscale, 7*(29), 12351–12355. https://doi.org/10.1039/c5nr02922h

Xue, J., Wu, T., Dai, Y., & Xia, Y. (2019). Electrospinning and electrospun nanofibers: Methods, materials, and applications. *Chemical Reviews, 119*, 5298–5415. https://doi.org/10.1021/acs.chemrev.8b00593

Xue, J., Yang, J., O'Connor, D. M., Zhu, C., Huo, D., Boulis, N. M., & Xia, Y. (2017). Differentiation of bone marrow stem cells into Schwann cells for the promotion of neurite outgrowth on electrospun fibers. *ACS Applied Materials and Interfaces, 9*(14), 12299–12310. https://doi.org/10.1021/acsami.7b00882

Yamaguchi, K., Prabakaran, M., Ke, M., Gang, X., Chung, I. M., Um, I. C., … Kim, I. S. (2016). Highly dispersed nanoscale hydroxyapatite on cellulose nanofibers for bone regeneration. *Materials Letters, 168*, 56–61. https://doi.org/10.1016/j.matlet.2016.01.010

Yang, F., Murugan, R., Wang, S., & Ramakrishna, S. (2005). Electrospinning of nano/micro scale poly(l-lactic acid) aligned fibers and their potential in neural tissue engineering. *Biomaterials, 26*(15), 2603–2610. https://doi.org/10.1016/j.biomaterials.2004.06.051

Yang, G., Rothrauff, B. B., & Tuan, R. S. (2013). Tendon and ligament regeneration and repair: Clinical relevance and developmental paradigm. *Birth Defects Research Part C—Embryo Today: Reviews, 99*, 203–222. https://doi.org/10.1002/bdrc.21041

Yang, Y., Xia, T., Zhi, W., Wei, L., Weng, J., Zhang, C., & Li, X. (2011). Promotion of skin regeneration in diabetic rats by electrospun core-sheath fibers loaded with basic fibroblast growth factor. *Biomaterials, 32*(18), 4243–4254. https://doi.org/10.1016/j.biomaterials.2011.02.042

Yannas, I. V., Tzeranis, D. S., & So, P. T. C. (2017a). Regeneration mechanism for skin and peripheral nerves clarified at the organ and molecular scales. *Current Opinion in Biomedical Engineering, 6*, 1–7. https://doi.org/10.1016/j.cobme.2017.12.002

Yannas, I. V., Tzeranis, D. S., & So, P. T. C. (2017b). Regeneration of injured skin and peripheral nerves requires control of wound contraction, not scar formation. *Wound Repair and Regeneration, 25*(2), 177–191. https://doi.org/10.1111/wrr.12516

Zamborsky, R., Kilian, M., Csobonyeiova, M., & Danisovic, L. (2018). Regenerative medicine in orthopaedics and trauma: Challenges, regulation and ethical issues. *Ortopedia Traumatologia Rehabilitacja, 20*(3), 173–180. https://doi.org/10.5604/01.3001.0012.0764

Zhang, L., Chen, S., Liang, R., Chen, Y., Li, S., Li, S., … Yang, Y. (2018). Fabrication of alignment polycaprolactone scaffolds by combining use of electrospinning and micromolding for regulating Schwann cells behavior. *Journal of Biomedical Materials Research—Part A, 106*(12), 3123–3134. https://doi.org/10.1002/jbm.a.36507

Zhou, Y., Yang, D., Chen, X., Xu, Q., Lu, F., & Nie, J. (2008). Electrospun water-soluble carboxyethyl chitosan/poly(vinyl alcohol) nanofibrous membrane as potential wound dressing for skin regeneration. *Biomacromolecules, 9*(1), 349–354. https://doi.org/10.1021/bm7009015

Zhu, W., Masood, F., O'Brien, J., & Zhang, L. G. (2015). Highly aligned nanocomposite scaffolds by electrospinning and electrospraying for neural tissue regeneration. *Nanomedicine: Nanotechnology, Biology, and Medicine, 11*(3), 693–704. https://doi.org/10.1016/j.nano.2014.12.001

Chapter 3
Nanomaterials for Wound Healing

Gozde Uzunalli

Abstract Tissue wounds cause a significant social and economic burden, which results in diminished quality of life and increased patient mortality. In the USA alone, the estimated cost of wound care ranges from $28.1 to $96.8 billion. The rapid rise of an aging population with chronic debilitating diseases, including obesity and diabetes, contributes to the unmet critical need for better therapeutics for impaired wound healing in patients. Conventional wound management fails to provide rapid tissue repair and restoration of function, particularly in these diseased states. Wound healing is a tightly orchestrated, complex, and dynamic physiological process occurring in a tightly controlled manner that requires more advanced alternative therapeutic approaches. An understanding of the basic biological mechanisms in wound repair underpins the development of novel therapeutic approaches for restoration of tissue function. The use of nanomaterials in wound management represents a unique tool that can be specifically designed to closely reflect the underlying physiological processes in damaged tissues. Harnessing this technology results in a cost-effective and individualized approach to wound management, which has the potential to revolutionize wound management. This chapter presents a basic overview of the pathophysiology of wound healing and an approach to utilizing nanomaterials for wound healing therapy.

3.1 Pathophysiology of the Wound Healing

3.1.1 Wounds

Wounds are defined as tissue injuries (e.g., trauma or surgical incisions) that disrupt the anatomic structure and function of an organ. Injury primarily affects the external structure of the skin and can also extend to any direction through underlying soft tissues including muscles, tendons, nerves, and blood vessels (Lazarus et al., 1994;

G. Uzunalli (✉)
Department of Comparative Pathobiology, College of Veterinary Medicine, Purdue University, West Lafayette, IN, USA
e-mail: guzunall@purdue.edu

A. B. Tekinay (ed.), *Nanomaterials for Regenerative Medicine*, Stem Cell Biology and Regenerative Medicine, https://doi.org/10.1007/978-3-030-31202-2_3

Velnar, Bailey, & Smrkolj, 2009). The skin is a protective barrier from the external environment, and is easily exposed to mild and severe external factors (Boateng & Catanzano, 2015). Thus, the structural and functional integrity of the damaged skin must be restored quickly to maintain the homeostasis.

Wounds are clinically classified as acute or chronic wounds based on the duration of repair. Acute wounds are usually healed within several weeks, accompanied by the formation of scar tissue, which may be functionally and aesthetically imperfect. The primary causes of acute wounds include mechanical forces (i.e., lacerations, punctures, abrasions, tears), chemical, and thermal damage. In contrast, chronic wounds often fail to heal due to pathological conditions such as movement, immune dysfunction (i.e., diabetes), ischemia, or infection and severely affect underlying tissues (Clark, 2013; Han & Ceilley, 2017; Lazarus et al., 1994).

3.1.2 Acute Wound Healing and Scar Formation

Wound healing is one the most complex, dynamic, and fundamental biological process of all living organisms, which is orchestrated by local and systemic cellular processes and closely regulated via intra and intercellular signaling. Interactions between biochemical mediators including cytokines and growth factors, extracellular matrix (ECM) components, and progenitor cells mediate inflammation and tissue repair to restore normal tissue function (Clark, 2013). The usual outcome of the wound healing process in the skin is scar tissue formation, or fibrosis, which is analogous to the intact skin. After injury, an ordered process of repair occurs beginning with inflammatory cells infiltrating the affected region, followed by wound contraction, and finally fibrosis (Wadman, 2005). The formation of fibrosis, or scar tissue, can cause severe and irreversible functional and cosmetic deficits in the skin, and possible physiological damage to the patients (Holavanahalli, Helm, & Kowalske, 2010; Sen et al., 2009). Although the scar tissue may be structurally and functionally adequate for acting as a physical barrier, it is not equivalent to skin regeneration with the exception of fetal wound healing (Atala, Irvine, Moses, & Shaunak, 2010; Singer & Clark, 1999).

Immediately after the injury, the body responds with a series of biochemical events known as the wound-healing cascade to promote healing. Wound healing is divided into four dynamic and overlapping phases that are either activated or suppressed to complete the tissue repair.

3.1.2.1 Hemostasis

Hemostasis is the process of controlling the amount of blood loss and hemorrhage from a wound, which occurs in many species, ranging from zebrafish to humans. The hemostasis phase of wound healing begins immediately following the injury with hemorrhage due to blood vessel damage and extravasation of the blood constituents to the wound bed. The primary function of hemostasis is to restrict blood loss to ensure survival, and form a mechanical plug to fill the wound bed. Hemostasis also serves to protect the wound from infection and provides a provisional matrix (Lawrence, 1998; Robson, 1997). Three crucial steps are present in hemostasis; vasoconstriction of the blood vessels to decrease the blood flow, rapid formation of a platelet plug (hemostatic plug), and the coagulation process to form a fibrin mesh clot (Clark, 2013; Kirsner & Eaglstein, 1993).

Within minutes, platelets are recruited to the site of injury where they are activated by the exposed ECM of the injured subendothelium. Activated platelets adhere and aggregate to the damaged vascular wall to form bridges with collagen and fibronectin (Hawiger, 1987). Platelets also release mediators including serotonin, thromboxane A2, and adenosine diphosphate to promote vasoconstriction of the injured vessels, while the secretion of fibrinogen and von Willebrand factor (vWF) trigger recruitment of additional platelets.

Fibrinogen, vWF, and thrombin mediate platelet aggregation and act as a bridge between platelets and crosslinking them (Nurden, Nurden, Sanchez, Andia, & Anitua, 2008; Steed, 1997). Platelet aggregation relies on both the intrinsic and extrinsic pathways of coagulation. The intrinsic system of coagulation is the contact activation pathway in which subendothelial tissues are exposed to blood and it is activated by a specific enzyme called Hageman Factor XII (and other factors VIII, IX, XI). That activates Factor X and results in the conversion of prothrombin to thrombin (Young & McNaught, 2011). In contrast, extrinsic pathway is the tissue factor pathway that initiates tissue factor release to the blood and the activation of Factor VII for thrombin formation (Versteeg, Heemskerk, Levi, & Reitsma, 2013). Both pathways eventually lead to a complex common pathway where soluble fibrinogen is converted into polymerized fibrin by thrombin (Palta, Saroa, & Palta, 2014). This fibrin clot becomes an early wound cover and also functions as a scaffold for cellular migration through the wound bed.

Spatial and temporal platelet activation and aggregation not only maintain hemostasis but also contributes to initiation of inflammation, proliferation, and tissue remodeling phases by releasing chemoattractants (Clark, 2013; Lawrence, 1998; Singer & Clark, 1999), transforming growth factor-beta (TGF-β) and platelet-derived growth factor (PDGF) (Derynck, 1988; Li, Chen, & Kirsner, 2007; Lynch, Nixon, Colvin, & Antoniades, 1987; Ross, Raines, & Bowen-Pope, 1986).

3.1.2.2 Inflammation

After hemostasis commences, an immune barrier against pathogens becomes necessary to ensure proper repair. Coagulation stimulates the release of inflammatory mediators including prostaglandin, histamine, leukotrienes, interleukins, and proteases (Wulff & Wilgus, 2013). Prostaglandins cause vasodilation and vascular permeability, which stimulates the migration of monocytes and fluid leakage to the wound bed (edema). Diminished prostaglandin expression results in impaired wound healing (Syeda et al., 2012). Edema formation and increased chemotaxis of immune cells are the characteristics of the inflammation phase. In the early phase of inflammation, neutrophils are recruited to site of injury by TGF-β, tumor necrosis factor-α (TNF-α), and formyl methionyl peptides. Neutrophilic migration occurs within 6 h and peaks at 24–48 h post-injury (Dorward et al., 2015; Robson, Steed, & Franz, 2001). The function of neutrophils is producing antimicrobial substances and proteases against bacteria to remove the damaged tissue debris. Neutrophil activation and migration to the site of injury begins with selectin-mediated rolling along the endothelium and then adhesion to the endothelial cells. The weak adhesion with selectins is followed by more enhanced ECM-cell interactions between integrins and their binding molecules (Schmidt, Lee, Zemans, Yamashita, & Downey, 2011). Finally, neutrophils migrate in between the endothelial cells to reach the site of injury. Resolution of early inflammation is achieved by elimination of the neutrophils from the wound bed via the process of apoptosis, programmable cell death.

Platelet activation and subsequent increased vascular permeability promotes the next phase of inflammation and repair, to recruit circulating monocytes into the wound bed. Monocytes are the circulating progenitor cells of the macrophages and undergo phenotypical changes and transform into wound healing macrophages within tissues, when they are activated by interferon-γ (IFN-γ) (Mosser & Edwards, 2008). Together, recruitment of neutrophils and monocytes are coordinated by certain cytokines such as PDGF, TGF-β, and leukotrienes. Initially, activated macrophages display inflammatory response at the wound bed and release cytokines like TNF, interleukin-1 (IL-1), and biologically active oxygen and nitrogen intermediates like superoxide and nitric oxide (NO), which have antimicrobial effects (Murray & Wynn, 2011; Serbina, Salazar-Mather, Biron, Kuziel, & Pamer, 2003). Macrophages are fundamental for the phagocytosis of the remaining bacteria, the damaged tissue, and the cellular debris. Following phagocytosis, macrophages switch to a regulatory phenotype by expressing immunoregulatory proteins (Pesce et al., 2009). The secretion of PDGF, TGF family proteins, vascular endothelial growth factor (VEGF) and fibroblast growth factor (FGF) from the regulatory macrophages stimulate cellular proliferation and migration of epithelial cells, fibroblasts, and vascular endothelial cells. In addition, they contribute to the ECM production to promote further repair. Hence, they play a pivotal role in wound healing, and their depletion attenuates the repair of the tissue.

3.1.2.3 Proliferation

The proliferation phase of wound healing begins at day 3 and lasts for 3–4 weeks. Once homeostasis is established with essential immune response, intense cellular activity starts to dominate the wound healing with focus on rebuilding the new ECM enriched with new vascular network and resurfacing the wounded area.

3.1.2.3.1 Reestablishing the Dermal Cover: Re-epithelialization

The denuded wound surface represents a potential threat to the skin itself and the underlying internal organs. Although a provisional fibrin-based and pathogen-free matrix is contructed during the hemostasis and inflammation phases, a rapid reestablishment of the epithelial layer of the wound is crucial for protection from the external environment. Re-epithelialization is defined as the shielding of the injury area with continuous epithelial tissue and constitutes a crucial step of successful healing. It begins within a few hours to 1 day following injury. The reestablishment of a functional epidermis depends on the migration and proliferation of the keratinocytes, the reconstitution of the desmosomal adhesions and finally terminal differentiation of the keratinocytes to form a protective barrier.

Adjacent epidermal keratinocytes and epithelial stem cells migrate to the skin surface and proliferate to complete the re-epithelialization process. Keratinocyte mobility requires dissolution of the cell–cell and cell–basement membrane interactions as well as retraction of the intercellular tonofilaments and growth of lamellipodia (Cavani et al., 1993; Martin, 1997). Hence, keratinocytes gain the ability to crawl over the new granulation tissue (Li et al., 2007). Migrating cells also deposit laminin-5 to guide the sequential cells (Nguyen, Ryan, Gil, & Carter, 2000). The keratinocytes undergo proliferation alongside migration to ensure a sufficient number of cells to cover the wound. Once migration terminates forming a monolayer of cells, focal adhesions and hemidesmosomes are reestablished. In addition, a basement membrane consisting of laminins, perlecan, heparan sulfate proteoglycans, fibronectin, and collagens starts to form with hemidesmosomal bridges (Behrens et al., 2012; Li et al., 2007; Martin, 1997). Concurrently, keratinocytes terminally differentiate to organized and stratified epidermis phenotype to restore epithelial integrity. All these steps of re-epithelialization are to be known controlled by the local release of epithelial growth factor (EGF), TGF, and keratinocyte growth factor (KGF) (reviewed in Martin, 1997).

3.1.2.3.2 Reconstitution of Dermis: Granulation Tissue Formation

Wound healing is completed by the formation of the granulation tissue, which is composed of activated fibroblasts that produce the new ECM (fibrosis) and cells that are required for angiogenesis. Angiogenesis is the formation of new blood vessels to provide nutrients and oxygen needed to support healing. Macroscopically, granulation tissue appears as a pink and granulated tissue, which, at the microscopic level, is composed of perpendicularly arranged collagen bundles, blood vessels, and fibroblasts surrounded by edema (Singer & Clark, 1999). The external appearance of the granulation tissue is indicative of its healing status. Dark red or pale color is an indicator of devitalized granulation tissue, which is indicative of poor healing. In patients, excessive granulation tissue formation, or hyper-granulation, frequently occurs due to problems in repair, including excessive inflammation, infection, movement, or immune dysfunction. Larger defects can produce islands of dead, or necrotic tissue detached from the underlying blood supply. Due to the high amount of debris, inflammation is more intense, thus, the wound healing requires a greater volume of granulation tissue, which eventually results in greater scar tissue.

3.1.2.3.2.1 Fibrosis

Fibrosis, instead of regeneration, is the pathopyhsiological response of the injured tissue that drives rapid healing for performing the fundamental functions of the skin. It involves fibroblast proliferation and migration into the fibrin-based provisional matrix and the synthesis of collagen and other ECM proteins. Dermal fibroblasts are the predominant cell type in fibroplasia, and proliferate within 3 days. Fibroblast proliferation and migration is induced by PDGF and TGF-β1 released from the macrophages (Enoch & Leaper, 2005). Once they reach the wound bed, they adopt a (proto-)myofibroblast phenotype and then mature to form a differentiated myofibroblast (Van De Water, Varney, & Tomasek, 2013). Proto-myofibroblasts are characterized by augmented bundles of actin-myosin stress fibers and fibronectin expression as well as larger adhesion sites (Dugina, Fontao, Chaponnier, Vasiliev, & Gabbiani, 2001). Fibroblasts are activated by structural, mechanical, and physical changes of ECM and local cytokines.

The fibrin-based provisional matrix, that is enriched with growth factors in the inflammation phase, has 10–100 Pa Young's modulus. It has been demonstrated that stress fiber formation appears when the elastic modulus reaches 3–6 kPa in 2D culture conditions. Moreover, after 20 kPa, fibroblasts start to transform into differentiated myofibroblasts with dramatically increased α-smooth muscle actin (α-SMA) expression, which is the actin isoform specific for vascular endothelial cells. Besides, differentiation of fibroblasts to myofibroblasts and expression of α-SMA are also modulated by TGF-β1 and fibronectin (Hinz, 2007; Van De Water et al., 2013). Myofibroblasts are the crucial cells of the scar tissue formation due to their highly contractile force generation features.

The provisional wound matrix consists of a fibrin and fibronectin matrix which is gradually replaced with polymerized collagen, elastin, proteoglycans, and glycosaminoglycans (GAGs) synthesized by the fibroblasts. Early granulation tissue contains hyaluronic acid and fibronectin that forms a woven scaffold to the proliferating cells at the wound bed. Later, collagen type III predominates the granulation tissue. Type III is the weaker type of the collagen, which can be synthesized rapidly.

3.1.2.3.2.2 Reconstruction of Blood Vessels: Angiogenesis

Angiogenesis is a crucial stage where the endothelial cells initiate the sprouting of the preexisting vessels adjacent to the wound bed for providing enough nutrients and oxygen to the regenerating tissue. Vessel sprouting occurs through a similar process to the epithelial cell migration in re-epithelialization. At the tip of the capillaries, the endothelial cells have two distinct phenotypes called the tip cells and the stalk cells. The tip cells lead the migration of the other cells by using their cytoplasmic pseudopod extensions. These cells do not contain a vascular lumen and have minimal proliferation capacity while the stalk cells are located behind the tip cells can proliferate and form lumens (Motherwell, Anderson, & Murfee, 2018). The phenotypic alteration of the EC and migration into the perivascular space are promoted by PDGF, FN, and heparin (Greaves, Ashcroft, Baguneid, & Bayat, 2013). Even when the endothelial cells are provided by proper signaling, collagen and proteoglycan-rich ECM might not allow cell migration due to its dense structure. Hence, cell migration also requires proteolytic dissolution of the basement membrane by MMPs (Li et al., 2007).

Once the wound healing begins, production of substances such as lactic acid and hypoxia-inducible factor-1 (HIF-1) following hypoxia—a condition in which tissues are inadequately oxygenated—are amplified and trigger cytokine release from macrophages to stimulate angiogenesis (Hong et al., 2014; Li et al., 2007). In addition to cytokines, the endothelial cells are responsive to the angiogenic growth factors including FGF-2, VEGF, angiopoietins, and TGFs. FGF-2 is released from the macrophages and the endothelial cells and functions to regulate the migration and proliferation of the endothelial cells through internal integrin signaling. Among several integrins on the endothelial cells, $\alpha v\beta 3$ integrin is expressed only on the granulation tissue vessels—not the healthy skin the endothelial cells—and regulates MMP-2 and, subsequently dissolution of the collagen matrix (Brooks et al., 1994). FGF-2 also binds to heparan sulfates on the proteoglycans as a tetramer to be able to augment the cell–receptor binding (Schultz & Wysocki, 2009). The depletion of FGF-2 triggers delayed wound healing (Ortega, Ittmann, Tsang, Ehrlich, & Basilico, 1998).

VEGF is another key growth factor that functions in wound angiogenesis, and it is highly expressed following injury in the skin. Its high affinity for fibrin facilitates the mitogenic activity regarding the EC migration and capillary sprouting which is mediated by integrins including $\alpha v\beta 3$, $\alpha 1\beta 1$ and $\alpha 2\beta 1$ (Senger et al., 1996, 1997). In the hypoxic wound microenvironment, VEGF gradient occurs alongside the oxygen gradient. Hence, the gradient directs EC migration towards the wound core to trigger vessel formation (Knighton, Silver, & Hunt, 1981). In addition to integrin-VEGF communication, re-establishment of the vascular network is also dependent on ECM organization. It has been shown that laminin 8 that is produced by human dermal endothelial cells stimulates EC attachment, spreading, migration, and capillary formation, which are mediated by $\alpha v\beta 3$ and $\beta 1$ integrins (Li et al., 2006).

VEGF is expressed from keratinocytes, platelets, the endothelial cells, fibroblasts, and vascular smooth muscle cells. In addition, macrophages also function as VEGF-secreting cells during the wound healing process. During the early stages of wound healing, an inflammatory subset of macrophage precursors—$Ly6C^+$ cells—recruit at the wound bed as a result of chemokine receptor-2 (CCR2) signaling and release proangiogenic VEGF to start vascularization (Willenborg et al., 2012). The studies on selective macrophage depletion (Lucas et al., 2010) and keratinocyte-restricted VEGF secretion (Rossiter et al., 2004) demonstrated that the initiation of angiogenesis by secreting VEGF is performed by monocyte/macrophages. Angiogenesis is continued by other cell types in the late stage.

3.1.3 Wound Contraction

Wound contraction is a process where contractile forces pull wound margins together recruiting the adjoining tissue to shrink the volume of the wound. The process is tuned by complex cell–ECM interactions influenced by a number of cytokines. Fibroblasts are the key cells for contraction.

In healthy tissue, fibroblasts reside in the collagen-rich ECM of the dermis and are responsible for the synthesis of collagen, proteoglycans, and other proteins which keep the fibroblasts in resting state. However, injury demolishes not only the gradient of signaling molecules and the composition of the ECM (discussed in Sect. 3.1.2.3.2.1) but also the mechanical strength of the microenvironment. Myofibroblasts sense these changes through mechanotransduction. Since migration and contraction require cell polarization, they arrange the assembly and disassembly of the ECM, integrins, and focal adhesions within a dynamic range (reviewed in Hoffman, Grashoff, & Schwartz, 2011). Contractile forces are formed by myosin motor proteins linked to the stress fibers. With the help of the actin–myosin complex, myofibroblasts extend, and their actin bundles bind the fibronectin in the extracellular space. Together, the binding and retraction of collagen result in the wound contraction. In conjunction with re-epithelialization, the wound contraction enables complete wound closure. The strength of the wound contraction varies depending on the rigidity, the type of the wound, and the cell number. If integrin-binding occurs on a rigid surface, contractile force transmission can be more adequate. Traction forces in the full-thickness wounds are more effective than partial-thickness wounds (Li et al., 2007). The contraction also contributes to the remodeling of the ECM by significantly deforming the granulation tissue.

3.1.4 Tissue Remodeling and Scar Formation

The last phase of wound healing, the remodeling, consists of deposition of the ECM, development of the new epithelium and maturation of the scar tissue. The remodeling starts almost 3 weeks after damage and may last up to a year or more. This final stage of wound healing is regulated by several cellular/metabolic degradation and synthesis events to be able to reach an equilibrium at the wound bed. Scar tissue is the regular response of the organisms and form after the complete wound healing.

The critical feature of the remodeling phase is the alteration of the ECM composition to recover the normal tissue structure and integrity. The collagen composition of the uninjured dermis is 80% collagen type I and 15% collagen type III, which are responsible for the mechanical strength and structure of the dermis. Healthy skin collagen organization is mainly reticular (rodents) (Whitby & Ferguson, 1991) or in basket weave formation (humans) while injured skin displays more cross-linked collagens that lie parallel to the skin (Ehrlich & Krummel, 1996). The proliferation phase demands a rapid granulation tissue formation to provide a temporary scaffold for the invading and proliferating cells. The loose and hydrated ECM that mainly consists of type III collagen is gradually replaced by denser collagen type I during scar tissue maturation (Xue & Jackson, 2015). Matrix turnover is the consequence of the degradation of collagen type III and synthesis of type I, which are controlled by MMPs. At the end, MMPs are no longer needed and tissue inhibitors of metalloproteinases are responsible for the fine tuning of MMP activity (TIMPs) (Visse & Nagase, 2003). Collagen replacement also increases the tensile strength of the injured tissue to almost 50% of the healthy skin within a month. However, the desired strength is never achieved as normal tissue, and it can reach up to 70% of uninjured skin (Li et al., 2007). In addition to the weaker structure, skin appendages including hair follicles and glands cannot grow back in the newly formed dermis following injury.

Upon wound closure, excessive blood vessels regress until few remain. All dominant cell types in granulation tissue undergo apoptosis and scar tissue becomes relatively avascular and acellular (Reinke & Sorg, 2012). In addition, the re-epithelialization is fully completed at the end of scarring without rete ridge formation.

3.1.5 Healing of Chronic Wounds

Wound healing process encompasses orderly and timely cellular and metabolic events. However, some wounds fail to complete the process and show no effective healing within 3 months of the injury (Nunan, Harding, & Martin, 2014). Preclinical and clinical data demonstrate chronic wounds do not follow a linear progression of classical wound healing cascade and subsequently result in structurally and functionally poor outcome. Chronic wounds cause major complications, despite many begin as a minor injury. Several etiological factors in chronic wound development result in structurally and functionally poor outcome. These can be local factors such as prolonged hypoxia and pressure or systemic problems such as age, ischemia, diabetes mellitus, obesity, medications or smoking (Guo & DiPietro, 2010). Chronic wounds are prone to result in severe pathologies and comorbidities including chronic pain, persistent infections, cancerogenous wound, and sometimes amputation (Menke, Ward, Witten, Bonchev, & Diegelmann, 2007).

The main difference between acute and chronic wounds is the healing rate. Many chronic wounds are characterized by an excessive and prolonged inflammatory phase (Xue & Jackson, 2015). As discussed before, MMPs are inhibited by TIMPs at the end of the wound healing to protect the wound bed from excessive degradation in health. On the contrary, increased MMP amount released by activated neutrophils in chronic wounds unbalances the MMP and TIMP ratio, which favors wound degradation (Lobmann et al., 2002). As a result of pathologically increased inflammatory response, more inflammation is observed by triggering the cytokine production such as TNFα and suppressing the mitogenic activity of the cells

(Goldman, 2004). Persistent infection of the wound can also be associated with the formation of bacterial biofilms, which are resistance to antibiotic treatment and delay re-epithelialization (Demidova-Rice, Hamblin, & Herman, 2012; Schierle, De la Garza, Mustoe, & Galiano, 2009).

Keratinocytes at the wound edges fail to get activated and differentiated which result in epidermal hyperproliferation in other words a thick epithelium formation. In addition, fibroblasts undergo senescence and become insensitive to growth factors particularly TGF-β (Pastar et al., 2010).

Chronic wounds are categorized into diabetic foot ulcers, venous stasis, arterial ulcers, and pressure ulcers. Diabetic ulcers result from both arterial and venous insufficiency and neuropathy due to the diabetes complications. Diabetes causes sensory neuropathy that decreases the ability to sense pain, pressure, and temperature as high blood glucose levels damage the nerves of the extremities. Pathological phenotypic abnormalities of the cells are present in chronic wounds, including a low mitotic activity (Demidova-Rice et al., 2012; Singer, Tassiopoulos, & Kirsner, 2017). Venous ulcers caused by damaged valves in the leg veins result in venous hypertension, which in turn results in edema, leakage of fibrin, and eventually leukocyte activation which triggers persistent inflammation (Gohel, Windhaber, Tarlton, Whyman, & Poskitt, 2008). Subsequently, MMPs induce matrix degradation and result in ulcer formation. Compared to venous ulcers, arterial ulcers are less common and are seen in patients with atherosclerosis, embolism, diabetes, or macro and microvascular diseases. They occur due to reduced arterial blood flow in the lower extremities, which leads to ischemia and subsequent necrosis (Demidova-Rice et al., 2012). Pressure ulcers, on the other hand, are skin and underlying tissue injuries due to the unrelieved pressure and shear force. Ulceration is caused by tissue necrosis following local ischemic zones, and it is common in patients with reduced mobility (Cox, 2011).

3.1.6 Insights from Regenerative Wound Healing

During the evolutionary process, our body is adapted to prioritize closing the wounded area before fine-tuning the structure and function of the tissues, which allows rapidly reunited wounds without having original strength and organization. However, scarring causes healthcare and economic burden worldwide. It has been estimated that scarring costs $12 billion annually while postsurgical wound management costs $38 billion (Sen et al., 2009). Moreover, scarring causes function loss of the injured tissue and subsequently increases the morbidity and even mortality. Therefore, investigating alternative methods of wound repair and rapid restoration of tissue function are of paramount importance to the field. In 1979, it was discovered that fetal wounds heal without scar tissue formation (Rowlatt, 1979). Subsequent studies in fetal animal models, including mice, rats, rabbits, sheep, and nonhuman primates have demonstrated the regenerative fetal wound healing (Colwell, Krummel, Longaker, & Lorenz, 2006; Ihara & Motobayashi, 1992;

Longaker et al., 1994; Lorenz, Whitby, Longaker, & Adzick, 1993; Somasundaram & Prathap, 1970). Fetal wounds heal by complete regeneration of the dermal tissue to the normal structure, strength, and function with differentiation of functional appendages, nerves, and blood vessels. Although the stages of wound healing are similar in both fetal and adult healing, there are some fundamental changes that cause different outcomes. A hallmark of fetal regeneration is minimal inflammatory response. Additionally, there is a decreased expression of leukocyte cytokines, resulting in minimal neutrophil migration and notably, and a lack of macrophages (Olutoye, Zhu, Cass, & Smith, 2005). Finally, platelets remain unbound and dissociated in fetal regeneration (Hopkinson-Woolley, Hughes, Gordon, & Martin, 1994). The characteristics of postnatal and fetal wound healing are summarized in Table 3.1.

Paucity of TGF-β1 and PDGF secretion from platelets come into prominence throughout the prenatal healing, whereas TGF-β3 expression is increased (O'Kane & Ferguson, 1997). In addition to the cytokine and inflammatory differences in prenatal regeneration, there are striking differences in the surrounding extracellular matrix which support regaining the function and structure. The pattern of collagen deposition in fetal wounds is replete with type III collagen in contrast to mature wound healing, which is predominately carried out by collagen type I (Burd, Longaker, Adzick, Harrison, & Ehrlich, 1990). Type III collagen a fine reticular pattern and is smaller compared to type I collagen.

It was previously hypothesized that the warm, sterile, and nutrient-rich intrauterine environment contributes to the intrauterine regeneration. However, studies in young opossums demonstrated that the fetal regeneration can occur in an extrauterine environment lacking amniotic fluid (Armstrong & Ferguson, 1995). Additionally, a xenograft model of human fetal skin to immunodeficient mice proved that human fetal skin can heal without scar in adult environment. Thus, these results suggest that fetal wound healing has intrinsic characteristics rather than environment-dependent feature (Lorenz et al., 1992).

3.2 Nanomaterials for Wound Healing Applications

3.2.1 Nanomaterials of Natural Origin

For the modern concept of wound management, it is critical to envisage the requirements of damaged tissue and fulfill the expectation for regeneration. Although there have been advances in developing suitable wound healing materials, there remains a critical need for rapid and efficient wound management methods. Natural polymers have been applied to a myriad of wound healing strategies due to their prominent biological functions. In the body, these nanobioblocks orchestrate growth, migration, tissue organization and function as a 3D structural scaffold. Benefits of utilizing natural biomaterials in wound healing development include a high

Table 3.1 Characteristics of postnatal and fetal wound healing

	Postnatal wound healing	Prenatal wound healing	References
Scarring	Liver and bone have regeneration abilities. Other tissues primarily rely on fibrosis	Regeneration prior to 24-weeks of gestation	Colwell et al. (2006), Ihara and Motobayashi (1992), Longaker et al. (1994), Lorenz et al. (1993), Rowlatt (1979)
Re-epithelialization	Slow	Rapid	Armstrong and Ferguson (1995), Whitby and Ferguson (1991)
Wound closure	Slow	Rapid	
Oxygen tension	Low	High	
Extracellular characteristics			
ECM structure	Rigid and densely arranged parallel bundles of collagen	Flexible and fine reticular woven collagen	Burd et al. (1990)
Collagen	Collagen type I	Collagen type III	Burd et al. (1990)
ECM deposition	Slow	Rapid	Khatib, Cass, and Adzick (2018)
Hyaluronic acid	Low levels, inhibits cell migration	High levels, enhances cell migration	Khatib et al. (2018), Lo, Zimmermann, Nauta, Longaker, and Lorenz (2012)
Adhesion proteins and cell surface receptors	Low levels	High levels, promotes faster migration	Lo et al. (2012), Longaker et al. (1991), Whitby and Ferguson (1991)
Cellular characteristics			
Platelets	Highly aggregated	Decreased aggregation	Olutoye et al. (1995)
Mast cells	High, mature, and degranulated	Few, immature, fail to degranulate	Wulff and Wilgus (2013)
Neutrophils	High infiltration	Rare infiltration	Olutoye et al. (2005)
Macrophages	Numerous	Few	Hopkinson-Woolley et al. (1994)
Fibroblasts	Present with scant surface hyaluronic acid receptors	Two- to fourfold more hyaluronic acid receptors	Alaish, Yager, Diegelmann, and Cohen (1994)
Myofibroblast	Abundant	Notably absent at day 14	Estes et al. (1994)
Biochemical composition of the inflammatory response			
TGF-β1 and 2	High	Low	Cowin, Holmes, Brosnan, and Ferguson (2001), Lo et al. (2012)
TGF-β3	Low	High	Lin et al. (1995), Nath, LaRegina, Markham, Ksander, and Weeks (1994)
PDGF	Transient	Sustained	Haynes et al. (1994)
Interleukins	IL-6 and 8	Increased IL-10, decreased IL-6 and 8	Gordon et al. (2008), Liechty, Adzick, and Crombleholme (2000), Liechty, Crombleholme, Cass, Martin, and Adzick (1998)
VEGF	Low	High	Colwell et al. (2005)

biological compatibility and minimal inflammatory response and degradation. However, low strength, limited supply, and high cost remain limitations of natural nanomaterials.

3.2.1.1 Protein-Based Polymers

3.2.1.1.1 Collagen

Collagen is the most abundant protein in the body and accounts for two-thirds of the dry weight of the skin. Collagen networks form the majority of the ECM and provide integrity and strength to skin. In addition to its structural functions, collagen is essential for cell migration and differentiation during granulation tissue development. It is composed of three helical polypeptide chains, which assemble into fibrils (Chattopadhyay & Raines, 2014; Shoulders & Raines, 2009). As previously mentioned (Sects. 3.1.4 and 3.1.6), type I and type III collagen are the principal constituents of the postnatal and fetal skin tissue, respectively. Type I collagen is composed of a single α2 chain and two α1 chains. In contrast, type III fetal collagen is a supercoiled homotrimer of α1 chains (Ricard-Blum, 2011). Collagen can be obtained from cows, horses, marine mammals, and rodents. A high degree of homology is present between different species: however, recombinant technologies can be also a great source of collagen production (Ashtikar & Wacker, 2018; Pinkas, Ding, Raines, & Barron, 2011; van der Rest & Garrone, 1991).

Collagen is regarded as an ideal healing scaffold due to its high water uptake abiliy (Malafaya, Silva, & Reis, 2007). In clinical practice, collagen-based materials have been widely used in a variety of forms, concentrations, and applications (Bakhshayesh et al., 2018; Willerth & Sakiyama-Elbert, 2007). Collagen sponges are effective scaffolds that provide a moist and protected environment to the tissues. Apligraf® is the first tissue-engineered commercial dressing material composed of a two-layered collagen hydrogel with human keratinocytes and fibroblasts, approved in 1998 and 2000 for venous ulcer treatment and diabetic ulcers in the skin, respectively (Zaulyanov & Kirsner, 2007). Taking the success of engineered dressing into account, more effort has been devoted to increasing the efficacy of the material. Helary et al. developed a more concentrated collagen hydrogel to improve the dermal part of Apligraf®. Results demonstrated that concentrated collagen hydrogels improved the mechanical resistance of collagen compared to Apligraf®. It also promoted fibroblast proliferation, stimulated KGF and inhibited TGF-β1 expression in vitro (Helary, Zarka, & Giraud-Guille, 2012).

Following Apligraf®, other collagen-based materials have been successfully developed to accelerate wound healing and they have also been used in clinic. AlloDerm®, a collagen-based skin equivalent, has been used in the management of full-thickness burn injuries and support fibroblast migration and neovascularization. Helistat®, an absorbable sponge, acts as a mechanical hemostat by forming a physical matrix to promote platelet aggregation (Gabay & Boucher, 2013; Zwischenberger & Robert, 1999). Enrichment of collagen within bioactive molecules has also been demonstrated to promote wound healing. Integra®, a cellular bilayered membrane enriched with glycosaminoglycans (GAGs) and chondroitin-sulphate-6, showed improved skin quality in reconstructive surgery and burn patients (Clayman, Clayman, & Mozingo, 2006; Heimbach et al., 2003; Muangman et al., 2006). Additionally, the silicone outer layer of Integra® provided a moist environment free from bacterial invasion.

Apart from skin wound healing, collagen-based materials have been intensively studied for cornea injuries in the eye, since collagen is the predominant protein in the cornea (Michelacci, 2003). Hydrogel implants have been shown to stimulate cornea regeneration with no evidence of rejection for over 4 years in human patients (Fagerholm et al., 2014). Hydrogels can also improve cell migration into the corneal pouch and accelerate wound healing in vivo (Watanabe et al., 2011).

Despite the excellent biological and relatively strong mechanical properties, collagen often loses its characteristic shape and stiffness without proper chemical modifications (Aramwit, 2016).

3.2.1.1.2 Fibrin

Fibrin is a biopolymer derived from fibrinogen, and is used to create a fibrin mesh at the site of the damage. The fibrin structure captures platelets to form a primary clot and reduce bleeding (Wolberg, 2007). Fibrin has been used clinically as a hemostatic agent and sealent. The versatility of fibrin, including being used as a delivery agent for inflammatory cells and growth factors, makes it a promising material for skin tissue regeneration (Ahmed, Dare, & Hincke, 2008; Parani, Lokhande, Singh, & Gaharwar, 2016).

Fibrin based materials can be utilized in the form of hydrogels, sheets, and fibrin nanoparticles (NPs). Fibrin gels and sheets are used for autologous transplantation of human keratinocytes to improve re-epithelialization and organization of fibroblasts in burn injuries (Hunyadi, Farkas, Bertényi, Oláh, & Dobozy, 1988; Ronfard et al., 1991; Sun, Haycock, & Macneil, 2006). Fibrin glue incorporated with NPs loaded with growth factors are produced as well (Losi et al., 2013; Zhou, Zhao, Zhao, & Mou, 2011). EGF-loaded chitosan NP containing fibrin-based scaffold increased the stability of EGF and resulted in its sustained-release for 2 weeks in vitro (Zhou et al., 2011). Furthermore, diabetic wounds treated with VEGF and FGF-2 loaded-NPs on a fibrin scaffold fabricated by the spray phase-inversion technique accelerated wound closure, improved re-epithelialization, and collagen organization in granulation tissue (Losi et al., 2013).

Fibrin NPs can be loaded with various drugs and growth factors to stimulate various signaling pathways in wound healing. Fibrin NPs loaded with angiogenic growth factors have been utilized to increase cell proliferation and differentiation in vitro (Praveen, Sreerekha, Menon, Nair, & Chennazhi, 2012). Fibrin NPs are also suitable for antimicrobial drug encapsulation and delivery to infected wounds. Fluconazole and ciprofloxacin delivery within fibrin nanoconstruction displayed selective cytotoxicity against certain bacteria species such as *E. coli*, *S. aureus*, and *C. albicans*, but not against human dermal fibroblasts (Alphonsa et al., 2014).

3.2.1.1.3 Other Protein-Based Nanomaterials

The skin's elasticity and resilience is attributed to elastin filaments present in the dermis (Rodríguez-Cabello, González de Torre, Ibañez-Fonseca, & Alonso, 2018). Elastin also triggers differentiation and migration of the fibroblasts, terminal differentiation of the keratinocytes, and vasculogenesis by modulating myofibrillar organization of the smooth muscle cells during wound healing (Almine et al., 2010; de Vries et al., 1994; Karnik et al., 2003). The soluble form of elastin, tropoelastin, is composed of alternating hydrophobic and hydrophilic domains which self-assemble into fibrils for the construction of nanomaterials. The complexity of the elastin scaffolding in the body can be created in the laboratory with a variety of techniques including electrospinning or sol-gel transition in combination with cross-linkers or alkali conditions (Mithieux, Tu, Korkmaz, Braet, & Weiss, 2009; Rodríguez-Cabello et al., 2018). Highly porous electrospun elastin fibers promoted dermal fibroblast attachment and spreading, proliferation and migration in vitro and new vessel formation in vivo (Rnjak-Kovacina et al., 2011). Elastin-based nanomaterials are often combined with other natural and synthetic materials to increase their tensile strength (Rnjak-Kovacina et al., 2012).

Silk fibroin is a natural protein from silkworms and it is used in biomedical applications due to its mechanical robustness. It has been established as a wound dressing in diverse form such as fibers, films, and sponges. Silk wound dressings improved keratinocyte and fibroblast adhesion and spreading (Min et al., 2004). Additionally, silk/elastin film increased dermal fibroblast proliferation and accelerated re-epithelialization in vitro burn model (Vasconcelos, Gomes, & Cavaco-Paulo, 2012). The sponge form of silk–elastin composite, which forms a hydrogel at body temperature in exudative wounds, improved granulation tissue formation in cutaneous wounds (Kawabata et al., 2017). In addition, compared to conventional hydrogels, silk/elastin hydrogel exhibited better healing capacity by promoting the reepithelization in diabetic mice (Kawabata et al., 2018).

Keratin is a filament-forming protein produced by keratinocytes, and it can be engineered to have optimal flexibility in wound healing applications due to its strength (Park, Kim, Shin, Park, & Kim, 2013). Keratin is often extracted from sheep wool or human hair. The Arg-Gly-Asp and Leu-Asp-Val motifs within the keratin promote cell adhesion, proliferation, and migration (Srinivasan et al., 2010). In allogenic mice full-thickness wound model, a hair-derived keratin dressing promoted re-epithelialization and accelerated wound closure (Konop et al., 2017).

3.2.1.2 Polysaccharide-Based Polymers

3.2.1.2.1 Chitosan

Chitin is the most abundant mucopolysaccharide in nature and contributes to the rigidity in the crustacean exoskeleton and fungal cell walls (Jayakumar, Reis, & Mano, 2006). The hydroxyl and amino groups of the backbone confer chemical modification to manipulate the physicochemical properties of chitin. Chitin is a non-immunogenic and nontoxic biomaterial. However, both forms of natural chitin, the α and β crystalline types, have poor solubility in solvents, which makes its utilization challenging. A deacetylated form of chitin, chitosan, exhibits improved solubility (Pillai, Paul, & Sharma, 2009), and therefore, is used in biomedical applications.

Chitosan, which is a linear copolymer of D-glucosamine and *N*-acetyl-D-glucosamine, exhibits highly biocompatible features. This versatile material can be used as a hydrogel, film, fiber, sponge, or NP with or without addition of other polymers and cells. It is rapidly dissolved in acidic conditions, such as the conditions during wound healing. Its polycationic structure limits microbial growth by forming polyelectrolyte complexes with the phospholipid-containing cell surfaces of pathogens, including fungi, algae, and bacteria. Therefore, it is a potent antimicrobial agent for wound healing (Ahmed & Ikram, 2016). A commercially available chitosan acetate bandage HemCom® decreased the number of *P. aeruginosa* and *P. mirabilis* colonies and increased survival of mice over 15 days post-wounding compared to alginate/silver sulfadiazine treatments on infected wounds (Burkatovskaya et al., 2006). Recently, chitosan loaded silver inlaid gold NPs were engineered and provided better porosity, swelling, and retention properties than chitosan, chitosan-gold, and chitosan-silver NPs alone. This nanocomposite exhibited enhanced antibacterial activity on *E. coli* and *S. aureus* in vitro and faster healing in vivo (Li et al., 2017). Besides being pathogen-resistant, chitosan enhances platelet attachment and activation (Lord, Cheng, McCarthy, Jung, & Whitelock, 2011) to potentiate wound healing early in hemostasis (Okamoto et al., 2003). Recently, it has been demonstrated that 80% deacetylated chitosan selectively stimulates leukotriene B_4 production which assists neutrophil chemotaxis into the wound bed (Hoemann, Marchand, Rivard, El-Gabalawy, & Poubelle, 2017).

The amino groups on the chitosan that are protonated at physiological pH confer the ability to interact and crosslink between chitosan and other natural molecules (Mao, Liu, Yin, & Yao, 2003). Therefore, it is preferable to use chitosan composites rather than chitosan alone to enhance stability and healing. The self-assembled chitosan-coated halloysite nanotubes formed by electrostatic interactions were recently created as a pourable powder. In this study, the chitosan-coated nanotubes supported angiogenesis, re-epithelialization, and regeneration of hair follicles in a cutaneous wound model (Sandri et al., 2017).

3.2.1.2.2 Alginate

Alginate is a commercially extracted anionic polymer obtained from brown algae. Alginate contains β-(1–4)-linked D-mannuronic acid (M-block) and α-(1–4)-linked L-glucuronic acid (G-block). It is typically used in hydrogels due to its ease of gelation, which occurs through dimerization of the G-residues through crosslinking of divalent cations (Kristiansen et al., 2009). The composition of the G-block in alginate determines the mechanical properties of the hydrogel. Increasing the molecular weight and length of the G-residue blocks can enhance the stiffness of the material, but also may increase the viscosity (Lee & Mooney, 2012), which limits its manufacturing and application potential.

Alginate has high water absorptivity (up to 20 times of its weight) and therefore, functions to create a moist wound environment and limits wound exudation (Mogoşanu & Grumezescu, 2014). Therefore, it is suitable for wounds that require abundant drainage, but may not be used for dry wounds, including burns (Dhivya, Padma, & Santhini, 2015). Alginate-based material activity is pH-dependent, which can enable a designed controlled release system. Alginate shrinks to prevent drug release in acidic conditions, while basophilic conditions accelerate its degradation and induces a rapid release of desired molecules (Kuo & Ma, 2001).

Some commercially available alginate-based dressings are used for surgical wounds, diabetic foot ulcers, and infected cutaneous wounds (listed in Aderibigbe & Buyana, 2018). For example, Kaltostat® is an absorbent fibrous alginate salt. Following application on exudative wounds, it forms a gel due to the ionic interactions. The endotoxin in alginates promoted wound healing by increasing TNF-α secretion from macrophages (Thomas, Harding, & Moore, 2000). Incorporation of zinc oxide (ZnO) NPs into alginate hydrogel resulted in rapid clotting and re-epithelialization compared to Kaltostat® (Mohandas, Kumar, Raja, Lakshmanan, & Jayakumar, 2015). Although alginate alone has positive effects on wound healing, composite materials are commonly investigated to enhance the healing properties. A hydrogel sheet composed of the chitosan-fucoidan-alginate blend improves granulation tissue formation and angiogenesis in mitomycin C treated wounds compared to a commercially available alginate dressing (Murakami et al., 2010). Another composite material utilized in wound healing is strontium loaded silk fibroin/sodium alginate film. This composite can potentially stimulate angiogenesis through VEGF and FGF-2 activation (Li, Li, Guo, Qin, & Yu, 2017). Sustained release properties of alginate hydrogels have been demonstrated by incorporating alginate with simvastatin-containing mesoporous hydroxyapatite microspheres. This combination triggered vascular tube formation VEGF and FGF-2 secretion of human umbilical vein endothelial cells in an in vitro model. Incorporated simvastatin also enhanced angiogenesis, collagen deposition and maturation, re-epithelialization, and wound contraction in a rodent full-thickness wound model (Yu et al., 2016).

3.2.1.2.3 Other Polysaccharide-Based Nanomaterials

Hyaluronic acid (HA) is a non-sulfated GAG that plays essential roles in acute and chronic cutaneous wound healing including early inflammation and angiogenesis (Dechert, Ducale, Ward, & Yager, 2006). In addition, it is involved in prenatal wound healing (Table 3.1). Due to its hygroscopicity, it regulates cell adhesion and attachment during wound healing (Voigt & Driver, 2012). Commercially available, bilayer hyaluronic acid autografts, Hyalograft® and Laserskin®, demonstrated improved healing on the patients with extensive burn wounds (Hollander, Soranzo, Falk, & Windolf, 2001). In addition, long-term outcomes of the HA-based matrix promoted connective tissue growth and dermal regeneration (Faga et al., 2013). Recently, HA-based materials, used as a stem cell carrier to full-thickness excisional wounds, exhibited improved healing by supporting the long-term release of stem cell-derived soluble factors (Skardal et al., 2017). A novel amniotic membrane-derived material composed of a combined UV crosslinked HA hydrogel system demonstrated great potential in wound healing by promoting angiogenesis, keratinocyte proliferation, and wound closure (Murphy et al., 2017).

Nanocellulose is a novel material for wound healing management obtained from woody plants and bacteria via chemical or enzymatic treatments. Nanocellulose contains lignocellulosic, which enables a remarkable strength, flexibility, and strength-to-weight ratio. Nanocellulose can be used as cellulose nanocrystals or cellulose nanofibers for wound healing (Lin & Dufresne, 2014). Nanocellulose is a hydrophilic material due to its hydroxyl groups and its high fiber surface area enables the formation of a strong hydrogel through chemical interactions with the surrounding environment (Klemm et al., 2009). As a xeno-free material, bacterial cellulose is particularly promising for wound applications due to its high swelling properties in neutral and alkaline conditions (Chinga-Carrasco & Syverud, 2014). Nanocellulose fibers induced rapid re-epithelialization in the treatment of skin graft donor sites in preclinical studies (Hakkarainen et al., 2016). Moreover, bacterial nanocellulose was shown to reduce the healing time on diabetic foot ulcers and skin tears in patients (Solway, Clark, & Levinson, 2011; Solway, Consalter, & Levinson, 2010). Application of a bacterial nanocellulose dressing incorporated with polyhexamethylene biguanide and sericin accelerated wound closure and collagen formation in a full-thickness wound healing rodent model within 14 days compared to standard care (Napavichayanun, Yamdech, & Aramwit, 2016).

3.2.2 Synthetic Nanomaterials

3.2.2.1 Synthetic Polymers

Synthetic polymers are highly useful in biomedical applications as they possess homogenous physical and chemical properties and higher stability compared to natural polymers (Zhong, Zhang, & Lim, 2010). They can be cheaply synthesized on a large scale and tailored for specific wound healing applications. These materials can be engineered with various physical and mechanical features such as high porosity, adhesiveness, biodegradability.

3.2.2.1.1 Polyethylene Glycol (PEG)

As a nonionic and hydrophilic polymer, polyethylene glycol is one of the most commonly used polymers in various medical applications. It is has been approved by Food and Drug Administration (FDA) for human use and has a well-characterized safety profile and biocompatibility.

A major challenge in utilizing nanomaterials for wound healing development is the possibility of a detrimental immune reaction, which degrades the nanomaterial and causes surface chemistry changes. The presence of polyethylene chains within the polymer resists protein adhesion and therefore, PEG had demonstrated low immunogenicity under in vivo conditions (Alcantar, Aydil, & Israelachvili, 2000). Although PEG can form thermosensitive hydrogels, it has poor mechanical strength. Cross-linking with other polymers increases its mechanical properties to produce proper biomaterials for wound management (Gokarneshan, Anitha Rachel, Rajendran, Lavanya, & Ghoshal, 2015; Peppas, Keys, Torres-Lugo, & Lowman, 1999). In PEG hydrogels, the hydroxyl ends of the chains are left uncovered, suitable for biological modifications (Zhu, 2010). Application of a PEG hydrogel system on a full-thickness cutaneous incision demonstrated improved wound closure and hemostasis compared to a standard wound dressing. PEG hydrogel can be degraded after 14 days of application, following cutaneous wound healing. In addition, rapid re-epithelialization and collagen production were observed on PEG hydrogel treated wounds (Chen et al., 2018). Dong et al. also engineered an injectable PEG-gelatin hydrogel system with high tunable cross-linking and hydrogel properties. It has been revealed that adipose-derived stem cell-encapsulated hydrogel provides better cell retention due to the 3D matrix support of the system. Furthermore, stem cells accelerate the wound closure, and enhance the reepithelization and granulation tissue formation when they are encapsulated in a PEG-based hydrogel system (Dong et al., 2017).

3.2.2.1.2 Polylactic Acid (PLA)

Polylactic acid, PLA, is another biocompatible and biodegradable thermoplastic polymer that has been investigated over the last several decades for tissue engineering and medical device applications (Vink et al., 2004). The FDA has approved the direct contact of PLA with biological fluids. Optically active and inactive forms of PLA offer an adjustable degradation rate and tensile strength (Farah, Anderson, & Langer, 2016). As an inactive biomolecule, PLA has many desirable features for wound healing including a high stiffness, hydrophobic charge to retain a moist healing environment, and close modeling of the ECM. PLA nanofibers are also favorable as drug delivery vehicles (Mohiti-Asli et al., 2017; Perumal et al., 2017). Recently, a skin graft made of bone marrow derived stem cells (BMSC) seeded on a highly porous 3D PLA nanofiber system via wet-electrospinning was evaluated on full-thickness skin wound healing. BMSC/PLA nanofiber treated wounds exhibited progressive healing including the formation of organized granulation tissue with compact collagen deposition surrounded by a differentiated epithelium (Ghorbani et al., 2018).

3.2.2.1.3 Poly(lactic-co-glycolic acid)

Polylactic-co-glycolic acid (PLGA) is an amorphous and biodegradable copolymer composed of polylactic acid and polyglycolic acid molecules. It has been approved by FDA for parenteral administration and has been studied in medical applications ranging from wound healing, drug delivery, and skeletal implants (Makadia & Siegel, 2011; Zhang, Wu, Jing, & Ding, 2005). Under in vivo conditions, ester bonds of PLGA can be easily hydrolyzed to produce lactic acid (Shenderova, Burke, & Schwendeman, 1999). Since lactic acid is produced by macrophages to kill bacteria (Hussain, Ghani, & Hunt, 1989) and plays a role in collagen synthesis and angiogenesis (Constant et al., 2000; Porporato et al., 2012), it has been considered as a promising wound dressing material.

PLGA can be easily tuned by changing the surface chemistry and the properties; for example, the degradation rate of PLGA depends on the ratio of lactic acid and glycolic acid. PLGA-based materials can form nanofibers, nanoparticles, or hydrogels for wound management in combination with drugs, especially hydrophobic drugs, or growth factors. Chen et al. developed a drug loaded PLGA/collagen blend using an electrospinning technique. The nanofibrous membrane provided a sustained release at the injury site and effectively accelerated wound healing in bacteria infected wounds (Chen, Liao, Liu, & Chan, 2012). The synergistic effect of GF delivery and the maintenance of bioactivity was reported in diabetic and nondiabetic animals using PLGA-NPs. VEGF or VEGF/FGF-2 loaded PLGA-NPs promote granulation tissue formation and complete re-epithelialization (Losi et al., 2013) and formation of an organized epithelium with aligned dermal collagen (Chereddy, Vandermeulen, & Préat, 2016). In addition, a thermosensitive PLGA/polyethylene glycol (PEG) hydrogel containing a TGF-β1 plasmid supported moisture retention at the injury site resulting in faster wound healing in a diabetic mouse model (Lee, Li, & Huang, 2003).

3.2.2.1.4 Poly(caprolactone)

Polycaprolactone (PCL) is a semi-crystalline synthetic polymer with high elasticity (Dash & Konkimalla, 2012). A nanofibrous form of PCL, largely similar to collagen fibers, had been widely investigated for wound healing applications (Saeed, Mirzadeh, Zandi, & Barzin, 2017). The structure of PCL is suitable for chemical modifications and blending with other polymers to obtain intended properties. Complete degradation of PCL can take 2–3 years with hydrolytic and enzymatic cleavage based on its molecular weight (Sun et al., 2006). Due to its stability at ambient conditions, it is a favorable material for nanofiber production using electrospinning. PCL was evaluated for wound and burn management where it was generally used as a supporting material. Chitosan/HA-coated electrospun PCL nanofibers has shown promise in wound healing advancement, as they promoted keratinocyte adhesion and proliferation in vitro (Croisier, Atanasova, Poumay, & Jérôme, 2014). Another PCL composite with gelatin and collagen type I remains a candidate for wound healing, as it enhanced fibroblast adhesion and proliferation (Gautam, Chou, Dinda, Potdar, & Mishra, 2014). In a study examining the combination of PCL with HA membrane, the material retained the exudate at the injury site compared to a commercially available dressing. However, no significant difference between control and PCL/HA membrane regarding the wound healing process was demonstrated in this study (Cai et al., 2014). Mussel adhesive protein blended PCL nanofibers have been demonstated to facilitate keratinocyte migration and increased FGF-2 binding in vitro and improved wound healing in vivo (Kim et al., 2017).

3.2.2.1.5 Other Synthetic Nanomaterials

Polyurethane (PU) has been extensively studied as a dressing material due to its biocompatibility, flexibility, and barrier properties. PU-based wound dressings have been shown to restrict bacterial contamination and wound dehydration, while allowing crucial gas exchange at the site of injury (Hinrichs, Lommen, Wildevuur, & Feijen, 1992). PU-based materials can be used as films or foams. Films are transparent, thin and self-adhesive materials suitable for uninfected and superficial wounds. Foams are highly absorbent and prevent injection and are utilized in burns and chronic wounds (Kamoun, Kenawy, & Chen, 2017). PU-based commercially available wound healing products include Pellethane®, Tegaderm®, and Opsite®.

3.2.2.2 Metallic Nanomaterials

Metal and metal oxide NPs have been extensively investigated in wound healing applications for decades. They have been proven to prevent wound infection with an antimicrobial action against bacteria, viruses, yeast, and fungi (Lara, Garza-Treviño, Ixtepan-Turrent, & Singh, 2011).

Silver is the most common metal used in wound management for centuries as a topical solution of silver nitrate (Parani et al., 2016; Tocco, Zavan, Bassetto, & Vindigni, 2012). In an aqueous environment, silver nanoparticles (AgNPs) release silver ions as a result of oxidization. Metallic silver is an inert material. In contrast, the ionized form, which forms in acidic conditions with exudative wounds, exhibits antimicrobial properties (Rai, Yadav, & Gade, 2009). It should be emphasized here the antimicrobial activity is attributed to the silver, not nanoparticle enhancement. The importance of utilizing a NP system is to create a larger surface area for improved cell surface interactions and microbial penetration. The antimicrobial activity is dependent on the size and shape of the NPs. Triangular shaped NPs size between 7 and 20 nm demonstrate a higher bactericidal activity due to the larger surface area and rapid release of silver ions (Tocco et al., 2012; Xiu, Zhang, Puppala, Colvin, & Alvarez, 2012) (Pal, Tak, & Song, 2007). AgNPs treated patients with second degree burns demonstrated wound healing 13 days earlier without any adverse effects compared to silver sulfadiazine, a standard topical antimicrobial treatment for burns (Adhya et al., 2014). Studies on diabetic ulcer treatment revealed that an antifouling hybrid hydrogel containing AgNPs have superior wound healing activity with increased fibroblast migration, granulation tissue formation, and angiogenesis (Shi et al., 2018). Additionally, mice treated with AgNPs showed increased keratinocyte proliferation and migration, and myofibroblast differentiation with improved wound contraction (Liu et al., 2010).

Another metallic material that has been used as an inorganic antibacterial agent is zinc oxide (ZnO). Utilization of ZnO resulted in reduced inflammation and bacterial growth and improved re-epithelialization on chronic wounds and burn injuries (Rajendran, Kumar, Houreld, & Abrahamse, 2018). Compared to AgNPs, ZnONPs show less cytotoxicity to mammalian cells in vitro (Bondarenko et al., 2013). Incorporation of ZnONPs within chitosan hydrogel showed enhanced blood clotting, and antibacterial activity. In addition, treatment of partial thickness wounds with ZnONPs/chitosan hydrogels promoted wound healing and faster collagen deposition and re-epithelialization (Sudheesh Kumar et al., 2012). Recently, ZnONPs incorporated within heparinized polyvinyl alcohol/chitosan hydrogels were shown to exhibit strong antimicrobial activity and are considered a promising dressing material (Khorasani, Joorabloo, Moghaddam, Shamsi, & MansooriMoghadam, 2018).

Gold NPs (AuNPs) also have been tested for wound healing applications due to their stability, optical, and chemical properties and ease of surface modifications. For AuNPs, surface modification or incorporation with other biomolecules is required prior to wound healing applications. Adding peptides, or polysaccharides, or GF modifications to AuNPs increased their potency to enhance healing (Akturk et al., 2016). AuNPs can inhibit free radicals including hydrogen peroxide, nitric oxide and hydroxyl, and function as a strong antioxidant (Medhe, Bansal, & Srivastava, 2014). The application of AuNPs on cutaneous wounds showed an increase in collagen, VEGF, and angiopoietin expression and a decrease in TGF-β1 and MMP levels (Kim et al., 2015). Furthermore, antibiotic-coated AuNP doped PCL/gelatin nanofiber mat was investigated on infected full-thickness wounds and showed decreased bacterial load and improved healing (Yang et al., 2017).

3.2.2.3 Self-Assembling Peptide Nanomaterials

Supramolecular self-assembly is the spontaneous organization of nanoblocks, which occurs in biological systems to build complex structures, including peptides and proteins. By using principles found in nature, self-assembling peptide nanomaterials can be designed using various combinations of amino acids with astonishing speed and efficiency compared to top-down conventional approaches. Fabrication of self-assembled peptide nanostructures create stable materials, which lack immunogenicity compared to other natural materials. Therefore, self-assembling peptide nanomaterials is an emerging field in synthetic biology.

Self-assembly occurs between individual building blocks through electrostatic interactions including hydrophobic and hydrophilic forces, hydrogen bonding, van der Waals interactions, and π–π stacking. In contrast to the covalent bonding system used in polymeric systems, the combination of these weak interactions in self-assembled peptide system generates non-energy driven and thermodynamically stable materials (Cui, Webber, & Stupp, 2010; Hosseinkhani, Hong, & Yu, 2013). Self-assembled peptide systems can mimic the natural ECM and interact with the cells and tissues (Cui et al., 2010). These materials have been investigated for various nanomedical applications due to their excellent biodegradability, biocompatibility, bioactivity and ease of bottom-up synthesis (Chen et al., 2015; Liu et al., 2009; Schneider, Garlick, & Egles, 2008; Yeboah et al., 2016).

Peptide amphiphiles (PA) are a promising class of self-assembling peptide-based materials inspired by the lipid structure of the cell membranes. These contain a hydrophobic alkyl tail and a hydrophilic head, which mediate self-assembled nanofibers (Toksoz, Mammadov, Tekinay, & Guler, 2011). The hydrophilic head consists of a short β-sheet forming peptide sequence capable of generating the non-covalent interactions to build one-dimensional nanofiber structures (Matson, Zha, & Stupp, 2011). The self-assembly process can occur as a result of neutralization of acidic or basic amino acid-containing PA (Hartgerink, Beniash, & Stupp, 2001), the addition of divalent cations (Hartgerink, Beniash, & Stupp, 2002), or combining two oppositely charged PA molecules (Niece, Hartgerink, Donners, & Stupp, 2003).

PA nanofibers can be designed to bear bioactive peptide epitopes on their surfaces to mimic biological conditions. In aqueous solutions, the hydrophobic collapse of the aliphatic tails due to the presence of the β-sheet domain exposes the bioactive signals, which stimulates the desired biological activities (Hendricks, Sato, Palmer, & Stupp, 2017). The cellular response to the bioactive epitopes are dependent on the local concentration of the available epitopes. In addition, physical entrapment of these epitopes on to the PA nanofibers induces rapid cellular differentiation compared to soluble epitope treatments (Silva et al., 2004).

Given the cellular, molecular, and metabolic complexity of wound healing, bioactive epitopes can be specifically designed to trigger a myriad of biological processes. Heparin-mimetic PA nanofibers have been shown to promote angiogenesis through binding to proangiogenic growth factors, VEGF and FGF-2 (Mammadov, Mammadov, Guler, & Tekinay, 2012). The presence of sulfonate, carboxylate, and hydroxyl groups was sufficient to induce the vascular tube formation of endothelial cells without additional growth factors. The self-assembled heparin-mimetic PA hydrogel demonstrated an effective bioactive wound dressing for rapid and functional repair of acute wounds and burns (Uzunalli et al., 2016; Yergoz et al., 2017). The excisional and burn wounds exhibited better granulation tissue organization with increased collagen deposition and angiogenesis. Cutaneous adnexa were also observed in wounds treated with heparin-mimetic PA. Furthermore, the presentation of the bioactive scaffolds is crucial for wound healing applications.

In the corneal stroma of the eye, the wound healing process is similar that previously discussed for the skin. In a corneal wound model, the fibronectin-derived peptide RGD coupled silk film has been shown to enhance cell attachment and proliferation (Gil et al., 2010), whereas RGD-bearing PA scaffold did not (Uzunalli et al., 2014). Laminin-derived YIGSR-sequence containing PA scaffold supported keratocyte proliferation and attachment in an in vitro system and corneal stromal healing in vivo (Uzunalli et al., 2014).

Beyond carrying biological cues, 3D nanofibrous PA scaffolds provide a wound specific topography to mirror the ECM. Fibroblasts have the ability to modulate their cellular and molecular behaviors depending on the matrix morphology (Andrews & Hunt, 2008). Furthermore, keratinocyte proliferation and migration depends on the stiffness of the underlying substrate (Zarkoob, Bodduluri, Ponnaluri, Selby, & Sander, 2015). In addition, a porous structure supports the development of granulation tissue and wound resolution (Doillon, 1988). PA nanofibers can be designed with a desired pore size and stiffness by modifying concentrations or adding divalent cations (Cui et al., 2010; Dagdas, Tombuloglu, Tekinay, Dana, & Guler, 2011).

In conclusion, this designable emerging technology represents an untapped resource, which offers great potential for wound regeneration and restoration of tissue function.

3.3 Conclusion and Future Directions

Wound healing is a dynamic physiological process with an end goal of restoring the function of the devitalized tissue. Ideally, complex cellular activities and signaling cascades interact in the body to facilitate rapid wound closure. However, natural healing through fibrosis results in loss of tissue structure and function. Therefore, as the mean age of the population and number of patients with chronic diseases increase, the economic burden of wound management continues to rise. The limitations of the conventional wound care strategies including dressing, debridement, and antibiotic treatment cause failure of the completely regeneration of the wounded tissues. In the past three decades, tremendous effort was made to promote tissue regeneration by therapeutically manipulating repair-based interventions for wound treatment. A variety of natural and synthetic nanomaterials and their combinations have been engineered for individual pathways in wound healing. Using nanomaterials has enormous potential to promote self-healing mechanisms that can mimic fetal regeneration.

However, there are still several challenges regarding the use of nanomaterials for wound healing. The heterogeneous nature of the wounded tissues requires a better understanding of the underlying mechanisms and cellular cascades to personalize nanomaterials for different wound healing applications. Additionally, limited data is present regarding the systemic toxicity of the nanomaterials that can interfere with the biological functions. Novel analytic techniques that can assure the quality and reproducibility of the nanomaterial synthesis are crucial. Although significant advances have been made using animal models of wound healing, they fail to recapitulate the physiology of human patients. Therefore, animal models in combination with organ-on-a-chip systems and clinical trials can provide a better insight and reveal their true potential. Further knowledge of cellular and molecular changes in adult and fetal wound healing including advances in technology, stem cell research, and -omics will play a bigger role for engineering personalized therapies for wound healing.

References

Aderibigbe, B. A., & Buyana, B. (2018). Alginate in wound dressings. *Pharmaceutics, 10*, E42.

Adhya, A., Bain, J., Ray, O., Hazra, A., Adhikari, S., Dutta, G., … Majumdar, B. K. (2014). Healing of burn wounds by topical treatment: A randomized controlled comparison between silver sulfadiazine and nano-crystalline silver. *Journal of Basic and Clinical Pharmacy, 6*, 29–34.

Ahmed, S., & Ikram, S. (2016). Chitosan based scaffolds and their applications in wound healing. *Achievements in the Life Sciences, 10*, 27–37.

Ahmed, T. A. E., Dare, E. V., & Hincke, M. (2008). Fibrin: A versatile scaffold for tissue engineering applications. *Tissue Engineering. Part B, Reviews, 14*, 199–215.

Akturk, O., Kismet, K., Yasti, A. C., Kuru, S., Duymus, M. E., Kaya, F., … Keskin, D. (2016). Collagen/gold nanoparticle nanocomposites: A potential skin wound healing biomaterial. *Journal of Biomaterials Applications, 31*, 283–301.

Alaish, S. M., Yager, D., Diegelmann, R. F., & Cohen, I. K. (1994). Biology of fetal wound healing: Hyaluronate receptor expression in fetal fibroblasts. *Journal of Pediatric Surgery, 29*, 1040–1043.

Alcantar, N. A., Aydil, E. S., & Israelachvili, J. N. (2000). Polyethylene glycol-coated biocompatible surfaces. *Journal of Biomedical Materials Research, 51*, 343–351.

Almine, J. F., Bax, D. V., Mithieux, S. M., Nivison-Smith, L., Rnjak, J., Waterhouse, A., … Weiss, A. S. (2010). Elastin-based materials. Chem Soc Rev, 39(9), 3371–3379. https://doi.org/10.1039/b919452p.

Alphonsa, B. M., Sudheesh Kumar, P. T., Praveen, G., Biswas, R., Chennazhi, K. P., & Jayakumar, R. (2014). Antimicrobial drugs encapsulated in fibrin nanoparticles for treating microbial infested wounds. *Pharmaceutical Research, 31*, 1338–1351.

Andrews, K. D., & Hunt, J. A. (2008). Upregulation of matrix and adhesion molecules induced by controlled topography. *Journal of Materials Science. Materials in Medicine, 19*, 1601–1608.

Aramwit, P. (2016). Introduction to biomaterials for wound healing. In *Wound healing biomaterials* (pp. 3–38). Amsterdam: Elsevier.

Armstrong, J. R., & Ferguson, M. W. (1995). Ontogeny of the skin and the transition from scar-free to scarring phenotype during wound healing in the pouch young of a marsupial, Monodelphis domestica. *Developmental Biology, 169*, 242–260.

Ashtikar, M., & Wacker, M. G. (2018). Nanopharmaceuticals for wound healing—Lost in translation? *Advanced Drug Delivery Reviews, 129*, 194–218.

Atala, A., Irvine, D. J., Moses, M., & Shaunak, S. (2010). Wound healing versus regeneration: Role of the tissue environment in regenerative medicine. *MRS Bulletin/Materials Research Society, 35*.

Bakhshayesh, A. R. D., Annabi, N., Khalilov, R., Akbarzadeh, A., Samiei, M., Alizadeh, E., … Montaseri, A. (2018). Recent advances on biomedical applications of scaffolds in wound healing and dermal tissue engineering. *Artificial Cells, Nanomedicine, and Biotechnology, 46*, 691–705.

Behrens, D. T., Villone, D., Koch, M., Brunner, G., Sorokin, L., Robenek, H., ... Hansen, U. (2012). The epidermal basement membrane is a composite of separate laminin- or collagen IV-containing networks connected by aggregated perlecan, but not by nidogens. *The Journal of Biological Chemistry, 287*, 18700–18709.
Boateng, J., & Catanzano, O. (2015). Advanced therapeutic dressings for effective wound healing—A review. *Journal of Pharmaceutical Sciences, 104*, 3653–3680.
Bondarenko, O., Juganson, K., Ivask, A., Kasemets, K., Mortimer, M., & Kahru, A. (2013). Toxicity of Ag, CuO and ZnO nanoparticles to selected environmentally relevant test organisms and mammalian cells in vitro: a critical review. *Archives of Toxicology, 87*, 1181–1200.
Brooks, P. C., Sanders, L. C., Aimes, R. T., Stetler-Stevenson, W. G., Quigley, J. P., & Cheresh, D. A. (1994). Localization of matrix metalloproteinase MMP-2 to the surface of invasive cells by interaction with integrin alpha v beta 3. *Science, 85*(5), 683–693.
Burd, D. A., Longaker, M. T., Adzick, N. S., Harrison, M. R., & Ehrlich, H. P. (1990). Foetal wound healing in a large animal model: The deposition of collagen is confirmed. *British Journal of Plastic Surgery, 43*, 571–577.
Burkatovskaya, M., Tegos, G. P., Swietlik, E., Demidova, T. N., Castano A, P., & Hamblin, M. R. (2006). Use of chitosan bandage to prevent fatal infections developing from highly contaminated wounds in mice. *Biomaterials, 27*, 4157–4164.
Cai, E. Z., Teo, E. Y., Jing, L., Koh, Y. P., Qian, T. S., Wen, F., ... Lim, T. C. (2014). Bio-conjugated polycaprolactone membranes: A novel wound dressing. *Archives of Plastic Surgery, 41*, 638–646.
Cavani, A., Zambruno, G., Marconi, A., Manca, V., Marchetti, M., & Giannetti, A. (1993). Distinctive integrin expression in the newly forming epidermis during wound healing in humans. *The Journal of Investigative Dermatology, 101*, 600–604.
Chattopadhyay, S., & Raines, R. T. (2014). Review collagen-based biomaterials for wound healing. *Biopolymers, 101*, 821–833.
Chen, D. W.-C., Liao, J.-Y., Liu, S.-J., & Chan, E.-C. (2012). Novel biodegradable sandwich-structured nanofibrous drug-eluting membranes for repair of infected wounds: An in vitro and in vivo study. *International Journal of Nanomedicine, 7*, 763–771.
Chen, S., Zhang, M., Shao, X., Wang, X., Zhang, L., Xu, P., ... Zhang, L. (2015). A laminin mimetic peptide SIKVAV-conjugated chitosan hydrogel promoting wound healing by enhancing angiogenesis, re-epithelialization and collagen deposition. *Journal of Materials Chemistry B, 3*, 6798–6804.
Chen, S.-L., Fu, R.-H., Liao, S.-F., Liu, S.-P., Lin, S.-Z., & Wang, Y.-C. (2018). A PEG-based hydrogel for effective wound care management. *Cell Transplantation, 27*, 275–284.
Chereddy, K. K., Vandermeulen, G., & Préat, V. (2016). PLGA based drug delivery systems: Promising carriers for wound healing activity. *Wound Repair and Regeneration, 24*, 223–236.
Chinga-Carrasco, G., & Syverud, K. (2014). Pretreatment-dependent surface chemistry of wood nanocellulose for pH-sensitive hydrogels. *Journal of Biomaterials Applications, 29*, 423–432.
Clark, R. A. F. (2013). *The molecular and cellular biology of wound repair*. Berlin: Springer Science & Business Media.
Clayman, M. A., Clayman, S. M., & Mozingo, D. W. (2006). The use of collagen-glycosaminoglycan copolymer (integra) for the repair of hypertrophic scars and keloids. *Journal of Burn Care & Research, 27*, 404–409.
Colwell, A. S., Beanes, S. R., Soo, C., Dang, C., Ting, K., Longaker, M. T., ... Lorenz, H. P. (2005). Increased angiogenesis and expression of vascular endothelial growth factor during scarless repair. *Plastic and Reconstructive Surgery, 115*, 204.
Colwell, A. S., Krummel, T. M., Longaker, M. T., & Lorenz, H. P. (2006). An in vivo mouse excisional wound model of scarless healing. *Plastic and Reconstructive Surgery, 117*, 2292.
Constant, J. S., Feng, J. J., Zabel, D. D., Yuan, H., Suh, D. Y., Scheuenstuhl, H., ... Hussain, M. Z. (2000). Lactate elicits vascular endothelial growth factor from macrophages: A possible alternative to hypoxia. *Wound Repair and Regeneration, 8*, 353–360.

Cowin, A. J., Holmes, T. M., Brosnan, P., & Ferguson, M. W. (2001). Expression of TGF-beta and its receptors in murine fetal and adult dermal wounds. *European Journal of Dermatology, 11*, 424–431.
Cox, J. (2011). Predictors of pressure ulcers in adult critical care patients. *American Journal of Critical Care, 20*, 364–375.
Croisier, F., Atanasova, G., Poumay, Y., & Jérôme, C. (2014). Polysaccharide-coated PCL nanofibers for wound dressing applications. *Advanced Healthcare Materials, 3*, 2032–2039.
Cui, H., Webber, M. J., & Stupp, S. I. (2010). Self-assembly of peptide amphiphiles: From molecules to nanostructures to biomaterials. *Peptide Science, 94*, 1–18.
Dagdas, Y. S., Tombuloglu, A., Tekinay, A. B., Dana, A., & Guler, M. O. (2011). Interfiber interactions alter the stiffness of gels formed by supramolecular self-assembled nanofibers. *Soft Matter, 7*, 3524.
Dash, T. K., & Konkimalla, V. B. (2012). Poly-є-caprolactone based formulations for drug delivery and tissue engineering: A review. *Journal of Controlled Release, 158*, 15–33.
Dechert, T. A., Ducale, A. E., Ward, S. I., & Yager, D. R. (2006). Hyaluronan in human acute and chronic dermal wounds. *Wound Repair and Regeneration, 14*, 252–258.
Demidova-Rice, T. N., Hamblin, M. R., & Herman, I. M. (2012). Acute and impaired wound healing: Pathophysiology and current methods for drug delivery, Part 1: Normal and chronic wounds: Biology, causes, and approaches to care. *Advances in Skin and Wound Care, 25*, 304–314.
Derynck, R. (1988). Transforming growth factor α. *Cell, 54*, 593–595.
de Vries, H.-J., Middlekoop, E., Mekkes, J.-R., Dutrieux, R.-P., Wildevuur, C.-H., & Westerhof, H. (1994). Dermal regeneration in native non-cross-linked collagen sponges with different extracellular matrix molecules. *Wound Repair and Regeneration, 2*(1), 37–47.
Dhivya, S., Padma, V. V., & Santhini, E. (2015). Wound dressings—A review. *Biomedicine, 5*(4), 22.
Doillon, C. J. (1988). Porous collagen sponge wound dressings: In vivo and in vitro studies. *Journal of Biomaterials Applications, 2*, 562–578.
Dong, Y., Sigen, A., Rodrigues, M., Li, X., Kwon, S. H., Kosaric, N., … Gurtner, G. C. (2017). Injectable and tunable gelatin hydrogels enhance stem cell retention and improve cutaneous wound healing. *Advanced Functional Materials, 27*, 1606619.
Dorward, D. A., Lucas, C. D., Chapman, G. B., Haslett, C., Dhaliwal, K., & Rossi, A. G. (2015). The role of formylated peptides and formyl peptide receptor 1 in governing neutrophil function during acute inflammation. *The American Journal of Pathology, 185*, 1172–1184.
Dugina, V., Fontao, L., Chaponnier, C., Vasiliev, J., & Gabbiani, G. (2001). Focal adhesion features during myofibroblastic differentiation are controlled by intracellular and extracellular factors. *Journal of Cell Science, 114*, 3285–3296.
Ehrlich, H. P., & Krummel, T. M. (1996). Regulation of wound healing from a connective tissue perspective. *Wound Repair and Regeneration, 4*, 203–210.
Enoch, S., & Leaper, D. J. (2005). Basic science of wound healing. *Surgery, 23*, 37–42.
Estes, J. M., Berg, J. S. V., Adzick, N. S., MacGillivray, T. E., Desmoulière, A., & Gabbiani, G. (1994). Phenotypic and functional features of myofibroblasts in sheep fetal wounds. *Differentiation, 56*, 173–181.
Faga, A., Nicoletti, G., Brenta, F., Scevola, S., Abatangelo, G., & Brun, P. (2013). Hyaluronic acid three-dimensional scaffold for surgical revision of retracting scars: A human experimental study. *International Wound Journal, 10*, 329–335.
Fagerholm, P., Lagali, N. S., Ong, J. A., Merrett, K., Jackson, W. B., Polarek, J. W., … Griffith, M. (2014). Stable corneal regeneration four years after implantation of a cell-free recombinant human collagen scaffold. *Biomaterials, 35*, 2420–2427.
Farah, S., Anderson, D. G., & Langer, R. (2016). Physical and mechanical properties of PLA, and their functions in widespread applications—A comprehensive review. *Advanced Drug Delivery Reviews, 107*, 367–392.
Gabay, M., & Boucher, B. A. (2013). An essential primer for understanding the role of topical hemostats, surgical sealants, and adhesives for maintaining hemostasis. *Pharmacotherapy, 33*, 935–955.

Gautam, S., Chou, C.-F., Dinda, A. K., Potdar, P. D., & Mishra, N. C. (2014). Surface modification of nanofibrous polycaprolactone/gelatin composite scaffold by collagen type I grafting for skin tissue engineering. *Materials Science and Engineering: C, 34*, 402–409.
Ghorbani, S., Eyni, H., Tiraihi, T., Asl, L. S., Soleimani, M., Atashi, A., … Warkiani, M. E. (2018). Combined effects of 3D bone marrow stem cell-seeded wet-electrospun poly lactic acid scaffolds on full-thickness skin wound healing. *International Journal of Polymeric Materials and Polymeric Biomaterials, 67*, 905–912.
Gil, E. S., Mandal, B. B., Park, S.-H., Marchant, J. K., Omenetto, F. G., & Kaplan, D. L. (2010). Helicoidal multi-lamellar features of RGD-functionalized silk biomaterials for corneal tissue engineering. *Biomaterials, 31*, 8953–8963.
Gohel, M. S., Windhaber, R. A. J., Tarlton, J. F., Whyman, M. R., & Poskitt, K. R. (2008). The relationship between cytokine concentrations and wound healing in chronic venous ulceration. *Journal of Vascular Surgery, 48*, 1272–1277.
Gokarneshan, N., Anitha Rachel, D., Rajendran, V., Lavanya, B., & Ghoshal, A. (2015). Smart textile wound dressings. In *Emerging research trends in medical textiles* (pp. 27–38). Singapore: Springer.
Goldman, R. (2004). Growth factors and chronic wound healing: Past, present, and future. *Advances in Skin and Wound Care, 17*, 24–35.
Gordon, A., Kozin, E. D., Keswani, S. G., Vaikunth, S. S., Katz, A. B., Zoltick, P. W., … Crombleholme, T. M. (2008). Permissive environment in postnatal wounds induced by adenoviral-mediated overexpression of the anti-inflammatory cytokine interleukin-10 prevents scar formation. *Wound Repair and Regeneration, 16*, 70–79.
Greaves, N. S., Ashcroft, K. J., Baguneid, M., & Bayat, A. (2013). Current understanding of molecular and cellular mechanisms in fibroplasia and angiogenesis during acute wound healing. *Journal of Dermatological Science, 72*, 206–217.
Guo, S., & DiPietro, L. A. (2010). Factors affecting wound healing. *Journal of Dental Research, 89*, 219–229.
Hakkarainen, T., Koivuniemi, R., Kosonen, M., Escobedo-Lucea, C., Sanz-Garcia, A., Vuola, J., … Kavola, H. (2016). Nanofibrillar cellulose wound dressing in skin graft donor site treatment. *Journal of Controlled Release, 244*, 292–301.
Han, G., & Ceilley, R. (2017). Chronic wound healing: A review of current management and treatments. *Advances in Therapy, 34*, 599–610.
Hartgerink, J. D., Beniash, E., & Stupp, S. I. (2001). Self-assembly and mineralization of peptide-amphiphile nanofibers. *Science, 294*, 1684–1688.
Hartgerink, J. D., Beniash, E., & Stupp, S. I. (2002). Peptide-amphiphile nanofibers: A versatile scaffold for the preparation of self-assembling materials. *Proceedings of the National Academy of Sciences of the United States of America, 99*, 5133–5138.
Hawiger, J. (1987). Formation and regulation of platelet and fibrin hemostatic plug. *Human Pathology, 18*, 111–122.
Haynes, J. H., Johnson, D. E., Mast, B. A., Diegelmann, R. F., Salzberg, D. A., Cohen, I. K., & Krummel, T. M. (1994). Platelet-derived growth factor induces fetal wound fibrosis. *Journal of Pediatric Surgery, 29*, 1405–1408.
Heimbach, D. M., Warden, G. D., Luterman, A., Jordan, M. H., Ozobia, N., Ryan, C. M., … Dimick, A. R. (2003). Multicenter postapproval clinical trial of Integra® dermal regeneration template for burn treatment. *The Journal of Burn Care & Rehabilitation, 24*, 42–48.
Helary, C., Zarka, M., & Giraud-Guille, M. M. (2012). Fibroblasts within concentrated collagen hydrogels favour chronic skin wound healing. *Journal of Tissue Engineering and Regenerative Medicine, 6*, 225–237.
Hendricks, M. P., Sato, K., Palmer, L. C., & Stupp, S. I. (2017). Supramolecular assembly of peptide amphiphiles. *Accounts of Chemical Research, 50*, 2440–2448.
Hinrichs, W. L., Lommen, E. J., Wildevuur, C. R., & Feijen, J. (1992). Fabrication and characterization of an asymmetric polyurethane membrane for use as a wound dressing. *Journal of Applied Biomaterials, 3*, 287–303.

Hinz, B. (2007). Formation and function of the myofibroblast during tissue repair. *The Journal of Investigative Dermatology, 127*, 526–537.
Hoemann, C. D., Marchand, C., Rivard, G.-E., El-Gabalawy, H., & Poubelle, P. E. (2017). Effect of chitosan and coagulation factors on the wound repair phenotype of bioengineered blood clots. *International Journal of Biological Macromolecules, 104*, 1916–1924.
Hoffman, B. D., Grashoff, C., & Schwartz, M. A. (2011). Dynamic molecular processes mediate cellular mechanotransduction. *Nature, 475*, 316–323.
Holavanahalli, R. K., Helm, P. A., & Kowalske, K. J. (2010). Long-term outcomes in patients surviving large burns: The skin. *Journal of Burn Care and Research, 31*, 631–639.
Hollander, D. A., Soranzo, C., Falk, S., & Windolf, J. (2001). Extensive traumatic soft tissue loss: Reconstruction in severely injured patients using cultured hyaluronan-based three-dimensional dermal and epidermal autografts. *Journal of Trauma and Acute Care Surgery, 50*, 1125.
Hong, W. X., Hu, M. S., Esquivel, M., Liang, G. Y., Rennert, R. C., McArdle, A., … Longaker, M. T. (2014). The role of hypoxia-inducible factor in wound healing. *Advances in Wound Care, 3*, 390–399.
Hopkinson-Woolley, J., Hughes, D., Gordon, S., & Martin, P. (1994). Macrophage recruitment during limb development and wound healing in the embryonic and foetal mouse. *Journal of Cell Science, 107*(Pt 5), 1159–1167.
Hosseinkhani, H., Hong, P.-D., & Yu, D.-S. (2013). Self-assembled proteins and peptides for regenerative medicine. *Chemical Reviews, 113*, 4837–4861.
Hunyadi, J., Farkas, B., Bertényi, C., Oláh, J., & Dobozy, A. (1988). Keratinocyte grafting: A new means of transplantation for full-thickness wounds. *The Journal of Dermatologic Surgery and Oncology, 14*, 75–78.
Hussain, M. Z., Ghani, Q. P., & Hunt, T. K. (1989). Inhibition of prolyl hydroxylase by poly(ADP-ribose) and phosphoribosyl-AMP. Possible role of ADP-ribosylation in intracellular prolyl hydroxylase regulation. *Journal of Biological Chemistry, 264*, 7850–7855.
Ihara, S., & Motobayashi, Y. (1992). Wound closure in foetal rat skin. *Development, 114*, 573–582.
Jayakumar, R., Reis, R. L., & Mano, J. F. (2006). Chemistry and applications of phosphorylated chitin and chitosan. *E-Polymers, 6*. https://doi.org/10.1515/epoly.2006.6.1.447
Kamoun, E. A., Kenawy, E.-R. S., & Chen, X. (2017). A review on polymeric hydrogel membranes for wound dressing applications: PVA-based hydrogel dressings. *Journal of Advanced Research, 8*, 217–233.
Karnik, S.-K., Brooke, B.-S., Bayes-Genis, A., Sorensen, L., Wythe, J.-D., Schwartz, R.-S., … Li, D.-Y. (2003). A critical role for elastin signaling in vascular morphogenesis and disease, *130*(2), 411–423.
Kawabata, S., Kanda, N., Hirasawa, Y., Noda, K., Matsuura, Y., Suzuki, S., & Kawai, K. (2018). The utility of silk-elastin hydrogel as a new material for wound healing. *Plastic and Reconstructive Surgery. Global Open, 6*, e1778.
Kawabata, S., Kawai, K., Somamoto, S., Noda, K., Matsuura, Y., Nakamura, Y., & Suzuki, S. (2017). The development of a novel wound healing material, silk-elastin sponge. *Journal of Biomaterials Science. Polymer Edition, 28*, 2143–2153.
Khatib, L., Cass, D. L., & Adzick, N. S. (2018). Updates in fetal wound healing and scar prevention. In J. A. Katowitz & W. R. Katowitz (Eds.), *Pediatric oculoplastic surgery* (pp. 45–58). Cham: Springer International Publishing.
Khorasani, M. T., Joorabloo, A., Moghaddam, A., Shamsi, H., & MansooriMoghadam, Z. (2018). Incorporation of ZnO nanoparticles into heparinised polyvinyl alcohol/chitosan hydrogels for wound dressing application. *International Journal of Biological Macromolecules, 114*, 1203–1215.
Kim, B. J., Cheong, H., Choi, E.-S., Yun, S.-H., Choi, B.-H., Park, K.-S., … Cha, H. J. (2017). Accelerated skin wound healing using electrospun nanofibrous mats blended with mussel adhesive protein and polycaprolactone. *Journal of Biomedical Materials Research. Part A, 105*, 218–225.

Kim, J. E., Lee, J., Jang, M., Kwak, M. H., Go, J., Kho, E. K., … Hwang, D. Y. (2015). Accelerated healing of cutaneous wounds using phytochemically stabilized gold nanoparticle deposited hydrocolloid membranes. *Biomaterials Science, 3*, 509–519.

Kirsner, R. S., & Eaglstein, W. H. (1993). The wound healing process. *Dermatologic Clinics, 11*, 629–640.

Klemm, D., Schumann, D., Kramer, F., Heßler, N., Koth, D., & Sultanova, B. (2009). Nanocellulose materials—Different cellulose, different functionality. *Macromolecular Symposia, 280*, 60–71.

Knighton, D. R., Silver, I. A., & Hunt, T. K. (1981). Regulation of wound-healing angiogenesis-effect of oxygen gradients and inspired oxygen concentration. *Surgery, 90*, 262–270.

Konop, M., Sulejczak, D., Czuwara, J., Kosson, P., Misicka, A., Lipkowski, A. W., & Rudnicka, L. (2017). The role of allogenic keratin-derived dressing in wound healing in a mouse model. *Wound Repair and Regeneration, 25*, 62–74.

Kristiansen, K. A., Schirmer, B. C., Aachmann, F. L., Skjåk-Bræk, G., Draget, K. I., & Christensen, B. E. (2009). Novel alginates prepared by independent control of chain stiffness and distribution of G-residues: Structure and gelling properties. *Carbohydrate Polymers, 77*, 725–735.

Kuo, C. K., & Ma, P. X. (2001). Ionically crosslinked alginate hydrogels as scaffolds for tissue engineering: Part 1. Structure, gelation rate and mechanical properties. *Biomaterials, 22*, 511–521.

Lara, H. H., Garza-Treviño, E. N., Ixtepan-Turrent, L., & Singh, D. K. (2011). Silver nanoparticles are broad-spectrum bactericidal and virucidal compounds. *Journal of Nanobiotechnology, 9*, 30.

Lawrence, W. T. (1998). Physiology of the acute wound. *Clinics in Plastic Surgery, 25*, 321–340.

Lazarus, G. S., Cooper, D. M., Knighton, D. R., Margolis, D. J., Pecoraro, R. E., Rodeheaver, G., & Robson, M. C. (1994). Definitions and guidelines for assessment of wounds and evaluation of healing. *Archives of Dermatology, 130*, 489–493.

Lee, K. Y., & Mooney, D. J. (2012). Alginate: Properties and biomedical applications. *Progress in Polymer Science, 37*, 106–126.

Lee, P.-Y., Li, Z., & Huang, L. (2003). Thermosensitive hydrogel as a Tgf-β1 gene delivery vehicle enhances diabetic wound healing. *Pharmaceutical Research, 20*, 1995–2000.

Li, J., Chen, J., & Kirsner, R. (2007). Pathophysiology of acute wound healing. *Clinics in Dermatology, 25*, 9–18.

Li, J. (2006). Overexpression of Laminin-8 in Human Dermal Microvascular Endothelial Cells Promotes Angiogenesis-Related Functions. *The Journal of investigative dermatology, 2*, 434–440.

Li, Q., Lu, F., Zhou, G., Yu, K., Lu, B., Xiao, Y., … Lan, G. (2017). Silver inlaid with gold nanoparticle/chitosan wound dressing enhances antibacterial activity and porosity, and promotes wound healing. *Biomacromolecules, 18*, 3766–3775.

Li, S., Li, L., Guo, C., Qin, H., & Yu, X. (2017). A promising wound dressing material with excellent cytocompatibility and proangiogenesis action for wound healing: Strontium loaded Silk fibroin/Sodium alginate (SF/SA) blend films. *International Journal of Biological Macromolecules, 104*, 969–978.

Liechty, K. W., Adzick, N. S., & Crombleholme, T. M. (2000). Diminished interleukin 6 (IL-6) production during scarless human fetal wound repair. *Cytokine, 12*, 671–676.

Liechty, K. W., Crombleholme, T. M., Cass, D. L., Martin, B., & Adzick, N. S. (1998). Diminished interleukin-8 (IL-8) production in the fetal wound healing response. *The Journal of Surgical Research, 77*, 80–84.

Lin, N., & Dufresne, A. (2014). Nanocellulose in biomedicine: Current status and future prospect. *European Polymer Journal, 59*, 302–325.

Lin, R. Y., Sullivan, K. M., Argenta, P. A., Meuli, M., Lorenz, H. P., & Adzick, N. S. (1995). Exogenous transforming growth factor-beta amplifies its own expression and induces scar formation in a model of human fetal skin repair. *Annals of Surgery, 222*, 146–154.

Liu, L., Xu, K., Wang, H., Jeremy Tan, P. K., Fan, W., Venkatraman, S. S., ... Yang, Y.-Y. (2009). Self-assembled cationic peptide nanoparticles as an efficient antimicrobial agent. *Nature Nanotechnology, 4*, 457–463.

Liu, X., Lee, P., Ho, C., Lui, V. C. H., Chen, Y., Che, C., ... Wong, K. K. Y. (2010). Silver nanoparticles mediate differential responses in keratinocytes and fibroblasts during skin wound healing. *ChemMedChem, 5*, 468–475.

Lo, D. D., Zimmermann, A. S., Nauta, A., Longaker, M. T., & Lorenz, H. P. (2012). Scarless fetal skin wound healing update. *Birth Defects Research Part C Embryo Today Reviews, 96*, 237–247.

Lobmann, R., Ambrosch, A., Schultz, G., Waldmann, K., Schiweck, S., & Lehnert, H. (2002). Expression of matrix-metalloproteinases and their inhibitors in the wounds of diabetic and non-diabetic patients. *Diabetologia, 45*, 1011–1016.

Longaker, M. T., Whitby, D. J., Ferguson, M. W., Lorenz, H. P., Harrison, M. R., & Adzick, N. S. (1994). Adult skin wounds in the fetal environment heal with scar formation. *Annals of Surgery, 219*, 65–72.

Longaker, M. T., Whitby, D. J., Jennings, R. W., Duncan, B. W., Ferguson, M. W. J., Harrison, M. R., & Adzick, N. S. (1991). Fetal diaphragmatic wounds heal with scar formation. *The Journal of Surgical Research, 50*, 375–385.

Lord, M. S., Cheng, B., McCarthy, S. J., Jung, M., & Whitelock, J. M. (2011). The modulation of platelet adhesion and activation by chitosan through plasma and extracellular matrix proteins. *Biomaterials, 32*, 6655–6662.

Lorenz, H. P., Longaker, M. T., Perkocha, L. A., Jennings, R. W., Harrison, M. R., & Adzick, N. S. (1992). Scarless wound repair: A human fetal skin model. *Development, 114*, 253–259.

Lorenz, H. P., Whitby, D. J., Longaker, M. T., & Adzick, N. S. (1993). Fetal wound healing. The ontogeny of scar formation in the non-human primate. *Annals of Surgery, 217*, 391–396.

Losi, P., Briganti, E., Errico, C., Lisella, A., Sanguinetti, E., Chiellini, F., & Soldani, G. (2013). Fibrin-based scaffold incorporating VEGF- and bFGF-loaded nanoparticles stimulates wound healing in diabetic mice. *Acta Biomaterialia, 9*, 7814–7821.

Lucas, T., Waisman, A., Ranjan, R., Roes, J., Krieg, T., Müller, W., ... Eming, S. A. (2010). Differential roles of macrophages in diverse phases of skin repair. *Journal of Immunology, 184*, 3964–3977.

Lynch, S. E., Nixon, J. C., Colvin, R. B., & Antoniades, H. N. (1987). Role of platelet-derived growth factor in wound healing: Synergistic effects with other growth factors. *Proceedings of the National Academy of Sciences of the United States of America, 84*, 7696–7700.

Makadia, H. K., & Siegel, S. J. (2011). Poly lactic-co-glycolic acid (PLGA) as biodegradable controlled drug delivery carrier. *Polymers, 3*, 1377–1397.

Malafaya, P. B., Silva, G. A., & Reis, R. L. (2007). Natural–origin polymers as carriers and scaffolds for biomolecules and cell delivery in tissue engineering applications. *Advanced Drug Delivery Reviews, 59*, 207–233.

Mammadov, R., Mammadov, B., Guler, M. O., & Tekinay, A. B. (2012). Growth factor binding on heparin mimetic peptide nanofibers. *Biomacromolecules, 13*, 3311–3319.

Mao, J. S., Liu, H. F., Yin, Y. J., & Yao, K. D. (2003). The properties of chitosan–gelatin membranes and scaffolds modified with hyaluronic acid by different methods. *Biomaterials, 24*, 1621 1629.

Martin, P. (1997). Wound healing—Aiming for perfect skin regeneration. *Science, 276*, 75–81.

Matson, J. B., Zha, R. H., & Stupp, S. I. (2011). Peptide self-assembly for crafting functional biological materials. *Current Opinion in Solid State & Materials Science, 15*, 225–235.

Medhe, S., Bansal, P., & Srivastava, M. M. (2014). Enhanced antioxidant activity of gold nanoparticle embedded 3,6-dihydroxyflavone: A combinational study. *Applied Nanoscience, 4*, 153–161.

Menke, N. B., Ward, K. R., Witten, T. M., Bonchev, D. G., & Diegelmann, R. F. (2007). Impaired wound healing. *Clinics in Dermatology, 25*, 19–25.

Michelacci, Y. M. (2003). Collagens and proteoglycans of the corneal extracellular matrix. *Brazilian Journal of Medical and Biological Research, 36*, 1037–1046.

Min, B.-M., Lee, G., Kim, S. H., Nam, Y. S., Lee, T. S., & Park, W. H. (2004). Electrospinning of silk fibroin nanofibers and its effect on the adhesion and spreading of normal human keratinocytes and fibroblasts in vitro. *Biomaterials, 25*, 1289–1297.
Mithieux, S. M., Tu, Y., Korkmaz, E., Braet, F., & Weiss, A. S. (2009). In situ polymerization of tropoelastin in the absence of chemical cross-linking. *Biomaterials, 30*, 431–435.
Mogoşanu, G. D., & Grumezescu, A. M. (2014). Natural and synthetic polymers for wounds and burns dressing. *International Journal of Pharmaceutics, 463*, 127–136.
Mohandas, A., Kumar, P. T. S., Raja, B., Lakshmanan, V.-K., & Jayakumar, R. (2015). Exploration of alginate hydrogel/nano zinc oxide composite bandages for infected wounds. *International Journal of Nanomedicine, 10*, 53–66.
Mohiti-Asli, M., Saha, S., Murphy, S. V., Gracz, H., Pourdeyhimi, B., Atala, A., & Loboa, E. G. (2017). Ibuprofen loaded PLA nanofibrous scaffolds increase proliferation of human skin cells in vitro and promote healing of full thickness incision wounds in vivo. *Journal of Biomedical Materials Research. Part B, Applied Biomaterials, 105*, 327–339.
Mosser, D. M., & Edwards, J. P. (2008). Exploring the full spectrum of macrophage activation. *Nature Reviews. Immunology, 8*, 958–969.
Motherwell, J. M., Anderson, C. R., & Murfee, W. L. (2018). Endothelial cell phenotypes are maintained during angiogenesis in cultured microvascular networks. *Scientific Reports, 8*, 5887.
Muangman, P., Deubner, H., Honari, S., Heimbach, D. M., Engrav, L. H., Klein, M. B., & Gibran, N. S. (2006). Correlation of clinical outcome of Integra application with microbiologic and pathological biopsies. *Journal of Trauma and Acute Care Surgery, 61*, 1212.
Murakami, K., Aoki, H., Nakamura, S., Nakamura, S., Takikawa, M., Hanzawa, M., … Ishihara, M. (2010). Hydrogel blends of chitin/chitosan, fucoidan and alginate as healing-impaired wound dressings. *Biomaterials, 31*, 83–90.
Murphy, S. V., Skardal, A., Song, L., Sutton, K., Haug, R., Mack, D. L., … Atala, A. (2017). Solubilized amnion membrane hyaluronic acid hydrogel accelerates full-thickness wound healing. *Stem Cells Translational Medicine, 6*, 2020–2032.
Murray, P. J., & Wynn, T. A. (2011). Protective and pathogenic functions of macrophage subsets. *Nature Reviews. Immunology, 11*, 723–737.
Napavichayanun, S., Yamdech, R., & Aramwit, P. (2016). The safety and efficacy of bacterial nanocellulose wound dressing incorporating sericin and polyhexamethylene biguanide: In vitro, in vivo and clinical studies. *Archives of Dermatological Research, 308*, 123–132.
Nath, R. K., LaRegina, M., Markham, H., Ksander, G. A., & Weeks, P. M. (1994). The expression of transforming growth factor type beta in fetal and adult rabbit skin wounds. *Journal of Pediatric Surgery, 29*, 416–421.
Nguyen, B. P., Ryan, M. C., Gil, S. G., & Carter, W. G. (2000). Deposition of laminin 5 in epidermal wounds regulates integrin signaling and adhesion. *Current Opinion in Cell Biology, 12*, 554–562.
Niece, K. L., Hartgerink, J. D., Donners, J. J. J. M., & Stupp, S. I. (2003). Self-assembly combining two bioactive peptide-amphiphile molecules into nanofibers by electrostatic attraction. *Journal of the American Chemical Society, 125*, 7146–7147.
Nunan, R., Harding, K. G., & Martin, P. (2014). Clinical challenges of chronic wounds: Searching for an optimal animal model to recapitulate their complexity. *Disease Models & Mechanisms, 7*, 1205–1213.
Nurden, A. T., Nurden, P., Sanchez, M., Andia, I., & Anitua, E. (2008). Platelets and wound healing. *Frontiers in Bioscience, 13*, 3525–3548.
O'Kane, S., & Ferguson, M. W. (1997). Transforming growth factor beta s and wound healing. *The International Journal of Biochemistry & Cell Biology, 29*, 63–78.
Okamoto, Y., Yano, R., Miyatake, K., Tomohiro, I., Shigemasa, Y., & Minami, S. (2003). Effects of chitin and chitosan on blood coagulation. *Carbohydrate Polymers, 53*, 337–342.
Olutoye, O. O., Alaish, S. M., Carr, M. E., Paik, M., Yager, D. R., Cohen, I. K., & Diegelmann, R. F. (1995). Aggregatory characteristics and expression of the collagen adhesion receptor in fetal porcine platelets. *Journal of Pediatric Surgery, 30*, 1649–1653.

Olutoye, O. O., Zhu, X., Cass, D. L., & Smith, C. W. (2005). Neutrophil recruitment by fetal porcine endothelial cells: Implications in scarless fetal wound healing. *Pediatric Research, 58*, 1290–1294.
Ortega, S., Ittmann, M., Tsang, S. H., Ehrlich, M., & Basilico, C. (1998). Neuronal defects and delayed wound healing in mice lacking fibroblast growth factor 2. *Proceedings of the National Academy of Sciences of the United States of America, 95*, 5672–5677.
Pal, S., Tak, Y. K., & Song, J. M. (2007). Does the antibacterial activity of silver nanoparticles depend on the shape of the nanoparticle? A study of the gram-negative bacterium Escherichia coli. *Applied and Environmental Microbiology, 73*, 1712–1720.
Palta, S., Saroa, R., & Palta, A. (2014). Overview of the coagulation system. *Indian Journal of Anaesthesia, 58*, 515–523.
Parani, M., Lokhande, G., Singh, A., & Gaharwar, A. K. (2016). Engineered nanomaterials for infection control and healing acute and chronic wounds. *ACS Applied Materials & Interfaces, 8*, 10049–10069.
Park, M., Kim, B.-S., Shin, H. K., Park, S.-J., & Kim, H.-Y. (2013). Preparation and characterization of keratin-based biocomposite hydrogels prepared by electron beam irradiation. *Materials Science & Engineering. C, Materials for Biological Applications, 33*, 5051–5057.
Pastar, I., Stojadinovic, O., Krzyzanowska, A., Barrientos, S., Stuelten, C., Zimmerman, K., ... Tomic-Canic, M. (2010). Attenuation of the transforming growth factor β-signaling pathway in chronic venous ulcers. *Molecular Medicine, 16*, 92–101.
Peppas, N. A., Keys, K. B., Torres-Lugo, M., & Lowman, A. M. (1999). Poly(ethylene glycol)-containing hydrogels in drug delivery. *Journal of Controlled Release, 62*, 81–87.
Perumal, G., Pappuru, S., Chakraborty, D., Maya Nandkumar, A., Chand, D. K., & Doble, M. (2017). Synthesis and characterization of curcumin loaded PLA-Hyperbranched polyglycerol electrospun blend for wound dressing applications. *Materials Science & Engineering. C, Materials for Biological Applications, 76*, 1196–1204.
Pesce, J. T., Ramalingam, T. R., Mentink-Kane, M. M., Wilson, M. S., El Kasmi, K. C., Smith, A. M., ... Wynn, T. A. (2009). Arginase-1-expressing macrophages suppress Th2 cytokine-driven inflammation and fibrosis. *PLoS Pathogens, 5*, e1000371.
Pillai, C. K. S., Paul, W., & Sharma, C. P. (2009). Chitin and chitosan polymers: Chemistry, solubility and fiber formation. *Progress in Polymer Science, 34*, 641–678.
Pinkas, D. M., Ding, S., Raines, R. T., & Barron, A. E. (2011). Tunable, post-translational hydroxylation of collagen Domains in Escherichia coli. *ACS Chemical Biology, 6*, 320–324.
Porporato, P. E., Payen, V. L., De Saedeleer, C. J., Préat, V., Thissen, J.-P., Feron, O., & Sonveaux, P. (2012). Lactate stimulates angiogenesis and accelerates the healing of superficial and ischemic wounds in mice. *Angiogenesis, 15*, 581–592.
Praveen, G., Sreerekha, P. R., Menon, D., Nair, S. V., & Chennazhi, K. P. (2012). Fibrin nanoconstructs: A novel processing method and their use as controlled delivery agents. *Nanotechnology, 23*, 095102.
Rai, M., Yadav, A., & Gade, A. (2009). Silver nanoparticles as a new generation of antimicrobials. *Biotechnology Advances, 27*, 76–83.
Rajendran, N. K., Kumar, S. S. D., Houreld, N. N., & Abrahamse, H. (2018). A review on nanoparticle based treatment for wound healing. *Journal of Drug Delivery Science and Technology, 44*, 421–430.
Reinke, J. M., & Sorg, H. (2012). Wound repair and regeneration. *European Surgical Research, 49*, 35–43.
Ricard-Blum, S. (2011). The collagen family. *Cold Spring Harbor Perspectives in Biology, 3*, a004978.
Rnjak-Kovacina, J., Wise, S. G., Li, Z., Maitz, P. K. M., Young, C. J., Wang, Y., & Weiss, A. S. (2011). Tailoring the porosity and pore size of electrospun synthetic human elastin scaffolds for dermal tissue engineering. *Biomaterials, 32*, 6729–6736.
Rnjak-Kovacina, J., Wise, S. G., Li, Z., Maitz, P. K. M., Young, C. J., Wang, Y., & Weiss, A. S. (2012). Electrospun synthetic human elastin:collagen composite scaffolds for dermal tissue engineering. *Acta Biomaterialia, 8*, 3714–3722.

Robson, M. C. (1997). Wound infection. A failure of wound healing caused by an imbalance of bacteria. *The Surgical Clinics of North America, 77*, 637–650.
Robson, M. C., Steed, D. L., & Franz, M. G. (2001). Wound healing: Biologic features and approaches to maximize healing trajectories. *Current Problems in Surgery, 38*, 72–140.
Rodríguez-Cabello, J. C., González de Torre, I., Ibañez-Fonseca, A., & Alonso, M. (2018). Bioactive scaffolds based on elastin-like materials for wound healing. *Advanced Drug Delivery Reviews, 129*, 118–133.
Ronfard, V., Broly, H., Mitchell, V., Galizia, J. P., Hochart, D., Chambon, E., ... Huart, J. J. (1991). Use of human keratinocytes cultured on fibrin glue in the treatment of burn wounds. *Burns, 17*, 181–184.
Ross, R., Raines, E. W., & Bowen-Pope, D. F. (1986). The biology of platelet-derived growth factor. *Cell, 46*, 155–169.
Rossiter, H., Barresi, C., Pammer, J., Rendl, M., Haigh, J., Wagner, E. F., & Tschachler, E. (2004). Loss of vascular endothelial growth factor A activity in murine epidermal keratinocytes delays wound healing and inhibits tumor formation. *Cancer Research, 64*, 3508–3516.
Rowlatt, U. (1979). Intrauterine wound healing in a 20 week human fetus. *Virchows Archiv. A, Pathological Anatomy and Histology, 381*, 353–361.
Saeed, S. M., Mirzadeh, H., Zandi, M., & Barzin, J. (2017). Designing and fabrication of curcumin loaded PCL/PVA multi-layer nanofibrous electrospun structures as active wound dressing. *Progress in Biomaterials, 6*, 39–48.
Sandri, G., Aguzzi, C., Rossi, S., Bonferoni, M. C., Bruni, G., Boselli, C., ... Ferrari, F. (2017). Halloysite and chitosan oligosaccharide nanocomposite for wound healing. *Acta Biomaterialia, 57*, 216–224.
Schierle, C. F., De la Garza, M., Mustoe, T. A., & Galiano, R. D. (2009). Staphylococcal biofilms impair wound healing by delaying reepithelialization in a murine cutaneous wound model. *Wound Repair and Regeneration, 17*, 354–359.
Schmidt, E. P., Lee, W. L., Zemans, R. L., Yamashita, C., & Downey, G. P. (2011). On, around, and through: Neutrophil-endothelial interactions in innate immunity. *Physiology, 26*, 334–347.
Schneider, A., Garlick, J. A., & Egles, C. (2008). Self-assembling peptide nanofiber scaffolds accelerate wound healing. *PLoS One, 3*, e1410.
Schultz, G. S., & Wysocki, A. (2009). Interactions between extracellular matrix and growth factors in wound healing. *Wound Repair and Regeneration, 17*, 153–162.
Sen, C. K., Gordillo, G. M., Roy, S., Kirsner, R., Lambert, L., Hunt, T. K., ... Longaker, M. T. (2009). Human skin wounds: A major and snowballing threat to public health and the economy. *Wound Repair and Regeneration, 17*, 763–771.
Senger, D. R., Claffey, K. P., Benes, J. E., Perruzzi, C. A., Sergiou, A. P., & Detmar, M. (1997). Angiogenesis promoted by vascular endothelial growth factor: Regulation through $\alpha 1\beta 1$ and $\alpha 2\beta 1$ integrins. *Proceedings of the National Academy of Sciences of the United States of America, 94*, 13612–13617.
Senger, D. R., Ledbetter, S. R., Claffey, K. P., Papadopoulos-Sergiou, A., Peruzzi, C. A., & Detmar, M. (1996). Stimulation of endothelial cell migration by vascular permeability factor/vascular endothelial growth factor through cooperative mechanisms involving the alphavbeta3 integrin, osteopontin, and thrombin. *The American Journal of Pathology, 149*, 293–305.
Serbina, N. V., Salazar-Mather, T. P., Biron, C. A., Kuziel, W. A., & Pamer, E. G. (2003). TNF/iNOS-producing dendritic cells mediate innate immune defense against bacterial infection. *Immunity, 19*, 59–70.
Shenderova, A., Burke, T. G., & Schwendeman, S. P. (1999). The acidic microclimate in poly(lactide-co-glycolide) microspheres stabilizes camptothecins. *Pharmaceutical Research, 16*, 241–248.
Shi, G., Chen, W., Zhang, Y., Dai, X., Zhang, X., & Wu, Z. (2018). An antifouling hydrogel containing silver nanoparticles for modulating the therapeutic immune response in chronic wound healing. *Langmuir, 35*(5), 1837–1845.

Shoulders, M. D., & Raines, R. T. (2009). Collagen structure and stability. *Annual Review of Biochemistry, 78*, 929–958.

Silva, G. A., Czeisler, C., Niece, K. L., Beniash, E., Harrington, D. A., Kessler, J. A., & Stupp, S. I. (2004). Selective differentiation of neural progenitor cells by high-epitope density nanofibers. *Science, 303*, 1352–1355.

Singer, A. J., & Clark, R. A. F. (1999). Cutaneous wound healing. *The New England Journal of Medicine, 341*, 738–746.

Singer, A. J., Tassiopoulos, A., & Kirsner, R. S. (2017). Evaluation and management of lower-extremity ulcers. *The New England Journal of Medicine, 377*, 1559–1567.

Skardal, A., Murphy, S. V., Crowell, K., Mack, D., Atala, A., & Soker, S. (2017). A tunable hydrogel system for long-term release of cell-secreted cytokines and bioprinted in situ wound cell delivery. *Journal of Biomedical Materials Research. Part B, Applied Biomaterials, 105*, 1986–2000.

Solway, D. R., Clark, W. A., & Levinson, D. J. (2011). A parallel open-label trial to evaluate microbial cellulose wound dressing in the treatment of diabetic foot ulcers. *International Wound Journal, 8*, 69–73.

Solway, D. R., Consalter, M., & Levinson, D. J. (2010). Microbial cellulose wound dressing in the treatment of skin tears in the frail elderly. *Wounds, 22*, 17–19.

Somasundaram, K., & Prathap, K. (1970). Intra-uterine healing of skin wounds in rabbit foetuses. *The Journal of Pathology, 100*, 81–86.

Srinivasan, B., Kumar, R., Shanmugam, K., Sivagnam, U. T., Reddy, N. P., & Sehgal, P. K. (2010). Porous keratin scaffold-promising biomaterial for tissue engineering and drug delivery. *Journal of Biomedical Materials Research. Part B, Applied Biomaterials, 92*, 5–12.

Steed, D. L. (1997). The role of growth factors in wound healing. *The Surgical Clinics of North America, 77*, 575–586.

Sudheesh Kumar, P. T., Lakshmanan, V.-K., Anilkumar, T. V., Ramya, C., Reshmi, P., Unnikrishnan, A. G., … Jayakumar, R. (2012). Flexible and microporous chitosan hydrogel/nano ZnO composite bandages for wound dressing: In vitro and in vivo evaluation. *ACS Applied Materials & Interfaces, 4*, 2618–2629.

Sun, T., Haycock, J., & Macneil, S. (2006). In situ image analysis of interactions between normal human keratinocytes and fibroblasts cultured in three-dimensional fibrin gels. *Biomaterials, 27*, 3459–3465.

Syeda, M. M., Jing, X., Mirza, R. H., Yu, H., Sellers, R. S., & Chi, Y. (2012). Prostaglandin transporter modulates wound healing in diabetes by regulating prostaglandin-induced angiogenesis. *The American Journal of Pathology, 181*, 334–346.

Thomas, A., Harding, K. G., & Moore, K. (2000). Alginates from wound dressings activate human macrophages to secrete tumour necrosis factor-α. *Biomaterials, 21*, 1797–1802.

Tocco, I., Zavan, B., Bassetto, F., & Vindigni, V. (2012). Nanotechnology-based therapies for skin wound regeneration. *Journal of Nanomaterials, 2012*, 1–11.

Toksoz, S., Mammadov, R., Tekinay, A. B., & Guler, M. O. (2011). Electrostatic effects on nanofiber formation of self-assembling peptide amphiphiles. *Journal of Colloid and Interface Science, 356*, 131–137.

Uzunalli, G., Mammadov, R., Yesildal, F., Alhan, D., Ozturk, S., Ozgurtas, T., … Tekinay, A. B. (2016). Angiogenic heparin-mimetic peptide nanofiber gel improves regenerative healing of acute wounds. *ACS Biomaterials Science and Engineering*. https://doi.org/10.1021/acsbiomaterials.6b00165

Uzunalli, G., Soran, Z., Erkal, T. S., Dagdas, Y. S., Dinc, E., Hondur, A. M., … Tekinay, A. B. (2014). Bioactive self-assembled peptide nanofibers for corneal stroma regeneration. *Acta Biomaterialia, 10*, 1156–1166.

Van De Water, L., Varney, S., & Tomasek, J. J. (2013). Mechanoregulation of the myofibroblast in wound contraction, scarring, and fibrosis: Opportunities for new therapeutic intervention. *Advances in Wound Care, 2*, 122–141.

van der Rest, M., & Garrone, R. (1991). Collagen family of proteins. *FASEB Journal, 5*, 2814–2823.

Vasconcelos, A., Gomes, A. C., & Cavaco-Paulo, A. (2012). Novel silk fibroin/elastin wound dressings. *Acta Biomaterialia, 8*, 3049–3060.
Velnar, T., Bailey, T., & Smrkolj, V. (2009). The wound healing process: An overview of the cellular and molecular mechanisms. *The Journal of International Medical Research, 37*, 1528–1542.
Versteeg, H. H., Heemskerk, J. W. M., Levi, M., & Reitsma, P. H. (2013). New fundamentals in hemostasis. *Physiological Reviews, 93*, 327–358.
Vink, E. T. H., Rábago, K. R., Glassner, D. A., Springs, B., O'Connor, R. P., Kolstad, J., & Gruber, P. R. (2004). The sustainability of NatureWorks™ polylactide polymers and Ingeo™ polylactide fibers: An update of the future. *Macromolecular Bioscience, 4*, 551–564.
Visse, R., & Nagase, H. (2003). Matrix metalloproteinases and tissue inhibitors of metalloproteinases: Structure, function, and biochemistry. *Circulation Research, 92*, 827–839.
Voigt, J., & Driver, V. R. (2012). Hyaluronic acid derivatives and their healing effect on burns, epithelial surgical wounds, and chronic wounds: A systematic review and meta-analysis of randomized controlled trials. *Wound Repair and Regeneration, 20*, 317–331.
Wadman, M. (2005). Scar prevention: The healing touch. *Nature, 436*(7054), 1079–1080.
Watanabe, R., Hayashi, R., Kimura, Y., Tanaka, Y., Kageyama, T., Hara, S., … Nishida, K. (2011). A novel gelatin hydrogel carrier sheet for corneal endothelial transplantation. *Tissue Engineering. Part A, 17*, 2213–2219.
Whitby, D. J., & Ferguson, M. W. (1991). The extracellular matrix of lip wounds in fetal, neonatal and adult mice. *Development, 112*, 651–668.
Willenborg, S., Lucas, T., van Loo, G., Knipper, J. A., Krieg, T., Haase, I., … Eming, S. A. (2012). CCR2 recruits an inflammatory macrophage subpopulation critical for angiogenesis in tissue repair. *Blood, 120*, 613–625.
Willerth, S. M., & Sakiyama-Elbert, S. E. (2007). Approaches to neural tissue engineering using scaffolds for drug delivery. *Advanced Drug Delivery Reviews, 59*, 325–338.
Wolberg, A. S. (2007). Thrombin generation and fibrin clot structure. *Blood Reviews, 21*, 131–142.
Wulff, B. C., & Wilgus, T. A. (2013). Mast cell activity in the healing wound: More than meets the eye? *Experimental Dermatology, 22*, 507–510.
Xiu, Z., Zhang, Q., Puppala, H. L., Colvin, V. L., & Alvarez, P. J. J. (2012). Negligible particle-specific antibacterial activity of silver nanoparticles. *Nano Letters, 12*, 4271–4275.
Xue, M., & Jackson, C. J. (2015). Extracellular matrix reorganization during wound healing and its impact on abnormal scarring. *Advances in Wound Care, 4*, 119–136.
Yang, X., Yang, J., Wang, L., Ran, B., Jia, Y., Zhang, L., … Jiang, X. (2017). Pharmaceutical intermediate-modified gold nanoparticles: Against multidrug-resistant bacteria and wound-healing application *via* an electrospun scaffold. *ACS Nano, 11*, 5737–5745.
Yeboah, A., Cohen, R. I., Faulknor, R., Schloss, R., Yarmush, M. L., & Berthiaume, F. (2016). The development and characterization of SDF1α-elastin-like-peptide nanoparticles for wound healing. *Journal of Controlled Release, 232*, 238–247.
Yergoz, F., Hastar, N., Cimenci, C. E., Ozkan, A. D., Tekinay, T., Guler, M. O., & Tekinay, A. B. (2017). Heparin mimetic peptide nanofiber gel promotes regeneration of full thickness burn injury. *Biomaterials, 134*, 117–127.
Young, A., & McNaught, C.-E. (2011). The physiology of wound healing. *Surgery, 29*, 475–479.
Yu, W., Jiang, Y.-Y., Sun, T.-W., Qi, C., Zhao, H., Chen, F., … He, Y. (2016). Design of a novel wound dressing consisting of alginate hydrogel and simvastatin-incorporated mesoporous hydroxyapatite microspheres for cutaneous wound healing. *RSC Advances, 6*, 104375–104387.
Zarkoob, H., Bodduluri, S., Ponnaluri, S. V., Selby, J. C., & Sander, E. A. (2015). Substrate stiffness affects human keratinocyte colony formation. *Cellular and Molecular Bioengineering, 8*, 32–50.
Zaulyanov, L., & Kirsner, R. S. (2007). A review of a bi-layered living cell treatment (Apligraf®) in the treatment of venous leg ulcers and diabetic foot ulcers. *Clinical Interventions in Aging, 2*, 93–98.
Zhang, J., Wu, L., Jing, D., & Ding, J. (2005). A comparative study of porous scaffolds with cubic and spherical macropores. *Polymer, 46*, 4979–4985.

Zhong, S. P., Zhang, Y. Z., & Lim, C. T. (2010). Tissue scaffolds for skin wound healing and dermal reconstruction. *Wiley Interdisciplinary Reviews. Nanomedicine and Nanobiotechnology, 2*, 510–525.

Zhou, W., Zhao, M., Zhao, Y., & Mou, Y. (2011). A fibrin gel loaded with chitosan nanoparticles for local delivery of rhEGF: Preparation and in vitro release studies. *Journal of Materials Science. Materials in Medicine, 22*, 1221.

Zhu, J. (2010). Bioactive modification of poly(ethylene glycol) hydrogels for tissue engineering. *Biomaterials, 31*, 4639–4656.

Zwischenberger, J. B., & Robert, L. (1999). Comparison of two topical collagen-based hemostatic sponges during cardiothoracic procedures. *Journal of Investigative Surgery, 12*, 101–106.

Chapter 4
Immunomodulatory Nanomaterials

Turgay Tekinay

Abstract The immune system provides protection against infections and toxins and serves as a guard for all tissues against pathogens and malfunctioning cells. Modulation of the immune system to use immune cells as therapeutic agents has been an important target for the treatment of many diseases that could not be cured otherwise such as cancer and autoimmune diseases. In addition, modulation of the immune system is crucial for tissue regeneration, since tissue regeneration and wound healing processes consist of a complicated and ordered array of events, a considerable part of which include the involvement of immune cells. Nanomaterials in the form of nanoparticles and nanofibers provide a wide array of tools for modulation of the immune system. Different types of nanomaterials have been developed to be used for effective targeting and treatment of cancer, as vaccines, and for the treatment of autoimmune disorders. In this chapter, different nanomaterials with immunomodulatory effects will be reviewed with an emphasis on cancer, autoimmune diseases, and vaccine development. In addition, future perspectives for developing materials with more refined immunomodulatory characteristics will be discussed.

4.1 Introduction

The immune system protects against infections and toxins, and plays crucial roles in the regeneration of the tissues after injury, including fighting against pathogens that might attack the regenerating tissue; removal of debris of the wounded and scar tissues; and degradation of the materials that are used for regenerative purposes. It is composed of many different types of cells and lymphoid organs that are tightly regulated and plays an invaluable role in the defense against invading microbes (Moon, Huang, & Irvine, 2012). Spleen, nasal-associated lymphoid tissue, Peyer's patches in the gut, and lymph nodes distributed throughout the body are all parts of the immune system and immune cells are generated at the thymus and bone marrow.

T. Tekinay (✉)
Algitek Biotechnology, Ankara, Turkey

A. B. Tekinay (ed.), *Nanomaterials for Regenerative Medicine*, Stem Cell Biology and Regenerative Medicine, https://doi.org/10.1007/978-3-030-31202-2_4

There are two major types of immune response which are innate and adaptive immunity. Innate immunity is nonspecific and acts as the initial defense against invading microorganisms and foreign particles. Macrophages, dendritic cells, neutrophils, and mast cells perform phagocytosis to destroy the foreign cells and release cytokines. Adaptive immunity plays a role as the second line of defense. In that response, antigen presenting cells (APCs) bring specific antigens to highly specialized T cells and B cells. Antigen is recognized and an immune response that is specific to the invading microorganism is elicited and the target is cleared (Hussain, Vanoirbeek, & Hoet, 2012; Norman, 2005).

The immune system is important in the fight against diseases; however, incorrect regulation of the immune response, immunosuppression and immunostimulation may also lead to pathological conditions (Norman, 2005). Immunosuppression is the state in which the immune system functions are decreased and the response is weakened, which may result in invasion by the pathogens or rapid growth of tumor cells. Immunostimulation enhances the immune response, overactivity of which may lead to a strong adverse response and may result in autoimmune disorders. Since immune system is a very complex mechanism and is very tightly regulated, any effect, inhibition or activation of a pathway may cause unexpected side effects to other pathways or different cells or tissues. Thus, the inflammatory agents are used in a controlled manner due to possible side effects (Chou et al., 2013).

Vaccines are great examples of protection against diseases in which the immune system is stimulated to protect individuals from infectious microbial organisms (Pulendran & Ahmed, 2011). Although there are very successful vaccines against various diseases, there is still a need for vaccines against many other severely infectious pathogens, such as HIV, malaria, tuberculosis, and hepatitis C. Current vaccine development approaches are centered on rational design to have more potency and less immunogenicity, and the number and types of vaccine candidates are increasing rapidly (Mamo & Poland, 2012; Oberg, Kennedy, Li, Ovsyannikova, & Poland, 2011; Rappuoli, Mandl, Black, & De Gregorio, 2011).

Immune system may also provide protection from tumors by using cancer vaccines to stimulate the immune system to inhibit tumors (DeMaria & Bilusic, 2019; Hodi et al., 2010; Kalos et al., 2011; Lollini, Cavallo, Nanni, & Forni, 2006; Williams et al., 2011). Cancer immunotherapy is a popular field that aims to develop novel cancer therapy approaches by understanding and utilizing immune pathways (DeMaria & Bilusic, 2019). On the other side of the spectrum, over-stimulation of the immune system may cause autoimmune disorders that require suppression of the immune system (Feldmann & Steinman, 2005).

Immunomodulatory drugs have been highly coveted for balancing the immune system for autoimmune diseases or specifically enhancing certain immune cells for protection against or treatment of cancer or other infectious diseases and for aiding tissue regeneration. However, due to the complexity of the immune system, using a single-agent immunomodulatory drug may result in severe side effects. Nanomaterials are suggested to be ideal for selective delivery of immune regulating molecules.

4.2 Nanomaterials and Immune System

Nanotechnology gives the scientists ability to design nanomaterials at different sizes, shapes, chemical and physical properties and composition for use in the treatment of different diseases (Couvreur & Vauthier, 2006; Hubbell, Thomas, & Swartz, 2009; Mamo & Poland, 2012; Moghimi, Hunter, & Murray, 2005; L. Zhao et al., 2014). Nanomaterials have been the subject of research for being used as drug delivery vehicles or adjuvants for vaccines (Hubbell et al., 2009; Kasturi et al., 2011; Moon et al., 2011; Reddy et al., 2007). There is also interest in use of them in diagnostic systems (Cho et al., 2011; Noh, Jang, Ahn, Lim, & Chung, 2011). Moreover, they may be used for the treatment of cancer (Hellstrom et al., 2001; Kalos et al., 2011; Steenblock & Fahmy, 2008) and as delivery vehicles for immunotherapy drugs (Ali & Mooney, 2010; Peer et al., 2007; Svenson, 2012). The shape, size and surface properties of the nanoparticles can be engineered to aid in cell specific responses (Treuel, Jiang, & Nienhaus, 2013). An important advantage of the nanoparticles is their ease of functionalization through adding or removing properties.

Nanomaterials with immunomodulatory properties are highly coveted for regenerative medicine applications to enable the immune cells to work in a controlled manner. Depending on the material that they are made of, nanoparticles might be perceived as foreign materials when they enter the body, eliciting an immune response in the form of either suppression or stimulation (Dobrovolskaia & McNeil, 2007). Immunomodulatory properties of nanoparticles may be used in development of vaccines or anti-allergy drugs (Parween, Gupta, & Chauhan, 2011; Ryan et al., 2007).

There is ongoing research on targeted delivery of nanoparticles to specific tissues and cells for treatment or diagnosis. Nanoparticles should be relatively stable in the body and should not elicit immune response to prevent degradation, but they should also reach the target site and stay there. Nanoparticles may accumulate in the desired tissue through blood vessels or bind to specific biological structures at the target site (Pelaz et al., 2017). Also, nanoparticles may be tailored to prevent nonspecific binding. Recent advances in nanoengineering methods allowed production of nanoparticles that may be used to target diseased tissues, in diagnostic applications or to regulate the immune system to produce vaccines, inhibit tumors or prevent autoimmunity.

In the following sections, we will describe how nanomaterials can be used for different immunomodulatory purposes and for aiding immunotherapeutics. We will also summarize recent developments in the design and implementation of immunomodulatory particles.

4.3 Nanomaterials for Immunotherapy of Cancer

According to World Health Organization data, cancer is the second leading cause of death in the world, and approximately 1 in 6 deaths is caused by cancer. The economic burden of cancer among both developed and under-developed countries is also devastating, and even though an incredible amount of effort has been spent on cancer research and therapy, the currently available methods are still limited, which causes the high lethality rate of cancer.

Main methods for treating cancer are surgery, radiation and chemotherapy. Immunotherapy has recently been developed and it is suggested to be a promising approach for the treatment of various diseases including cancer, because it regulates the immune system to target and destroy cancer cells and tumors (Song, Musetti, & Huang, 2017). Immunotherapy is the regulation of the immune system (activation or suppression) for treatment of diseases. Immunosuppressive approaches may be used to reduce allergic reactions and reduce excessive inflammation in organ transplants. In contrast, in cancer patients, immune system is aimed to be more active and new methods are developed to activate the system against malignant cells. There are different approaches and materials used in cancer immunotherapy, which is expected to result in more durable antitumor responses and reduce metastasis and recurrence compared to previous treatment methods. Currently, immunotherapy is one of the most highly coveted treatment methods; however, it is also a very expensive treatment method and is not suitable for all cancer types.

4.3.1 Nanomaterials for Cancer Immunotherapy

Tumor microenvironment prevents the activity of the immune system, and tumor cells inhibit the activity of tumor-specific T-cells (Munn & Bronte, 2016; Vasievich & Huang, 2011). New research focuses on regulating the tumor microenvironment by inhibiting immunosuppressor molecules or activating soluble mediators, which in turn may elicit an immune response. Nanomaterials have the ability to activate or suppress immune system depending on their type, size, side groups, and functionalization and use of nanomaterials may reduce these side effects by achieving specific delivery to target tumors. Nanomaterials may be used to selectively distribute immune checkpoint regulators to the tumors to inhibit their growth. For example, PLGA nanoparticles that were conjugated with anti-OX40 mAb resulted in stronger cytokine production and increased overall antitumor cytotoxic cell response when injected into the tumor compared to the administration of free antibody (M. Chen, Ouyang, Zhou, Li, & Ye, 2014). Targeted inhibition of cytokines using various approaches is also important for preventing the immunosuppressive environment of the tumor. Park et al. developed nanoscale liposomal polymeric gels (nanolipogels, nLGs) for co-delivery of IL-2 and small molecule TGF-β receptor-I inhibitor SB505124 for this purpose (Park et al., 2012).

Targeted delivery to tumors may also allow the use of highly potent drugs with reduced side effects. Intratumoral injection of anticancer antibodies from functionalized nanoporous silica inhibited tumor growth more efficiently and at a longer duration than systemic antibody injections (Lei et al., 2010). In another study, anti-CD137 mAb and engineered IL-2Fc fusion protein that were anchored to PEGylated liposomes prevented lethal toxicity and increased systemic antitumor immunity compared to free anti-CD137 antibodies. In this study, it was also determined that the size of the nanomaterials is important. The nanoparticles that are small enough to reach into the tumor but large enough to join the systemic circulation were found to be highly efficient (Kwong, Gai, Elkhader, Wittrup, & Irvine, 2013). Heo et al. found that after injection with immunomodulating oligodeoxynucleotides or siRNA encapsulated PLGA's in addition to chemotherapy, dendritic cells become active and migrate to the tumor-draining lymph nodes. This combination therapy using immunomodulatory nanomaterials inhibited tumor growth and increased the survival rate (Heo, Kim, Yun, & Lim, 2015).

Cell therapy is an alternative for treatment of many diseases; however, it has major limitations such as loss of function of the transplanted cells. A novel approach uses nanoparticles, created from liposomes and liposome-like synthetic nanoparticles 100–300 nm in diameter with a drug-loaded core and phospholipid surface layer, which carry adjuvants. These nanoparticles continuously stimulate donor cells and were shown to increase the efficiency of tumor elimination (Stephan, Moon, Um, Bersthteyn, & Irvine, 2010).

4.3.2 Cancer Vaccines

There are two types of cancer vaccines, therapeutic and prophylactic. Prophylactic vaccines are used to prevent cancers such as the hepatocellular carcinoma secondary to hepatitis B virus and squamous cell carcinoma secondary to human papillomavirus (HPV) (DeMaria & Bilusic, 2019). Therapeutic vaccines, on the other hand, are aimed to treat cancer.

Tumor cells express a variety of antigens, some of which are specific to them, and are not produced by healthy cells. In earlier studies, the cancer vaccines were developed with whole-cells together with adjuvants to target tumor cells. However, because of a need to develop more potent vaccines with fewer side effects, current research focuses on antigens that are specific to tumors (Herlyn & Birebent, 1999). Some of these targets are products of mutated oncogenes (p53, ras, PSA, GP-100, MART-1, B-raf). For cancer to occur, the cells have to have many mutations and escape through many checkpoints, thus every tumor has a different composition. Therefore, a personalized approach may be more effective for cancer treatment, and new developments in genomics allow scientists to determine the specific mutations in the patients. Together with the developments in nanotechnology, this information may be used to customize a specific therapeutic approach for each patient. Initial clinical trials of personalized cancer vaccines have shown the feasibility, safety, and immunotherapeutic activity of targeting individual tumors (Sahin & Türeci, 2018).

Cancer vaccines may not only target different antigens and immune adjuvants, but also use different vaccine platforms, such as peptides/proteins, whole tumor cells, recombinant vectors, dendritic cells (DCs), gangliosides, and genes (DeMaria & Bilusic, 2019). Cancer cells may also be coated with nanoparticles conjugated with tumor antigens to elicit an immune response (Fang et al., 2014). There are already approved cancer vaccines for treatment against early-stage bladder cancer, mCRPC, and metastatic melanoma, such as TheraCys® and TICE®, PROVENGE, and IMLYGIC®, and several new vaccines are undergoing clinical trials.

Cancer vaccines are promising; however, their impact in metastatic carcinomas is not very high. Cancer vaccines need to work in synergy with other therapeutic approaches to inhibit local (tumor-site) immune response and act as an immunosuppressive agent while stimulating antitumor response. Nanoparticles are very important in that respect, since they may be tailored for different functions.

There is an increased focus on vaccine development by using nanoparticles and nanotechnology methods. Tailored nanoparticles may have fewer side effects and can be more effective. Nanoparticle use in vaccines may protect antigens, allow for more specific targeted delivery and result in sustained slow release. The nanoparticle vaccine field is developing very rapidly and promising; however, there are some problems that need to be addressed, an important one of which is the lack of understanding of the behavior of nanoparticles in the body, when they are used as a shuttle system or an adjuvant in vaccines (L. Zhao et al., 2014).

Nanomaterials may be used for encapsulating antigen and adjuvants, protecting them from degradation and to increase the efficiency of T-cell response (Irvine, Hanson, Rakhra, & Tokatlian, 2015; Irvine, Swartz, & Szeto, 2013; Zhu, Zhang, Ni, Niu, & Chen, 2017). Some carbon-based nanomaterials, such as carbon nanotubes and graphene, might affect immune cells by specifically activating them and initiate an antitumor immune response (Orecchioni et al., 2014; Pescatori et al., 2013; Xu et al., 2013). In one example, PC7A nanoparticles were shown to deliver tumor antigens to cytosol and activating the stimulator of interferon genes (STING) pathway (Luo et al., 2017).

4.4 Nanomaterials for Development of Vaccines

Vaccines are microbial antigens or attenuated/killed microbes administered together with an adjuvant to induce antigen-specific immune responses that result in long-lasting immune memory against specific pathogens. In autoimmune disorders, the vaccine should work in an opposite manner: to inhibit immune responses against self-cells without affecting the capability of the immune system to act against foreign organisms or materials or cancer cells (Clemente-Casares, Tsai, Yang, & Santamaria, 2011).

The size of the particle effects the time required for drainage of that particle into the lymph node (Manolova et al., 2008). When antigens were covalently bound to nano-beads, the immune response differed with respect to the size of the nano-beads (Fifis et al., 2004). Nanoparticle size is also important in delivery. Amorphous silica nanoparticles that are greater than 100 nm have more difficulty in entering into cytosol compared to smaller nanoparticles (between 70 and 10 nm) (Hirai et al., 2012).

Dendritic cells ingest nanoparticles at varying efficiency. The rate of uptake differs by size, charge and hydrophobicity. Nanoparticles bigger than 500 nm are ingested at low rates (Foged, Brodin, Frokjaer, & Sundblad, 2005). Nanoparticles with positive charge are ingested more efficiently compared to neutral or negatively charged particles (Wischke, Borchert, Zimmermann, Siebenbrodt, & Lorenzen, 2006).

An advantage of nanoparticles is that they may be functionalized through addition of receptors on their surface to increase efficiency of delivery. Use of Fc receptors to deliver antigens to human dendritic cells have been suggested to be successful, by using intact antibodies or engineered fragments (Cruz et al., 2011; Mi et al., 2008).

Another advantage of nanoparticle vaccines is that they enter the APC via phagocytosis, unlike soluble antigens which enter by micropinocytosis. Thus nanoparticle vaccines present antigens more efficiently, which may also result in stronger immune responses (H. Shen et al., 2006).

Delivery of vaccines is another important subject. Several nanocarrier systems have been investigated for vaccine delivery, such as liposomes. Liposomes are important because of their adaptability and flexibility for use in different applications (Schwendener, 2014). Liposomes may be used to encapsulate different types of nanoparticles. Hydrophilic particles may be carried inside the liposome, while hydrophobic particles may be incorporated into the lipid bilayer. Liposomes may also be adjusted regarding their components, size and charge.

As a delivery system, nanoparticles may activate immune system by directly delivering antigen to the immune system cells or perform targeted delivery (Girija & Balasubramanian, 2018; Mody et al., 2013). To function as immunomodulators, nanoparticles may be tailored to activate immune pathways which might then enhance or inhibit antigen processing and immunogenicity. Gold and silica nanoparticles have been analyzed for their potential for use as cargo delivery system (Brito & O'Hagan, 2014; Shah, O'hagan, Amiji, & Brito, 2014). They may be tailored to strongly bind to antigens, co-deliver adjuvants and multi epitope antigens into lymphoid organs and into antigen-presenting cells (Zhu et al., 2017). The antigens can be attached to nanoparticles by encapsulation, conjugation, or adsorption (L. Zhao et al., 2014). Adsorption is not very strong, and relies on electrical charge or hydrophobicity, which may lead to rapid dissociation of the antigen and the nanoparticle in vivo. Encapsulation and chemical conjugation results in stronger binding of the nanoparticle to the antigen (Pati, Shevtsov, & Sonawane, 2018). The antigen is released when the carrier particle is ingested and degraded by the cell. In chemical conjugation, antigen is coupled irreversibly to the nanoparticle (Andersson, Buldun, Pattinson, Draper, & Howarth, 2019). New research may focus on soft-matter nanoparticles that are based on emulsions which work as adjuvants when they are given into the body independently of the antigen, that is, no attachment (Morel et al., 2011; O'Hagan, Ott, & Van Nest, 1997). Further studies are required to fully understand the effects of these molecules on each other and to the immune system.

4.4.1 Nanoparticle Interactions with Antigen Presenting Cells

For the vaccines to work efficiently, antigens must be delivered to antigen presenting cells, dendritic cells and macrophages, so that these cells are activated and an immune response is elicited (Jones, 2008; Reddy, Swartz, & Hubbell, 2006). Interactions of these cells with nanoparticles and mechanisms of delivery into the cells are important for rational design of vaccines (Kumari & Yadav, 2011). Also, scientists have to be aware of the possible changes to the nanoparticle behavior with the changing size, shape, side-group, and overall functionalization (Khong et al., 2018; Xiang et al., 2006). For example, PLGA particles that are 300 nm in diameter were transported into the cells at a higher rate compared to 1, 7 and 17 μm particles (Joshi, Geary, & Salem, 2013). Internalization of positively charged polystyrene nanoparticles to the dendritic cells was also higher, possibly due to the interactions with the anionic cell membranes (Foged et al., 2005; Y. Shen, Hao, Ou, Hu, & Chen, 2018). However, it was not shown that this increased internalization is correlated with more potent immune response.

In addition, nanoparticles have been increasingly used to deliver not only antigen of interest but also co-adjuvants, such as poly(I:C), CpG and MPL (De Temmerman et al., 2011; Hafner, Corthésy, & Merkle, 2013). However, more work is needed in nano-vaccine research to overcome challenges including synthesizing nanoparticles that are cheap, uniform, and consistent, that are functionalized with desired properties, ability to target the particle to desired location, with fewer side effects. Thus, rational design of these nanoparticles with respect to the disorder is imperative. New technologies such as micro- and nanofluidics are also important for more efficient analyses of the effects of nanoparticles on different cell types (Hong, Lu, Liu, & Chen, 2019).

4.4.2 Polymeric Nanomaterials for Vaccine Development

Synthetic polymers have different compositions and properties and some of the most widely used ones to prepare nanoparticles are poly(D,L-lactide-co-glycolide) (PLG) (Kim et al., 1999; Köping-Höggård, Sánchez, & Alonso, 2005; C. Thomas, Rawat, Hope-Weeks, & Ahsan, 2011), poly(D,L-lactic-coglycolic acid) (PLGA) (Demento et al., 2012; Lü et al., 2009; Manish, Rahi, Kaur, Bhatnagar, & Singh, 2013; Silva et al., 2013), poly(g-glutamic acid) (g-PGA) (Akagi, Baba, & Akashi, 2012; Akagi, Kaneko, Kida, & Akashi, 2005), poly(ethylene glycol) (PEG) (Köping-Höggård et al., 2005), and polystyrene (Kalkanidis et al., 2006; Minigo et al., 2007). PLG and PLGA nanoparticles have high biocompatibility and biodegradability which are needed for use in drug delivery (D'souza et al., 2014; Danhier et al., 2012).

Importance of polymeric nanoparticles comes from their ability to deliver antigens to target cells or provide sustained release of their cargo antigens. These polymeric nanoparticles entrap antigens for delivery to certain cells or sustain antigen release by virtue of their slow biodegradation rate (Danhier et al., 2012). PLGA nanoparticles were shown to shuttle antigens against *Plasmodium vivax* with monophosphoryl lipid A as adjuvant (Moon, Suh, et al., 2012), hepatitis B virus (HBV) (C. Thomas, Rawat, et al., 2011), *Bacillus anthracis* (Manish et al., 2013), and model antigens such as ovalbumin and tetanus toxoid (Demento et al., 2012; Diwan, Tafaghodi, & Samuel, 2002).

Cationic alginate-polyethylenimine (PEI) nanogels have been proposed to be used as a vaccine delivery system (Li et al., 2013). Compared with the empty nanogels, nanoparticle loaded nanogels enhanced vaccine-induced antibody production more efficiently, showing the potential of this approach to be used to enhance vaccine-elicited humoral and cellular immune responses.

4.4.3 Metal Nanoparticles

Metal nanoparticles are widely used for different applications. They have well developed and controllable synthesis methods with precise sizes and shapes, and they can also be readily functionalized (Kalkanidis et al., 2006; L. Zhao et al., 2014).

Gold nanoparticles have been used in several studies for vaccine development (L. Zhao et al., 2014). The size of gold nanoparticles usually ranges from 2 to 150 nm, and they may be given spherical, rod or cubic shapes (Gregory, Titball, & Williamson, 2013; Niikura et al., 2013). In one study, the highly conserved extracellular region of the matrix 2 protein of influenza A virus was conjugated to gold nanoparticles to treat influenza. Intranasal administration of this nanoparticle system induced matrix 2 protein specific IgG serum antibodies (Tao, Ziemer, & Gill, 2014). Stone et al. fabricated gold nanorods which were attached to the respiratory syncytial virus by covalent binding of a viral protein. This gold nanorod construct contained the major protective antigen of the virus, the fusion protein (F), and was able to successfully induce immune response (Stone et al., 2013). In another study, gold nanoparticles were attached to a synthetic peptide resembling foot-and-mouth disease virus protein, with sizes ranging from 2 to 50 nm. These nanoparticles activated the antibody response at different strengths (Y. S. Chen, Hung, Lin, & Huang, 2010). Xu et al. conjugated the surface of the gold nanorods with cetyltrimethylammonium bromide (CTAB), poly(diallydimethylammonium chloride) (PDDAC), and PEI. PDDAC- or PEI-attached gold nanoparticles with DNA stimulated stronger immune response compared DNA only approach (Xu et al., 2012).

Surfaces of gold or iron oxide may also be functionalized by coating them with sugar molecules (glyconanoparticles). Carbohydrates and other molecules may be attached to metal nanoparticles, and the core may be filled with magnetic or fluorescence molecules (Marradi, Chiodo, García, & Penadés, 2013).

Some inorganic materials such as silica-based nanoparticles are biocompatible and are being investigated as nanovaccine constituents. Mesoporous silica nanoparticles may be tailored for targeted release of antigens by changing properties such as the shape, size and surface functionalization (Manzano et al., 2008). These mesoporous silica nanoparticles can carry more cargo compared to solid silica nanoparticles, and could be regulated to release the cargo in a controlled manner by changing mesoporous structures. Mesoporous silica nanoparticles are promising candidates for use in vaccines due to their controlled-release abilities.

4.4.4 Carbon-Based Nanomaterials

Carbon nanoparticles have been studied for their use in vaccine delivery (Gregory et al., 2013; T. Wang et al., 2011). Particles which were 450 nm in size and with 50 nm mesopores on the particle surface were produced. These pockets would be used to transport protein antigen, protecting the antigen and allowing oral administration (T. Wang et al., 2011). They are tolerated well inside the body and were given different shapes including mesoporous spheres (Bianco, Kostarelos, & Prato, 2005; Gupta et al., 2015). Carbon nanotubes are pure carbon molecules and are generally 0.8–2 nm in diameter with a length of 100–1000 nm (Parra, Abad-Somovilla, Mercader, Taton, & Abad-Fuentes, 2013; Villa et al., 2011).

4.4.5 Biological Materials for Vaccine Development

Proteins, peptides, carbohydrates, nucleic acids, and liposomal systems have all been proposed to be used for developing effective vaccines for various purposes. They can be used in the form of either nanoparticles and nanofibers or hydrogels. Recently, nanoparticles are proposed to be synthesized by self-assembling of proteins that assemble to form higher level quaternary structures. By using protein structure information, self-assembling nanoparticles that resulted in stronger immune response than influenza vaccines were designed (Kanekiyo et al., 2013). In this study, the viral hemagglutinin was genetically fused to ferritin, which is a protein composed of 24 identical polypeptides that can self-assemble into spherical 10 nm particles. This nanoparticle vaccine elicited hemagglutination inhibition antibody titers that were more than tenfold higher compared to the traditional vaccine. The advantages of protein nanoparticles in vaccine development include having highly organized structures and symmetry, biodegradability, and tailorability at three different interfaces and size (Neek, Kim, & Wang, 2019).

Vault nanoparticles are self-assembled to form an ellipsoid that contains an empty region inside (Buehler et al., 2014). Recent work which focused on the use of vault nanoparticles to carry antigens showed that these nanoparticles elicited potent immune response, that may also be useful in cancer vaccines (Kar et al., 2012). A new class of monodispersed, self-assembling vault nanoparticles comprises a protein shell exterior with a lipophilic core interior and these recombinant vaults contained a small amphipathic α-helix obtained from hepatitis C virus. This design resulted in a small area in which lipophilic compounds would be encapsulated and delivered to target cells (Buehler et al., 2014).

Liposomes are spherical structures composed of phospholipid bilayers with a core that may be used for antigen delivery. Liposomes are biodegradable, and range from 20 nm to 1 μm in size. Liposomes may be used as delivery systems for vaccines by encapsulation, adsorption, or surface coupling of antigens (Giddam, Zaman, Skwarczynski, & Toth, 2012). Liposomes do not have immunostimulatory or immunosuppression activity, so they are combined with adjuvants for vaccine development. An in-depth analysis of liposomal interaction with the immune system would allow scientists to design more potent vaccines. Liposomes may be used to deliver DNA vaccines or virosomes that contain viral envelope glycoproteins (Glück, Moser, & Metcalfe, 2004; Khatri et al., 2008). Liposomes also allow the use of intranasal approach for delivery of vaccines (S. Sharma, Mukkur, Benson, & Chen, 2009).

Recombinant vaccines using proteins have very little toxicity but their potency is not high. In one study, an interbilayer-cross-linked multilamellar vesicle system was produced by cross-linking headgroups of adjacent lipid bilayers. These vesicles internalize antigens in the core and immunostimulatory adjuvants on the vesicle walls. These vesicles were shown to have potency, eliciting very strong endogenous T-cell and antibody responses (Moon et al., 2011).

4.5 Nanomaterials for the Treatment of Autoimmune Diseases

Autoimmune disease is the name of around 80 disorders that share a common etiology: an immune attack on the body's own cells. There are self-reactive B-cells and T-cells that recognize and attack the self-antigens. During T-cell development, self-reactive cells are normally deleted, and problems in removal or inhibition of these self-reactive cells may lead to autoimmunity (Notarangelo, Gambineri, & Badolato, 2006).

The prevalence of autoimmune diseases is increasing in industrialized countries, which may be due to environmental changes, such as air quality. There are no successful cures for these diseases. The disease progresses slowly and organ and tissue damage occur before diagnosis. To battle against autoimmune diseases, inhibitors against immunostimulatory molecules may be used, such as monoclonal antibodies or small receptor blockers. These inhibitors downregulate or result in the degradation of immunostimulatory agents.

Scientists have tried to develop techniques to transport anti-inflammatory drugs to target immune cells in affected tissues and inhibit or limit their pathological effects. For the treatment of autoimmune diseases, it would be extremely important to inhibit the T-cells that attack the body's own tissues, but not effect T-cells in other tissues. A wide and very strong immunosuppression would result in infections. In the immune system, T-cells play a very critical role in the defense against diseases or degradation of tumor cells. However, they may be responsible from self-tissue damage in autoimmune disorders. Thus, it is imperative that T-cell activity is regulated to develop successful strategies against autoimmune diseases. This may be done by converting T-cells into other types, redirecting their program, such as conversion of effector cells to regulatory T-cells (Moon, Huang, & Irvine, 2012; O'Shea & Paul, 2010; Rose, 2016). Nanoparticles would be very important in these types of altering function activities. By specifically tailoring the abilities of the drugs, nanoparticles may regulate these cells (Park et al., 2011). Metal nanoparticles, liposomal systems, biomaterial-based nanomaterials and carbon-based nanoparticles have been widely studied for this purpose.

4.5.1 Carbon-Based Nanoparticles

Carbon nanotubes are formed by carbon atoms that arrange forming a two-dimensional hollow cylinder. Carbon nanotubes were shown to induce systemic immunosuppression in mice (L. A. Mitchell et al., 2007; Leah A. Mitchell, Lauer, Burchiel, & McDonald, 2009; Tkach et al., 2011; X. Wang, Podila, Shannahan, Rao, & Brown, 2013).

Fullerene is also formed by carbon atoms and forms a closed structure that may be a hollow sphere, ellipsoid, tube, or many other shapes and sizes. It has anti-inflammatory and antioxidant effects (Magoulas et al., 2012). Fullerenes inhibit the allergic response against Ag-driven type I hypersensitivity by decreasing the level of reactive oxygen species (ROS) (Ryan et al., 2007). Fullerene derivatives may defend against oxidative stress in ischemia-reperfused lungs (Y.-W. Chen, 2004). Fullerenes were also shown to inhibit the development of arthritis in a rat model (Yudoh, Karasawa, Masuko, & Kato, 2009). In another study, hydroxylated fullerenes inhibited neutrophil function in fathead minnows (Jovanović, Anastasova, Rowe, & Palić, 2011).

4.5.2 Metal Nanoparticles

Gold nanoparticles have diverse properties and their size can be tailored to modulate their immune response and biodistribution. In gold nanoparticles, gold core is inert and nontoxic, and the nanoparticles can be manufactured with a very wide size distribution ranging from 1 to 150 nm and can be easily synthesized by various methods (C. P. Sharma, 2010). Twenty-one nanometer spherical gold nanoparticles caused no apparent organ or cell toxicity in mice, but resulted in inhibition of inflammatory effects (H. Chen et al., 2013). Five and 15 nm gold nanoparticles reduce pro-inflammatory responses induced by interleukin-1 (Sumbayev et al., 2013).

Iron oxide nanoparticles have also been used for immunomodulatory purposes. When the mice were exposed to ovalbumin and to varying doses of iron oxide submicron- or nanoparticles, allergic response was significantly inhibited.

Interestingly, low doses of submicron particles had no significant effect on the allergic response while the same dose of nanoparticles had an adjuvant effect on the response to ovalbumin. This study clearly showed that the particle dose and size affect the allergic response (Ban, Langonné, Huguet, Guichard, & Goutet, 2013). Administration of iron oxide nanoparticles (58.7 nm) also suppressed T-helper 1 cell-mediated immunity in ovalbumin sensitized mice (Shen, Liang, Wang, Liao, & Jan, 2012). Compared to single instillation, repeated instillations resulted in a reduction of inflammatory cell numbers in both bronchoalveolar lavages and pulmonary parenchyma (Ban, Langonné, Huguet, & Goutet, 2012).

Cerium oxide nanoparticles have the potential to reduce reactive oxygen species production and may be used to battle chronic inflammation (Hirst et al., 2009; Schanen et al., 2013). Also, cerium oxide nanoparticles (5–8 nm) protected the cardiac progenitor cells from H_2O_2-induced cytotoxicity (Pagliari et al., 2012).

Quantum dots are artificial semiconductor particles that are a few nanometers in size and their distinctive conductive properties are usually determined by their size. Cadmium telluride quantum dot nanoparticles suppressed the immune responses of macrophages to *Pseudomonas aeruginosa* by reducing NO, TNF, KC/CXCL-1, and IL-8 levels (Nguyen, Seligy, & Tayabali, 2013). Immunosuppression was also observed in Juvenile rainbow trout. When Juvenile rainbow trout were exposed to 5, 10, and 20 nM cadmium tellurium quantum dots, each form of dots resulted in a different pattern of gene expression and lowered fish immune response (Gagné et al., 2010). Sub-toxic levels of cadmium telluride quantum dot nanoparticles were also shown to suppress immune responses against bacteria in macrophages and epithelial cells (Nguyen et al., 2013).

4.5.3 Polymeric Nanoparticles

The size, shape and material of polymeric nanoparticles have also been shown to be important for suppression of immune responses. Twenty nanometer polystyrene particles decreased the efficiency of dendritic cells to degrade soluble antigens, without affecting their ability to induce antigen-specific CD4+ T-cell proliferation. Thousand nanometer polystyrene particles did not have such effects, while 20 nm particles accumulated in the lysosomes. Size-dependent accumulation of particles in lysosomes modulates dendritic cell function through impaired antigen degradation (Seydoux et al., 2014).

When dendritic cells (DCs) were treated with model biomedical poly(vinyl alcohol)-coated super-paramagnetic iron oxide nanoparticles (PVA-SPIONs), they were observed to exhibit decreased antigen processing capacity and CD4+ T-cell stimulation capacity (Blank et al., 2011).

Inert 50 nm polystyrene nanoparticles were also shown to inhibit allergic lung inflammation by modification of pulmonary dendritic cell function (Hardy et al., 2012). Particles composed of PLGA or PEG did not result in production of pro-inflammatory cytokines or inflammasome activation in macrophages. When instilled into the lungs, particle composition and size may increase the number and type of innate immune cells in the lungs without triggering inflammatory responses (Roberts et al., 2013). In another study, polystyrene or biodegradable poly(lactide-co-glycolide) microparticles encapsulating encephalitogenic peptides (500 nm diameter) were observed to induce long-term T-cell tolerance and ameliorated experimental autoimmune encephalomyelitis (Getts et al., 2012).

4.5.4 *Biomaterial-Based Nanoparticles*

Peptide, carbohydrate, nucleic acid, and liposomal formulations have also been used for immunomodulation for the treatment of autoimmune diseases. Nanoparticles carrying disease-related peptide-major histocompatibility complexes were shown to reduce polyclonal autoimmunity by activating the selective expansion of memory-like autoregulatory CD8+ T-cells (Clemente-Casares et al., 2011). These antigens interacted with CD8+ T-cells only in diseased but not healthy individuals, thus these nanoparticles coated with any relevant pMHC may be used as vaccines to inhibit polyclonal autoimmune responses in a disease and organ-specific manner (Tsai et al., 2010).

Liposomal encapsulation of glucocorticoids both enhances their efficacy in the treatment of encephalomyelitis and alters their target cell specificity and their mode of action compared to free glucocorticoids (Schweingruber et al., 2011). A study with folate-targeted nanoparticles showed promising results in the treatment of inflammatory arthritis (T. P. Thomas, Goonewardena, et al., 2011). Liposomal glucocorticoids inhibited proinflammatory macrophage functions and upregulate anti-inflammatory genes, but had little effect on T-cell apoptosis and function.

Injection of DNA plasmids encoding immunomodulatory proteins (OX40-TRAIL), and a cationic lipid was shown to ameliorate experimental autoimmune encephalomyelitis (Yellayi et al., 2011). Controlled release of immunomodulating peptide antigens from PLGA suppressed production of inflammatory cytokines and helped to reduce dosing without increased frequency (H. Zhao, Kiptoo, Williams, Siahaan, & Topp, 2010). A study in siRNA silencing in inflammatory monocytes to suppress expression of the chemokine receptor CCR2 showed that this approach prevented accumulation of inflammatory monocytes and their differentiation into highly activated antigen-presenting macrophages at the sites of inflammation. This therapeutic approach reduced inflammation in atherosclerotic plaques, decreased infarct size after coronary artery occlusion, prolonged survival of pancreatic islet allografts after transplantation, and suppressed tumor growth (Leuschner et al., 2011). Polylactide-cyclosporine A nanoparticles were produced at sub-100 nm sizes and narrow particle size distributions, and released cyclosporine A continuously for targeted immunosuppression. This study showed that polylactide-cyclosporine A nanoparticles were internalized into the dendritic cells with a continuous release to the culture medium, suppressing proliferation of T-cells without any systemic release of the drug (Azzi et al., 2010).

In another study, when leukemia inhibitory factor-loaded nanoparticles were directed to T-cells, vascularized heart grafts survived longer, Foxp3+ cells were induced and Th17 cells were expanded. These results show that engineered nanoparticles regulate immune pathways to elicit wanted response, thus enabling a new therapeutic approach for autoimmune disorders (Park et al., 2011).

4.6 Conclusions and Future Perspectives

Strategies to develop therapies based on inherent properties of different types of nanoparticles, which may be tailored for different diseases, should also employ methods to understand the mechanisms of the immunomodulatory action of these nanoparticles. There are many different types of nanoparticles with various chemical structures and sizes, and they can be functionalized with different side units. In addition to the size, particle composition, surface chemistry, ability to bind to plasma proteins, and drug excretion time and route is important for the immunomodulatory characteristics of the nanomaterials. These responses may be beneficial or harmful depending on the disease type. Research on immunomodulating nanomaterials to fight against cancer focuses on avoiding side effects and enhancing the tumor degrading ability of these nanoparticles. On the other hand, nanoparticles should have immune suppressing ability to treat autoimmune disorders.

It is extremely important that the correlation between the nanoparticle properties and immune response is elucidated for development of treatment and diagnosis methods in medicine. However, further detailed studies are required for tissue regeneration and diagnosis, prevention and treatment of diseases through immunomodulatory nanoparticles.

References

Akagi, T., Baba, M., & Akashi, M. (2012). Biodegradable nanoparticles as vaccine adjuvants and delivery systems: Regulation of immune responses by nanoparticle-based vaccine. *Advances in Polymer Science, 247*(1), 31–64. https://doi.org/10.1007/12_2011_150

Akagi, T., Kaneko, T., Kida, T., & Akashi, M. (2005). Preparation and characterization of biodegradable nanoparticles based on poly(γ-glutamic acid) with L-phenylalanine as a protein carrier. *Journal of Controlled Release, 108*(2–3), 226–236. https://doi.org/10.1016/j.jconrel.2005.08.003

Ali, O. A., & Mooney, D. J. (2010). Immunologically active biomaterials for cancer therapy. *Current Topics in Microbiology and Immunology, 344*(1), 279–297. https://doi.org/10.1007/82-2010-69

Andersson, A. M. C., Buldun, C. M., Pattinson, D. J., Draper, S. J., & Howarth, M. (2019). SnoopLigase peptide-peptide conjugation enables modular vaccine assembly. *Scientific Reports, 9*(1), 4625. https://doi.org/10.1038/s41598-019-40985-w

Azzi, J., Tang, L., Moore, R., Tong, R., El Haddad, N., Akiyoshi, T., … Abdi, R. (2010). Polylactide-cyclosporin A nanoparticles for targeted immunosuppression. *FASEB Journal, 24*(10), 3927–3938. https://doi.org/10.1096/fj.10-154690

Ban, M., Langonné, I., Huguet, N., & Goutet, M. (2012). Effect of submicron and nano-iron oxide particles on pulmonary immunity in mice. *Toxicology Letters, 210*(3), 267–275. https://doi.org/10.1016/j.toxlet.2012.02.004
Ban, M., Langonné, I., Huguet, N., Guichard, Y., & Goutet, M. (2013). Iron oxide particles modulate the ovalbumin-induced Th2 immune response in mice. *Toxicology Letters, 216*(1), 31–39. https://doi.org/10.1016/j.toxlet.2012.11.003
Bianco, A., Kostarelos, K., & Prato, M. (2005). Applications of carbon nanotubes in drug delivery. *Current Opinion in Chemical Biology, 9*(6), 674–679. https://doi.org/10.1016/j.cbpa.2005.10.005
Blank, F., Gerber, P., Rothen-Rutishauser, B., Sakulkhu, U., Salaklang, J., De Peyer, K., … Von Garnier, C. (2011). Biomedical nanoparticles modulate specific CD4 + T cell stimulation by inhibition of antigen processing in dendritic cells. *Nanotoxicology, 5*(4), 606–621. https://doi.org/10.3109/17435390.2010.541293
Brito, L. A., & O'Hagan, D. T. (2014). Designing and building the next generation of improved vaccine adjuvants. *Journal of Controlled Release, 190*, 563–579. https://doi.org/10.1016/j.jconrel.2014.06.027
Buehler, D. C., Marsden, M. D., Shen, S., Toso, D. B., Wu, X., Loo, J. A., … Rome, L. H. (2014). Bioengineered vaults: Self-assembling protein shell-lipophilic core nanoparticles for drug delivery. *ACS Nano, 8*(8), 7723–7732. https://doi.org/10.1021/nn5002694
Chen, H., Dorrigan, A., Saad, S., Hare, D. J., Cortie, M. B., & Valenzuela, S. M. (2013). In vivo study of spherical gold nanoparticles: Inflammatory effects and distribution in mice. *PLoS One, 8*(2), e58208. https://doi.org/10.1371/journal.pone.0058208
Chen, M., Ouyang, H., Zhou, S., Li, J., & Ye, Y. (2014). PLGA-nanoparticle mediated delivery of anti-OX40 monoclonal antibody enhances anti-tumor cytotoxic T cell responses. *Cellular Immunology, 287*(2), 91–99. https://doi.org/10.1016/j.cellimm.2014.01.003
Chen, Y. S., Hung, Y. C., Lin, W. H., & Huang, G. S. (2010). Assessment of gold nanoparticles as a size-dependent vaccine carrier for enhancing the antibody response against synthetic foot-and-mouth disease virus peptide. *Nanotechnology, 21*(19), 195101. https://doi.org/10.1088/0957-4484/21/19/195101
Chen, Y.-W. (2004). Fullerene derivatives protect against oxidative stress in RAW 264.7 cells and ischemia-reperfused lungs. *AJP: Regulatory, Integrative and Comparative Physiology, 287*(1), R21–R26. https://doi.org/10.1152/ajpregu.00310.2003
Cho, N. H., Cheong, T. C., Min, J. H., Wu, J. H., Lee, S. J., Kim, D., … Seong, S. Y. (2011). A multifunctional core-shell nanoparticle for dendritic cell-based cancer immunotherapy. *Nature Nanotechnology, 6*(10), 675–682. https://doi.org/10.1038/nnano.2011.149
Chou, S. H., Shetty, A. V., Geng, Y., Xu, L., Munirathinam, G., Pipathsouk, A., … Zheng, G. (2013). Palmitate-derivatized human IL-2: A potential anticancer immunotherapeutic of low systemic toxicity. *Cancer Immunology, Immunotherapy, 62*(3), 597–603. https://doi.org/10.1007/s00262-012-1364-8
Clemente-Casares, X., Tsai, S., Yang, Y., & Santamaria, P. (2011). Peptide-MHC-based nanovaccines for the treatment of autoimmunity: A "one size fits all" approach? *Journal of Molecular Medicine, 89*(8), 733–742. https://doi.org/10.1007/s00109-011-0757-z
Couvreur, P., & Vauthier, C. (2006). Nanotechnology: Intelligent design to treat complex disease. *Pharmaceutical Research, 23*(7), 1417–1450. https://doi.org/10.1007/s11095-006-0284-8
Cruz, L. J., Rueda, F., Cordobilla, B., Simón, L., Hosta, L., Albericio, F., & Domingo, J. C. (2011). Targeting nanosystems to human DCs via Fc receptor as an effective strategy to deliver antigen for immunotherapy. *Molecular Pharmaceutics, 8*(1), 104–116. https://doi.org/10.1021/mp100178k
D'souza, M. J., Tawde, S. A., Akalkotkar, A., Chablani, L., D'souza, M., & Chiriva-Internati, M. (2014). Nanotechnology in vaccine delivery. In *Molecular vaccines: From prophylaxis to therapy* (Vol. 2, pp. 727–741). Wien: Springer. https://doi.org/10.1007/978-3-319-00978-0_19
Danhier, F., Ansorena, E., Silva, J. M., Coco, R., Le Breton, A., & Préat, V. (2012). PLGA-based nanoparticles: An overview of biomedical applications. *Journal of Controlled Release, 161*(2), 505–522. https://doi.org/10.1016/j.jconrel.2012.01.043

De Temmerman, M. L., Rejman, J., Demeester, J., Irvine, D. J., Gander, B., & De Smedt, S. C. (2011). Particulate vaccines: On the quest for optimal delivery and immune response. *Drug Discovery Today, 16*(13–14), 569–582. https://doi.org/10.1016/j.drudis.2011.04.006

DeMaria, P. J., & Bilusic, M. (2019). Cancer vaccines. *Hematology/Oncology Clinics of North America, 33*(2), 199–214. https://doi.org/10.1016/j.hoc.2018.12.001

Demento, S. L., Cui, W., Criscione, J. M., Stern, E., Tulipan, J., Kaech, S. M., & Fahmy, T. M. (2012). Role of sustained antigen release from nanoparticle vaccines in shaping the T cell memory phenotype. *Biomaterials, 33*(19), 4957–4964. https://doi.org/10.1016/j.biomaterials.2012.03.041

Diwan, M., Tafaghodi, M., & Samuel, J. (2002). Enhancement of immune responses by co-delivery of a CpG oligodeoxynucleotide and tetanus toxoid in biodegradable nanospheres. *Journal of Controlled Release, 85*(1–3), 247–262. https://doi.org/10.1016/S0168-3659(02)00275-4

Dobrovolskaia, M. A., & McNeil, S. E. (2007). Immunological properties of engineered nanomaterials. *Nature Nanotechnology, 2*(8), 469–478. https://doi.org/10.1038/nnano.2007.223

Fang, R. H., Hu, C.-M. J., Luk, B. T., Gao, W., Copp, J. A., Tai, Y., … Zhang, L. (2014). Cancer cell membrane-coated nanoparticles for anticancer vaccination and drug delivery. *Nano Letters, 14*(4), 2181–2188. https://doi.org/10.1021/nl500618u

Feldmann, M., & Steinman, L. (2005). Design of effective immunotherapy for human autoimmunity. *Nature, 435*(7042), 612–619. https://doi.org/10.1038/nature03727

Fifis, T., Gamvrellis, A., Crimeen-Irwin, B., Pietersz, G. A., Li, J., Mottram, P. L., … Plebanski, M. (2004). Size-dependent immunogenicity: Therapeutic and protective properties of nano-vaccines against tumors. *The Journal of Immunology, 173*(5), 3148–3154. https://doi.org/10.4049/jimmunol.173.5.3148

Foged, C., Brodin, B., Frokjaer, S., & Sundblad, A. (2005). Particle size and surface charge affect particle uptake by human dendritic cells in an in vitro model. *International Journal of Pharmaceutics, 298*(2), 315–322. https://doi.org/10.1016/j.ijpharm.2005.03.035

Gagné, F., Fortier, M., Yu, L., Osachoff, H. L., Skirrow, R. C., Van Aggelen, G., … Fournier, M. (2010). Immunocompetence and alterations in hepatic gene expression in rainbow trout exposed to CdS/CdTe quantum dots. *Journal of Environmental Monitoring, 12*(8), 1556–1565. https://doi.org/10.1039/c0em00031k

Getts, D. R., Martin, A. J., Mccarthy, D. P., Terry, R. L., Hunter, Z. N., Yap, W. T., … Miller, S. D. (2012). Microparticles bearing encephalitogenic peptides induce T-cell tolerance and ameliorate experimental autoimmune encephalomyelitis. *Nature Biotechnology, 30*(12), 1217–1224. https://doi.org/10.1038/nbt.2434

Giddam, A. K., Zaman, M., Skwarczynski, M., & Toth, I. (2012). Liposome-based delivery system for vaccine candidates: Constructing an effective formulation. *Nanomedicine, 7*(12), 1877–1893. https://doi.org/10.2217/nnm.12.157

Girija, A. R., & Balasubramanian, S. (2018). Theragnostic potentials of core/shell mesoporous silica nanostructures. *Nanotheranostics, 3*(1), 1–40. https://doi.org/10.7150/ntno.27877

Glück, R., Moser, C., & Metcalfe, I. C. (2004). Influenza virosomes as an efficient system for adjuvanted vaccine delivery. *Expert Opinion on Biological Therapy, 4*(7), 1139–1145. https://doi.org/10.1517/14712598.4.7.1139

Gregory, A. E., Titball, R., & Williamson, D. (2013). Vaccine delivery using nanoparticles. *Frontiers in Cellular and Infection Microbiology, 3*, 13. https://doi.org/10.3389/fcimb.2013.00013

Gupta, A., Liberati, T. A., Verhulst, S. J., Main, B. J., Roberts, M. H., Potty, A. G. R., … El-Amin, S. F., III. (2015). Biocompatibility of single-walled carbon nanotube composites for bone regeneration. *Bone & Joint Research, 4*(5), 70–77. https://doi.org/10.1302/2046-3758.45.2000382

Hafner, A. M., Corthésy, B., & Merkle, H. P. (2013). Particulate formulations for the delivery of poly(I: C) as vaccine adjuvant. *Advanced Drug Delivery Reviews, 65*(10), 1386–1399. https://doi.org/10.1016/j.addr.2013.05.013

Hardy, C. L., LeMasurier, J. S., Belz, G. T., Scalzo-Inguanti, K., Yao, J., Xiang, S. D., … Plebanski, M. (2012). Inert 50-nm polystyrene nanoparticles that modify pulmonary dendritic cell function and inhibit allergic airway inflammation. *The Journal of Immunology, 188*(3), 1431–1441. https://doi.org/10.4049/jimmunol.1100156

Hellstrom, I., Ledbetter, J. A., Scholler, N., Yang, Y., Ye, Z., Goodman, G., … Hellstrom, K. E. (2001). CD3-mediated activation of tumor-reactive lymphocytes from patients with advanced cancer. *Proceedings of the National Academy of Sciences of the United States of America, 98*(12), 6783–6788. https://doi.org/10.1073/pnas.021557498

Heo, M. B., Kim, S. Y., Yun, W. S., & Lim, Y. T. (2015). Sequential delivery of an anticancer drug and combined immunomodulatory nanoparticles for efficient chemoimmunotherapy. *International Journal of Nanomedicine, 10*, 5981–5993. https://doi.org/10.2147/IJN.S90104

Herlyn, D., & Birebent, B. (1999). Advances in cancer vaccine development. *Annals of Medicine, 31*(1), 66–78.

Hirai, T., Yoshioka, Y., Takahashi, H., Ichihashi, K., Yoshida, T., Tochigi, S., … Tsutsumi, Y. (2012). Amorphous silica nanoparticles enhance cross-presentation in murine dendritic cells. *Biochemical and Biophysical Research Communications, 427*(3), 553–556. https://doi.org/10.1016/j.bbrc.2012.09.095

Hirst, S. M., Karakoti, A. S., Tyler, R. D., Sriranganathan, N., Seal, S., & Reilly, C. M. (2009). Anti-inflammatory properties of cerium oxide nanoparticles. *Small, 5*(24), 2848–2856. https://doi.org/10.1002/smll.200901048

Hodi, F. S., O'Day, S. J., McDermott, D. F., Weber, R. W., Sosman, J. A., Haanen, J. B., … Urba, W. J. (2010). Improved survival with ipilimumab in patients with metastatic melanoma. *New England Journal of Medicine, 363*(8), 711–723. https://doi.org/10.1056/nejmoa1003466

Hong, T., Lu, A., Liu, W., & Chen, C. (2019). Microdroplet synthesis of silver nanoparticles with controlled sizes. *Micromachines, 10*(4), 274. https://doi.org/10.3390/mi10040274

Hubbell, J. A., Thomas, S. N., & Swartz, M. A. (2009). Materials engineering for immunomodulation. *Nature, 462*(7272), 449–460. https://doi.org/10.1038/nature08604

Hussain, S., Vanoirbeek, J. A. J., & Hoet, P. H. M. (2012). Interactions of nanomaterials with the immune system. *Wiley Interdisciplinary Reviews: Nanomedicine and Nanobiotechnology, 4*(2), 169–183. https://doi.org/10.1002/wnan.166

Irvine, D. J., Hanson, M. C., Rakhra, K., & Tokatlian, T. (2015). Synthetic nanoparticles for vaccines and immunotherapy. *Chemical Reviews, 115*(19), 11109–11146. https://doi.org/10.1021/acs.chemrev.5b00109

Irvine, D. J., Swartz, M. A., & Szeto, G. L. (2013). Engineering synthetic vaccines using cues from natural immunity. *Nature Materials, 12*(11), 978–990. https://doi.org/10.1038/nmat3775

Jones, K. S. (2008). Biomaterials as vaccine adjuvants. *Biotechnology Progress, 24*(4), 807–814. https://doi.org/10.1002/btpr.10

Joshi, V. B., Geary, S. M., & Salem, A. K. (2013). Biodegradable particles as vaccine delivery systems: Size matters. *The AAPS Journal, 15*(1), 85–94. https://doi.org/10.1208/s12248-012-9418-6

Jovanović, B., Anastasova, L., Rowe, E. W., & Palić, D. (2011). Hydroxylated fullerenes inhibit neutrophil function in fathead minnow (Pimephales promelas Rafinesque, 1820). *Aquatic Toxicology, 101*(2), 474–482. https://doi.org/10.1016/j.aquatox.2010.11.002

Kalkanidis, M., Pietersz, G. A., Xiang, S. D., Mottram, P. L., Crimeen-Irwin, B., Ardipradja, K., & Plebanski, M. (2006). Methods for nano-particle based vaccine formulation and evaluation of their immunogenicity. *Methods, 40*(1), 20–29. https://doi.org/10.1016/j.ymeth.2006.05.018

Kalos, M., Levine, B. L., Porter, D. L., Katz, S., Grupp, S. A., Bagg, A., & June, C. H. (2011). T cells with chimeric antigen receptors have potent antitumor effects and can establish memory in patients with advanced leukemia. *Science Translational Medicine, 3*(95), 95ra73. https://doi.org/10.1126/scitranslmed.3002842

Kanekiyo, M., Wei, C. J., Yassine, H. M., McTamney, P. M., Boyington, J. C., Whittle, J. R. R., … Nabel, G. J. (2013). Self-assembling influenza nanoparticle vaccines elicit broadly neutralizing H1N1 antibodies. *Nature, 499*(7456), 102–106. https://doi.org/10.1038/nature12202

Kar, U. K., Jiang, J., Champion, C. I., Salehi, S., Srivastava, M., Sharma, S., … Kelly, K. A. (2012). Vault nanocapsules as adjuvants favor cell-mediated over antibody-mediated immune responses following immunization of mice. *PLoS One, 7*(7), e38553. https://doi.org/10.1371/journal.pone.0038553

Kasturi, S. P., Skountzou, I., Albrecht, R. A., Koutsonanos, D., Hua, T., Nakaya, H. I., … Pulendran, B. (2011). Programming the magnitude and persistence of antibody responses with innate immunity. *Nature, 470*(7335), 543–550. https://doi.org/10.1038/nature09737

Khatri, K., Goyal, A. K., Gupta, P. N., Mishra, N., Mehta, A., & Vyas, S. P. (2008). Surface modified liposomes for nasal delivery of DNA vaccine. *Vaccine, 26*(18), 2225–2233. https://doi.org/10.1016/j.vaccine.2008.02.058

Khong, H., Volmari, A., Sharma, M., Dai, Z., Imo, C. S., Hailemichael, Y., … Overwijk, W. W. (2018). Peptide vaccine formulation controls the duration of antigen presentation and magnitude of tumor-specific CD8 + T cell response. *The Journal of Immunology, 200*(10), 3464–3474. https://doi.org/10.4049/jimmunol.1700467

Kim, S. Y., Doh, H. J., Jang, M. H., Ha, Y. J., Chung, S. I. I., & Park, H. J. (1999). Oral immunization with Helicobacter pylori-loaded poly(D,L-Lactide-Co-Glycolide) nanoparticles. *Helicobacter, 4*(1), 33–39. https://doi.org/10.1046/j.1523-5378.1999.09046.x

Köping-Höggård, M., Sánchez, A., & Alonso, M. J. (2005). Nanoparticles as carriers for nasal vaccine delivery. *Expert Review of Vaccines, 4*(2), 185–196. https://doi.org/10.1586/14760584.4.2.185

Kumari, A., & Yadav, S. K. (2011). Cellular interactions of therapeutically delivered nanoparticles. *Expert Opinion on Drug Delivery, 8*(2), 141–151. https://doi.org/10.1517/17425247.2011.547934

Kwong, B., Gai, S. A., Elkhader, J., Wittrup, K. D., & Irvine, D. J. (2013). Localized immunotherapy via liposome-anchored anti- CD137 + IL-2 prevents lethal toxicity and elicits local and systemic antitumor immunity. *Cancer Research, 73*(5), 1547–1558. https://doi.org/10.1158/0008-5472.CAN-12-3343

Lei, C., Liu, P., Chen, B., Mao, Y., Engelmann, H., Shin, Y., … Hellstrom, K. E. (2010). Local release of highly loaded antibodies from functionalized nanoporous support for cancer immunotherapy. *Journal of the American Chemical Society, 132*(20), 6906–6907. https://doi.org/10.1021/ja102414t

Leuschner, F., Dutta, P., Gorbatov, R., Novobrantseva, T. I., Donahoe, J. S., Courties, G., … Nahrendorf, M. (2011). Therapeutic siRNA silencing in inflammatory monocytes in mice. *Nature Biotechnology, 29*(11), 1005–1010. https://doi.org/10.1038/nbt.1989

Li, P., Luo, Z., Liu, P., Gao, N., Zhang, Y., Pan, H., … Ma, Y. (2013). Bioreducible alginate-poly(ethylenimine) nanogels as an antigen-delivery system robustly enhance vaccine-elicited humoral and cellular immune responses. *Journal of Controlled Release, 168*, 271. https://doi.org/10.1016/j.jconrel.2013.03.025

Lollini, P. L., Cavallo, F., Nanni, P., & Forni, G. (2006). Vaccines for tumour prevention. *Nature Reviews Cancer, 6*(3), 204–216. https://doi.org/10.1038/nrc1815

Lü, J. M., Wang, X., Marin-Muller, C., Wang, H., Lin, P. H., Yao, Q., & Chen, C. (2009). Current advances in research and clinical applications of PLGA-based nanotechnology. *Expert Review of Molecular Diagnostics, 9*(4), 325–341. https://doi.org/10.1586/erm.09.15

Luo, M., Wang, H., Wang, Z., Cai, H., Lu, Z., Li, Y., … Gao, J. (2017). A STING-activating nanovaccine for cancer immunotherapy. *Nature Nanotechnology, 12*(7), 648–654. https://doi.org/10.1038/nnano.2017.52

Magoulas, G. E., Garnelis, T., Athanassopoulos, C. M., Papaioannou, D., Mattheolabakis, G., Avgoustakis, K., & Hadjipavlou-Litina, D. (2012). Synthesis and antioxidative/anti-inflammatory activity of novel fullerene-polyamine conjugates. *Tetrahedron, 68*(35), 7041–7049. https://doi.org/10.1016/j.tet.2012.06.066

Mamo, T., & Poland, G. A. (2012). Nanovaccinology: The next generation of vaccines meets 21st century materials science and engineering. *Vaccine, 30*(47), 6609–6611. https://doi.org/10.1016/j.vaccine.2012.08.023

Manish, M., Rahi, A., Kaur, M., Bhatnagar, R., & Singh, S. (2013). A single-dose PLGA encapsulated protective antigen domain 4 nanoformulation protects mice against bacillus anthracis spore challenge. *PLoS One, 8*(4), e61885. https://doi.org/10.1371/journal.pone.0061885

Manolova, V., Flace, A., Bauer, M., Schwarz, K., Saudan, P., & Bachmann, M. F. (2008). Nanoparticles target distinct dendritic cell populations according to their size. *European Journal of Immunology, 38*(5), 1404–1413. https://doi.org/10.1002/eji.200737984

Manzano, M., Aina, V., Areán, C. O., Balas, F., Cauda, V., Colilla, M., … Vallet-Regí, M. (2008). Studies on MCM-41 mesoporous silica for drug delivery: Effect of particle morphology and amine functionalization. *Chemical Engineering Journal, 137*(1), 30–37. https://doi.org/10.1016/j.cej.2007.07.078
Marradi, M., Chiodo, F., García, I., & Penadés, S. (2013). Glyconanoparticles as multifunctional and multimodal carbohydrate systems. *Chemical Society Reviews, 42*(11), 4728–4745. https://doi.org/10.1039/c2cs35420a
Mi, W., Wanjie, S., Lo, S.-T., Gan, Z., Pickl-Herk, B., Ober, R. J., & Ward, E. S. (2008). Targeting the neonatal fc receptor for antigen delivery using engineered fc fragments. *The Journal of Immunology, 181*(11), 7550–7561. https://doi.org/10.4049/jimmunol.181.11.7550
Minigo, G., Scholzen, A., Tang, C. K., Hanley, J. C., Kalkanidis, M., Pietersz, G. A., … Plebanski, M. (2007). Poly-L-lysine-coated nanoparticles: A potent delivery system to enhance DNA vaccine efficacy. *Vaccine, 25*(7), 1316–1327. https://doi.org/10.1016/j.vaccine.2006.09.086
Mitchell, L. A., Lauer, F. T., Burchiel, S. W., & McDonald, J. D. (2009). Mechanisms for how inhaled multiwalled carbon nanotubes suppress systemic immune function in mice. *Nature Nanotechnology, 4*(7), 451–456. https://doi.org/10.1038/nnano.2009.151
Mitchell, L. A., Gao, J., Wal, R. V., Gigliotti, A., Burchiel, S. W., & McDonald, J. D. (2007). Pulmonary and systemic immune response to inhaled multiwalled carbon nanotubes. *Toxicological Sciences, 100*(1), 203–214. https://doi.org/10.1093/toxsci/kfm196
Mody, K. T., Popat, A., Mahony, D., Cavallaro, A. S., Yu, C., & Mitter, N. (2013). Mesoporous silica nanoparticles as antigen carriers and adjuvants for vaccine delivery. *Nanoscale, 5*(12), 5167–5179. https://doi.org/10.1039/c3nr00357d
Moghimi, S. M., Hunter, A. C., & Murray, J. C. (2005). Nanomedicine: Current status and future prospects. *The FASEB Journal, 19*(3), 311–330. https://doi.org/10.1096/fj.04-2747rev
Moon, J. J., Huang, B., & Irvine, D. J. (2012). Engineering nano- and microparticles to tune immunity. *Advanced Materials, 24*(28), 3724–3746. https://doi.org/10.1002/adma.201200446
Moon, J. J., Suh, H., Bershteyn, A., Stephan, M. T., Liu, H., Huang, B., … Irvine, D. J. (2011). Interbilayer-crosslinked multilamellar vesicles as synthetic vaccines for potent humoral and cellular immune responses. *Nature Materials, 10*(3), 243–251. https://doi.org/10.1038/nmat2960
Moon, J. J., Suh, H., Polhemus, M. E., Ockenhouse, C. F., Yadava, A., & Irvine, D. J. (2012). Antigen-displaying lipid-enveloped PLGA nanoparticles as delivery agents for a Plasmodium vivax malaria vaccine. *PLoS One, 7*(2), e31472. https://doi.org/10.1371/journal.pone.0031472
Morel, S., Didierlaurent, A., Bourguignon, P., Delhaye, S., Baras, B., Jacob, V., … Van Mechelen, M. (2011). Adjuvant system AS03 containing α-tocopherol modulates innate immune response and leads to improved adaptive immunity. *Vaccine, 29*(13), 2461–2473. https://doi.org/10.1016/j.vaccine.2011.01.011
Munn, D. H., & Bronte, V. (2016). Immune suppressive mechanisms in the tumor microenvironment. *Current Opinion in Immunology, 39*, 1–6. https://doi.org/10.1016/j.coi.2015.10.009
Neek, M., Kim, T. I., & Wang, S. W. (2019). Protein-based nanoparticles in cancer vaccine development. *Nanomedicine: Nanotechnology, Biology, and Medicine, 15*(1), 164–174. https://doi.org/10.1016/j.nano.2018.09.004
Nguyen, K. C., Seligy, V. L., & Tayabali, A. F. (2013). Cadmium telluride quantum dot nanoparticle cytotoxicity and effects on model immune responses to Pseudomonas aeruginosa. *Nanotoxicology, 7*(2), 202–211. https://doi.org/10.3109/17435390.2011.648667
Niikura, K., Matsunaga, T., Suzuki, T., Kobayashi, S., Yamaguchi, H., Orba, Y., … Sawa, H. (2013). Gold nanoparticles as a vaccine platform: Influence of size and shape on immunological responses in vitro and in vivo. *ACS Nano, 7*(5), 3926–3938. https://doi.org/10.1021/nn3057005
Noh, Y.-W., Jang, Y.-S., Ahn, K.-J., Lim, Y. T., & Chung, B. H. (2011). Simultaneous in vivo tracking of dendritic cells and priming of an antigen-specific immune response. *Biomaterials, 32*(26), 6254–6263. https://doi.org/10.1016/j.biomaterials.2011.05.013
Norman, P. (2005). Immunobiology: The immune system in health and disease. *Journal of Allergy and Clinical Immunology, 96*(2), 274–274. https://doi.org/10.1016/s0091-6749(95)70025-0

Notarangelo, L. D., Gambineri, E., & Badolato, R. (2006). Immunodeficiencies with autoimmune consequences. *Advances in Immunology, 89*, 321–370. https://doi.org/10.1016/S0065-2776(05)89008-X

O'Hagan, D. T., Ott, G. S., & Van Nest, G. (1997). Recent advances in vaccine adjuvants: The development of MF59 emulsion and polymeric microparticles. *Molecular Medicine Today, 3*(2), 69–75. https://doi.org/10.1016/S1357-4310(96)10058-7

O'Shea, J., & Paul, W. E. (2010). Mechanisms underlying lineage commitment and plasticity of helper CD4 + T cells. *Science, 327*(5969), 1098–1102. https://doi.org/10.1126/science.1178334

Oberg, A. L., Kennedy, R. B., Li, P., Ovsyannikova, I. G., & Poland, G. A. (2011). Systems biology approaches to new vaccine development. *Current Opinion in Immunology, 23*(3), 436–443. https://doi.org/10.1016/j.coi.2011.04.005

Orecchioni, M., Bedognetti, D., Sgarrella, F., Marincola, F. M., Bianco, A., & Delogu, L. G. (2014). Impact of carbon nanotubes and graphene on immune cells. *Journal of Translational Medicine, 12*(1), 138. https://doi.org/10.1186/1479-5876-12-138

Pagliari, F., Mandoli, C., Forte, G., Magnani, E., Pagliari, S., Nardone, G., … Traversa, E. (2012). Cerium oxide nanoparticles protect cardiac progenitor cells from oxidative stress. *ACS Nano, 6*(5), 3767–3775. https://doi.org/10.1021/nn2048069

Park, J., Gao, W., Whiston, R., Strom, T. B., Metcalfe, S., & Fahmy, T. M. (2011). Modulation of CD4+ T lymphocyte lineage outcomes with targeted, nanoparticle-mediated cytokine delivery. *Molecular Pharmaceutics, 8*(1), 143–152. https://doi.org/10.1021/mp100203a

Park, J., Wrzesinski, S. H., Stern, E., Look, M., Criscione, J., Ragheb, R., … Fahmy, T. M. (2012). Combination delivery of TGF-β inhibitor and IL-2 by nanoscale liposomal polymeric gels enhances tumour immunotherapy. *Nature Materials, 11*(10), 895–905. https://doi.org/10.1038/nmat3355

Parra, J., Abad-Somovilla, A., Mercader, J. V., Taton, T. A., & Abad-Fuentes, A. (2013). Carbon nanotube-protein carriers enhance size-dependent self-adjuvant antibody response to haptens. *Journal of Controlled Release, 170*(2), 242–251. https://doi.org/10.1016/j.jconrel.2013.05.019

Parween, S., Gupta, P. K., & Chauhan, V. S. (2011). Induction of humoral immune response against PfMSP-119 and PvMSP-1 19 using gold nanoparticles along with alum. *Vaccine, 29*(13), 2451–2460. https://doi.org/10.1016/j.vaccine.2011.01.014

Pati, R., Shevtsov, M., & Sonawane, A. (2018). Nanoparticle vaccines against infectious diseases. *Frontiers in Immunology, 9*(OCT), 2224. https://doi.org/10.3389/fimmu.2018.02224

Peer, D., Karp, J. M., Hong, S., Farokhzad, O. C., Margalit, R., & Langer, R. (2007). Nanocarriers as an emerging platform for cancer therapy. *Nature Nanotechnology, 2*(12), 751–760. https://doi.org/10.1038/nnano.2007.387

Pelaz, B., Alexiou, C., Alvarez-Puebla, R. A., Alves, F., Andrews, A. M., Ashraf, S., … Parak, W. J. (2017). Diverse applications of nanomedicine. *ACS Nano, 11*(3), 2313–2381. https://doi.org/10.1021/acsnano.6b06040

Pescatori, M., Bedognetti, D., Venturelli, E., Ménard-Moyon, C., Bernardini, C., Muresu, E., … Delogu, L. G. (2013). Functionalized carbon nanotubes as immunomodulator systems. *Biomaterials, 34*(18), 4395–4403. https://doi.org/10.1016/j.biomaterials.2013.02.052

Pulendran, B., & Ahmed, R. (2011). Immunological mechanisms of vaccination. *Nature Immunology, 12*(6), 509–517.

Rappuoli, R., Mandl, C. W., Black, S., & De Gregorio, E. (2011). Vaccines for the twenty-first century society. *Nature Reviews Immunology, 11*(12), 865–872. https://doi.org/10.1038/nri3085

Reddy, S. T., Swartz, M. A., & Hubbell, J. A. (2006). Targeting dendritic cells with biomaterials: Developing the next generation of vaccines. *Trends in Immunology, 27*(12), 573–579. https://doi.org/10.1016/j.it.2006.10.005

Reddy, S. T., Van Der Vlies, A. J., Simeoni, E., Angeli, V., Randolph, G. J., O'Neil, C. P., … Hubbell, J. A. (2007). Exploiting lymphatic transport and complement activation in nanoparticle vaccines. *Nature Biotechnology, 25*(10), 1159–1164. https://doi.org/10.1038/nbt1332

Roberts, R. A., Shen, T., Allen, I. C., Hasan, W., DeSimone, J. M., & Ting, J. P. Y. (2013). Analysis of the murine immune response to pulmonary delivery of precisely fabricated nano- and microscale particles. *PLoS One, 8*(4), e62115. https://doi.org/10.1371/journal.pone.0062115

Rose, N. R. (2016). Prediction and prevention of autoimmune disease in the 21st Century: A review and preview. *American Journal of Epidemiology, 183*(5), 403–406. https://doi.org/10.1093/aje/kwv292

Ryan, J. J., Bateman, H. R., Stover, A., Gomez, G., Norton, S. K., Zhao, W., … Kepley, C. L. (2007). Fullerene nanomaterials inhibit the allergic response. *Journal of Immunology (Baltimore, Md. : 1950), 179*(1), 665–672.

Sahin, U., & Türeci, Ö. (2018). Personalized vaccines for cancer immunotherapy. *Science, 359*(6382), 1355–1360. https://doi.org/10.1126/science.aar7112

Schanen, B. C., Das, S., Reilly, C. M., Warren, W. L., Self, W. T., Seal, S., & Drake, D. R. (2013). Immunomodulation and T helper TH 1/TH 2 response polarization by CeO 2 and TiO 2 nanoparticles. *PLoS One, 8*(5), e62816. https://doi.org/10.1371/journal.pone.0062816

Schweingruber, N., Haine, A., Tiede, K., Karabinskaya, A., van den Brandt, J., Wüst, S., … Lühder, F. (2011). Liposomal encapsulation of glucocorticoids alters their mode of action in the treatment of experimental autoimmune encephalomyelitis. *The Journal of Immunology, 187*(8), 4310–4318. https://doi.org/10.4049/jimmunol.1101604

Schwendener, R. A. (2014). Liposomes as vaccine delivery systems: A review of the recent advances. *Therapeutic Advances in Vaccines, 2*(6), 159–182. https://doi.org/10.1177/2051013614541440

Seydoux, E., Rothen-Rutishauser, B., Nita, I. M., Balog, S., Gazdhar, A., Stumbles, P. A., … von Garnier, C. (2014). Size-dependent accumulation of particles in lysosomes modulates dendritic cell function through impaired antigen degradation. *International Journal of Nanomedicine, 9*(1), 3885–3902. https://doi.org/10.2147/IJN.S64353

Shah, R. R., O'hagan, D. T., Amiji, M. M., & Brito, L. A. (2014). The impact of size on particulate vaccine adjuvants. *Nanomedicine, 9*(17), 2671–2681. https://doi.org/10.2217/nnm.14.193

Sharma, C. P. (2010). *Biointegration of medical implant materials: Science and design* (pp. 1–412). Cambridge: Woodhead Publishing. https://doi.org/10.1533/9781845699802

Sharma, S., Mukkur, T. K. S., Benson, H. A. E., & Chen, Y. (2009). Pharmaceutical aspects of intranasal delivery of vaccines using particulate systems. *Journal of Pharmaceutical Sciences, 98*(3), 812–843. https://doi.org/10.1002/jps.21493

Shen, C. C., Liang, H. J., Wang, C. C., Liao, M. H., & Jan, T.-R. (2012). Iron oxide nanoparticles suppressed T helper 1 cell-mediated immunity in a murine model of delayed-type hypersensitivity. *International Journal of Nanomedicine, 7*, 2729. https://doi.org/10.2147/ijn.s31054

Shen, H., Ackerman, A. L., Cody, V., Giodini, A., Hinson, E. R., Cresswell, P., … Hanlon, D. J. (2006). Enhanced and prolonged cross-presentation following endosomal escape of exogenous antigens encapsulated in biodegradable nanoparticles. *Immunology, 117*(1), 78–88. https://doi.org/10.1111/j.1365-2567.2005.02268.x

Shen, Y., Hao, T., Ou, S., Hu, C., & Chen, L. (2018). Applications and perspectives of nanomaterials in novel vaccine development. *MedChemComm, 9*(2), 226–238. https://doi.org/10.1039/c7md00158d

Silva, A. L., Rosalia, R. A., Sazak, A., Carstens, M. G., Ossendorp, F., Oostendorp, J., & Jiskoot, W. (2013). Optimization of encapsulation of a synthetic long peptide in PLGA nanoparticles: Low-burst release is crucial for efficient CD8+ T cell activation. *European Journal of Pharmaceutics and Biopharmaceutics, 83*(3), 338–345. https://doi.org/10.1016/j.ejpb.2012.11.006

Song, W., Musetti, S. N., & Huang, L. (2017). Nanomaterials for cancer immunotherapy. *Biomaterials, 148*, 16–30. https://doi.org/10.1016/j.biomaterials.2017.09.017

Steenblock, E. R., & Fahmy, T. M. (2008). A comprehensive platform for ex vivo T-cell expansion based on biodegradable polymeric artificial antigen-presenting cells. *Molecular Therapy, 16*(4), 765–772. https://doi.org/10.1038/mt.2008.11

Stephan, M. T., Moon, J. J., Um, S. H., Bersthteyn, A., & Irvine, D. J. (2010). Therapeutic cell engineering with surface-conjugated synthetic nanoparticles. *Nature Medicine, 16*(9), 1035–1041. https://doi.org/10.1038/nm.2198

Stone, J. W., Thornburg, N. J., Blum, D. L., Kuhn, S. J., Wright, D. W., & Crowe, J. E. (2013). Gold nanorod vaccine for respiratory syncytial virus. *Nanotechnology, 24*(29), 295102. https://doi.org/10.1088/0957-4484/24/29/295102

Sumbayev, V. V., Yasinska, I. M., Garcia, C. P., Gilliland, D., Lall, G. S., Gibbs, B. F., … Calzolai, L. (2013). Gold nanoparticles downregulate interleukin-1β-induced pro-inflammatory responses. *Small, 9*(3), 472–477. https://doi.org/10.1002/smll.201201528

Svenson, S. (2012). Clinical translation of nanomedicines. *Current Opinion in Solid State and Materials Science, 16*(6), 287–294. https://doi.org/10.1016/j.cossms.2012.10.001

Tao, W., Ziemer, K. S., & Gill, H. S. (2014). Gold nanoparticle-M2e conjugate coformulated with CpG induces protective immunity against influenza A virus. *Nanomedicine, 9*(2), 237–251. https://doi.org/10.2217/nnm.13.58

Thomas, C., Rawat, A., Hope-Weeks, L., & Ahsan, F. (2011). Aerosolized PLA and PLGA nanoparticles enhance humoral, mucosal and cytokine responses to hepatitis B vaccine. *Molecular Pharmaceutics, 8*(2), 405–415. https://doi.org/10.1021/mp100255c

Thomas, T. P., Goonewardena, S. N., Majoros, I. J., Kotlyar, A., Cao, Z., Leroueil, P. R., & Baker, J. R. (2011). Folate-targeted nanoparticles show efficacy in the treatment of inflammatory arthritis. *Arthritis and Rheumatism, 63*(9), 2671–2680. https://doi.org/10.1002/art.30459

Tkach, A. V., Shurin, G. V., Shurin, M. R., Kisin, E. R., Murray, A. R., Young, S. H., … Shvedova, A. A. (2011). Direct effects of carbon nanotubes on dendritic cells induce immune suppression upon pulmonary exposure. *ACS Nano, 5*(7), 5755–5762. https://doi.org/10.1021/nn2014479

Treuel, L., Jiang, X., & Nienhaus, G. U. (2013). New views on cellular uptake and trafficking of manufactured nanoparticles. *Journal of the Royal Society Interface, 10*(82), 20120939. https://doi.org/10.1098/rsif.2012.0939

Tsai, S., Shameli, A., Yamanouchi, J., Clemente-Casares, X., Wang, J., Serra, P., … Santamaria, P. (2010). Reversal of autoimmunity by boosting memory-like autoregulatory T cells. *Immunity, 32*(4), 568–580. https://doi.org/10.1016/j.immuni.2010.03.015

Vasievich, E. A., & Huang, L. (2011). The suppressive tumor microenvironment: A challenge in cancer immunotherapy. *Molecular Pharmaceutics, 8*(3), 635–641. https://doi.org/10.1021/mp1004228

Villa, C. H., Dao, T., Ahearn, I., Fehrenbacher, N., Casey, E., Rey, D. A., … Scheinberg, D. A. (2011). Single-walled carbon nanotubes deliver peptide antigen into dendritic cells and enhance IgG responses to tumor-associated antigens. *ACS Nano, 5*(7), 5300–5311. https://doi.org/10.1021/nn200182x

Wang, T., Zou, M., Jiang, H., Ji, Z., Gao, P., & Cheng, G. (2011). Synthesis of a novel kind of carbon nanoparticle with large mesopores and macropores and its application as an oral vaccine adjuvant. *European Journal of Pharmaceutical Sciences, 44*(5), 653–659. https://doi.org/10.1016/j.ejps.2011.10.012

Wang, X., Podila, R., Shannahan, J. H., Rao, A. M., & Brown, J. M. (2013). Intravenously delivered graphene nanosheets and multiwalled carbon nanotubes induce site-specific Th2 inflammatory responses via the IL-33/ST2 axis. *International Journal of Nanomedicine, 8*, 1733–1748. https://doi.org/10.2147/IJN.S44211

Williams, S. B., Lay, A. H., Lau, C. S., Josephson, D. Y., Wilson, T. G., Choueiri, T. K., & Pal, S. K. (2011). New therapies for castrate-resistant prostate cancer. *Expert Opinion on Pharmacotherapy, 12*(13), 2069–2074. https://doi.org/10.1517/14656566.2011.590133

Wischke, C., Borchert, H. H., Zimmermann, J., Siebenbrodt, I., & Lorenzen, D. R. (2006). Stable cationic microparticles for enhanced model antigen delivery to dendritic cells. *Journal of Controlled Release, 114*(3), 359–368. https://doi.org/10.1016/j.jconrel.2006.06.020

Xiang, S. D., Scholzen, A., Minigo, G., David, C., Apostolopoulos, V., Mottram, P. L., & Plebanski, M. (2006). Pathogen recognition and development of particulate vaccines: Does size matter? *Methods, 40*(1), 1–9. https://doi.org/10.1016/j.ymeth.2006.05.016

Xu, L., Liu, Y., Chen, Z., Li, W., Liu, Y., Wang, L., … Chen, C. (2012). Surface-engineered gold nanorods: Promising DNA vaccine adjuvant for HIV-1 treatment. *Nano Letters, 12*(4), 2003–2012. https://doi.org/10.1021/nl300027p

Xu, L., Liu, Y., Chen, Z., Li, W., Liu, Y., Wang, L., … Chen, C. (2013). Morphologically virus-like fullerenol nanoparticles act as the dual-functional nanoadjuvant for HIV-1 vaccine. *Advanced Materials, 25*(41), 5928–5936. https://doi.org/10.1002/adma.201300583

Yellayi, S., Hilliard, B., Ghazanfar, M., Tsingalia, A., Nantz, M. H., Bollinger, L., … Hecker, J. G. (2011). A single intrathecal injection of DNA and an asymmetric cationic lipid as lipoplexes ameliorates experimental autoimmune encephalomyelitis. *Molecular Pharmaceutics, 8*(5), 1980–1984. https://doi.org/10.1021/mp2002413

Yudoh, K., Karasawa, R., Masuko, K., & Kato, T. (2009). Water-soluble fullerene (C60) inhibits the development of arthritis in the rat model of arthritis. *International Journal of Nanomedicine, 4*, 217–225.

Zhao, H., Kiptoo, P., Williams, T. D., Siahaan, T. J., & Topp, E. M. (2010). Immune response to controlled release of immunomodulating peptides in a murine experimental autoimmune encephalomyelitis (EAE) model. *Journal of Controlled Release, 141*(2), 145–152. https://doi.org/10.1016/j.jconrel.2009.09.002

Zhao, L., Seth, A., Wibowo, N., Zhao, C. X., Mitter, N., Yu, C., & Middelberg, A. P. J. (2014). Nanoparticle vaccines. *Vaccine, 32*(3), 327–337. https://doi.org/10.1016/j.vaccine.2013.11.069

Zhu, G., Zhang, F., Ni, Q., Niu, G., & Chen, X. (2017). Efficient nanovaccine delivery in cancer immunotherapy. *ACS Nano, 11*(3), 2387–2392. https://doi.org/10.1021/acsnano.7b00978

Chapter 5
Neuroregenerative Nanotherapeutics

Ayse Begum Tekinay, Brabu Balusamy, Anitha Senthamizhan, and Tamer Uyar

Abstract As one of the rare tissues where cell proliferation can be severely problematic, neural tissue damages due to accidents or diseases remain one of the major challenges in medicine. Neural tissue engineering offers great revolution in repairing neural injuries. Guidance for neurite extension and connectivity are considered as primary factors in designing neural scaffolds. Several types of nanomaterials have been proposed to be used for therapeutic purposes for neural regeneration. Among these materials electrospun nanofibers made of polymeric and natural biomaterials and self-assembled peptide/peptide amphiphile nanofibers are the most widely studied examples. Electrospinning is a versatile method in fabrication of nanofibers and garnering endless attention among various scientific disciplines including biomedical sciences for their potential use. The scaffolds made of electrospun nanofibers offer excellent environment for cell adhesion and proliferation that opens new avenue in the tissue regeneration application. Natural biomaterials, on the other hand, offer inherent biocompatibility and bioactivity, both of which are desirable qualities for biomaterials for neuroregeneration. Self-assembled peptide nanofibers are also quite advantageous nanomaterials with well-defined chemical structures, low immunogenicity, tailorable bioactivity, and inherent biocompatibility. Advancements in fabrication of neural scaffolds using abovementioned techniques have shed light on neural regeneration research into a new era as it offers the fabrication of tunable and diverse range of scaffolds. In this chapter, we briefly present nanotherapeutics that have been developed for neural regeneration and examples of various nanomaterials and their fabrication methods.

A. B. Tekinay (✉)
Eryigit Research & Development Center, Ivedik OSB, Ankara, Turkey
e-mail: abt@eryigit.com.tr

B. Balusamy · A. Senthamizhan
Fondazione Istituto Italiano di Tecnologia, Genoa, Italy

T. Uyar
Department of Fiber Science & Apparel Design, College of Human Ecology, Cornell University, Ithaca, NY, USA

A. B. Tekinay (ed.), *Nanomaterials for Regenerative Medicine*, Stem Cell Biology and Regenerative Medicine, https://doi.org/10.1007/978-3-030-31202-2_5

5.1 Introduction

A great deal of efforts have been made in the prospects of regenerative medicine over the decades that has set high expectations for discovering solutions of many human illnesses through numerous strategies (Atala, 2012; Mao & Mooney, 2015; Rosenthal & Badylak, 2016). The regenerative medicine aims the functional rehabilitation of organs and tissues injured through disease, trauma, aging and also to regularize congenital defects. In the recent past, rapid developments of nanotechnology opened new realm of advancements in biomedicine and the new fabrication methods have been extensively employed in various fields (Chan, Lee, Zhuo, & Ming, 2017; Gu, Wu, Chen, & Xiao, 2013; Nobile & Nobile, 2017; Pelaz et al., 2017; Ramos, Cruz, Tovani, & Ciancaglini, 2017; Sanna, Pala, & Sechi, 2014; A. Z. Wang & Tepper, 2014). Accordingly, various bottom-up and top-down approaches are applied in advancing the field of regenerative medicine and offer hope in culmination of several long-felt needs (Arora et al., 2012; Engel, Michiardi, Navarro, Lacroix, & Planell, 2008; A. Gupta et al., 2013; Solanki, Kim, & Lee, 2008; Tan et al., 2016; Verma, Domb, & Kumar, 2011). In essence, widespread nanomaterials including polymeric nanoparticles, inorganic nanoparticles, magnetic nanoparticles, nanofibers, dendrimers, and carbon nanotubes have been comprehensively explored for their potential in regenerative medicine (Verma et al., 2011). Neural tissue regeneration is especially problematic due to the inhibitory environment of the neural ECM. In the next subsections, we will describe the molecular characterizations and functions neural ECM molecules.

5.2 Neural ECM Components

The ECM in the nervous system is synthesized by both neural cells, glia and microglia. The neural ECM in the central nervous system of adult animals such as mice and humans not only guides axonal growth but also enables or inhibits neurite formation and extension and glia activation. Neural ECM is also a major determinant in establishing the differentiation paths of neural stem cells into neurons and glia. While ECM components such as laminin and fibronectin enable axonal or neurite growth, molecules such as semaphorins and chondroitin sulfate proteoglycans function as inhibitory components, which is required for proper balancing of neural synapse formation. In addition, the ECM signals also modulate the regenerative and degenerative processes by neurons and glia. The ECM of the peripheral and central nervous system are quite different from each other in terms of their inhibitory and permissive functions. Although the precise mechanisms of how the ECM molecules modulate the cellular activity are not completely known yet, some of the well-known ECM components have been widely studied for neuroregenerative applications and mimicking neural ECM has been one of the major approaches in developing nanomaterials for neuroregeneration.

5.2.1 *Collagens*

Collagens are the most widely found proteins in the ECMs of animal tissues. Today, there are 29 different types of collagen and they share a triple-helical structural formation. Different types of collagens can arrange into fibrillar or non-fibrillar formations and can form uniform or randomly-organized scaffolds. The collagens that are expressed in the peripheral nervous system (PNS) guide neurite extension and provide support for the neurons and the Schwann cells, while collagens are mostly found in the basement membrane structure in the central nervous system (CNS). In the PNS, the upregulation of collagen after injury results in delayed regeneration due to hindering of the neurite extension (Koopmans, Hasse, & Sinis, 2009). Collagen IV has also been associated with the pathogenesis of glaucoma (Wenbin Huang et al., 2013). Collagen IV, which is the main component of basal lamina in the CNS, is also expressed as a part of the basal lamina complex at the neuromuscular junction, where the other major collagens are Collagen Q and Collagen XIII (Singhal & Martin, 2011).

5.2.2 *Laminins*

Laminins are non-fibrous glycoproteins that are composed of α, β, and γ chains which combine together forming a coiled-coil structure. Laminins are found in almost all tissues and 16 isoforms have been identified with varied expression depending on the tissue and the age of the organism (Aumailley, 2013). Laminins are mostly recognized by integrins on the cell surface; however, several other types of proteins such as dystroglycans have also been identified as laminin receptors (Colognato et al., 2007). The binding of dystroglycans with the laminins are important for the regeneration of the PNS after injury. In CNS, laminin is primarily found in the basal lamina and provides guidance to axonal and neurite growth during development (Turney & Bridgman, 2005). Laminins have also been shown to induce proliferation and differentiation of neural stem cells in cell culture (Flanagan, Rebaza, Derzic, Schwartz, & Monuki, 2006; Hall, Lathia, Caldwell, & Ffrench-Constant, 2008; W. Ma et al., 2008). Laminin knockout studies have also shown the importance of laminins in embryogenesis and CNS development (Guldager Kring Rasmussen & Karsdal, 2016; Rasi et al., 2010). In humans, several mutations in laminin receptors such as dystroglycan and integrins as well as laminins have been associated with neurological problems (Hara et al., 2011; Longman et al., 2003; Matejas et al., 2010). Due to the ability of laminins to guide neurite outgrowth and axonal elongation, many of the regenerative medicine studies that aim to regenerate neural tissues have used either laminin or epitopes that are derived from laminin, and several of these studies will be explained in the nanomaterials for neuroregeneration section.

5.2.3 Tenascins

Among the four members of the Tenascin family, Tenascin C (TN-C) and Tenascin R (TN-R) are expressed in the CNS and both of their expression are upregulated after injury (Becker, Schweitzer, Feldner, Schachner, & Becker, 2004; Midwood & Orend, 2009; Tang, Davies, & Davies, 2003). While TN-C is expressed by radial glia, astrocytes and retinal and hippocampal neurons, TN-R is primarily expressed by neurons. On cell surface, TN-C interacts with integrins and was shown to modulate neuronal behavior in CNS during development in vivo (Garcion, Faissner, & Ffrench-Constant, 2001; Garcion, Halilagic, Faissner, & Ffrench-Constant, 2004; Tucker & Chiquet-Ehrismann, 2015). In cell culture, TN-C and TN-R have been both shown to modulate neuronal and stem cell behavior (Garcion et al., 2004; Gobaa et al., 2011; Wenhui Huang, Zhang, Niu, & Liao, 2009; Little, Healy, & Schaffer, 2008), thus, they provide novel ways of modulating stem cell behavior for neuroregeneration purposes as well. However, it should be noted that both TN-C and TN-R have varied effects on neural behavior depending on the presence of culture media and the cell type in vitro. In vivo, the inhibitory or induction effects of tenascins in neurite outgrowth and axonal elongation also depends on the cell type and other ECM elements.

5.2.4 Glycosaminoglycans

GAGs in the neural ECM are found either by themselves or as attached to proteoglycans. Chondroitin sulfate and chondroitin sulfate proteoglycans (CSPG) are the most widely found GAGs and proteoglycans in the CNS. Most of the CSPGs are expressed by astrocytes. The sulfation degree of these proteoglycans vary depending on the protein core and position where the sulfation occurs, that is, chondroitin-4-sulfate and chondroitin-6-sulfate, which results in a vast array of different molecules with various functions. CSPGs are mostly known for their inhibitory effects on neurite outgrowth and axonal elongation, and the expression levels of some of these CSPGs are highly upregulated after injury, inhibiting neuroregeneration (Shen et al., 2009; Yang, Kwok, & Fawcett, 2014). Hyaluronan-binding CSPGs such as aggrecan, brevican, neurocan, and versican, have been shown to inhibit neurite formation and axonal growth and are upregulated in glial scars after CNS injuries (Fawcett, Kwok, Afshari, & García-Alías, 2008; Yang et al., 2014). Chondroitinase ABC, which is an enzyme that breaks down chondroitin sulfate chains, was shown to reverse this inhibitory effect after injury (Fawcett et al., 2008; Ikegami et al., 2005; Suzuki et al., 2017).

The activity of proteoglycans on neural stem cells have also been investigated in vitro, where it was found that the presence of CSPGs in the culture environment causes disruptions in the development and differentiation of neural stem cells (Karumbaiah et al., 2015; Sirko, von Holst, Wizenmann, Gotz, & Faissner, 2007).

CSPGs were shown to be an important constituent of perineural nets (PNNs). PNNs are dense ECM structures that wrap the surface of the neurons and are composed of CSPGs, hyaluronan, tenascins and proteoglycan-link-proteins. Since PNNs are important for establishment of synaptic connections and plasticity, they have crucial functions in neuroregeneration after injury (Fawcett et al., 2008; Kwok, Dick, Wang, & Fawcett, 2011; Maroto et al., 2013).

5.2.5 *Netrins and Slits*

Netrins and slits are secreted proteins that function in axonal growth and movement in the nervous system and cellular functions such as adhesion and migration in other tissues (Lin, Rao, & Isacson, 2005; Stein & Tessier-Lavigne, 2001; K. L. W. Sun, Correia, & Kennedy, 2011; Yebra et al., 2003). Depending on the receptors on the growth cones of the neurons, they provide attractive or repellent signals for the axons and have binding sites for heparan sulfate proteoglycans and integrins (Stein & Tessier-Lavigne, 2001; Yebra et al., 2003). Netrins can have both short range and long range effects on axonal growth, which was also shown by gene knockout studies in mice (Dominici et al., 2017; Schwarting, Raitcheva, Bless, Ackerman, & Tobet, 2004).

Slits are glycoproteins and act as ligands for Robo receptors either for inhibiting axonal attraction or acting as axonal repellants (Blockus & Chédotal, 2016; Brose et al., 1999; Kidd, Bland, & Goodman, 1999). Both slits and robo proteins have been shown to have important effects on non-neuronal cellular functions such as cell adhesion and angiogenesis (Blockus & Chédotal, 2016). Thus, biological clues on these proteins can act as invaluable tools for developing nanomaterials for regenerative medicine applications.

5.2.6 *Reelin*

Reelin is a vital ECM glycoprotein for the development of the nervous system and mutations in this gene causes major CNS pathologies both in mice and humans (Baek et al., 2015; D'Arcangelo et al., 1995; Pujadas et al., 2014; Tissir & Goffinet, 2003). The major receptors for reelin are lipoprotein receptors VLDLR and ApoER2, expressed by migrating neurons and radial glia, but reelin was shown to bind to or effect activities of other receptors as well (Baek et al., 2015; Dulabon et al., 2000; Hiesberger et al., 1999; Weeber et al., 2002). Reelin modulates neuronal migration, synapse formation and function, which makes it an attractive therapeutic target for the treatment of CNS related pathologies (Durakoglugil, Chen, White, Kavalali, & Herz, 2009; Hashimoto-Torii et al., 2008; Sekine, Kubo, & Nakajima, 2014; Telese et al., 2015).

5.3 Electrospun Nanofibers for Neuroregeneration

Electrospinning is remarkably simple and versatile technique that enables production of nanofibers with diameters down to nanoscale using electrostatic repulsion of surface charges finally draw nanofibers from viscous fluid. The electrospinning setup typically consist of high voltage power supply, syringe pump, spinneret, and collector that has been extensively applied to fabricate fibers from wide range of materials including polymers, ceramics and their combinations. Interestingly, the electrospinning offers the possibilities to generate nanofibers in solid, hollow, porous, and core–shell structures. Furthermore, the electrospun nanofibers can be interior and/or surface functionalized using various active agents including nanoparticles during and following electrospinning process (J. Xue, Xie, Liu, & Xia, 2017). Nevertheless, morphology of the resultant electrospun nanofibers mainly governed by the electrospinning parameters determines their physicochemical characteristics and intended applications. Figure 5.1 illustrates the schematic representation of electrospinning process with various parameters influencing the morphology of fibers (Pelipenko, Kristl, Janković, Baumgartner, & Kocbek, 2013). Attributed to their exceptional characteristics, the electrospun nanofibers have tremendously applied in various sectors from environmental remediation to biomedicine (Senthamizhan, Balusamy, & Uyar, 2016; Shabafrooz, Mozafari, Vashaee, & Tayebi, 2014; Uyar & Kny, 2017).

Designing and fabrication of ideal scaffold for specific tissue regeneration is critical while the scaffold directly exposed to the cells, support cell proliferation and tissue development (Khademhosseini & Langer, 2016). Electrospun nanofibers primarily solve the major challenge in design and fabrication of suitable scaffold for regenerative medicine. Generally, in the tissue engineering perspective, the designed scaffold should mimic as native extracellular matrix (ECM) to offer temporary support in cell adhesion during formation of natural ECM by the cells. Since the electrospun nanofibers produced with high surface-to-volume ratio, porosity, and spatial interconnectivity offers better cellular communication and nutrient transport to serve as a better platform for cell attachment, differentiation, and proliferation, they qualify as an ideal scaffold in regeneration. Liu, Thomopoulos, and Xia (2012) stated that tissues including sciatic nerve, heart, tendon, and blood vessel can be recapitulated using nanofibrous scaffolds (W. Liu et al., 2012). Therefore, to date electrospun nanofibers has been immensely explored for their application in regenerative medicine (Bosworth & Downes, 2011; Braghirolli, Steffens, & Pranke, 2014; S. Chen, Li, Li, & Xie, 2018; W. Liu et al., 2012; B. Ma, Xie, Jiang, Shuler, & Bartlett, 2013; Pham, Sharma, & Mikos, 2006; Xie, MacEwan, Schwartz, & Xia, 2010).

Besides the electrospun nanofibrous scaffold mimicking ECM characteristics, alignment of nanofibers plays key role in the successful application of tissue regeneration as it greatly influences the cell adhesion, migration and differentiation. The alignment of electrospun fibers can be controlled through mechanical, electrostatical and magnetic force (W. Liu et al., 2012). As most of the tissues in human body

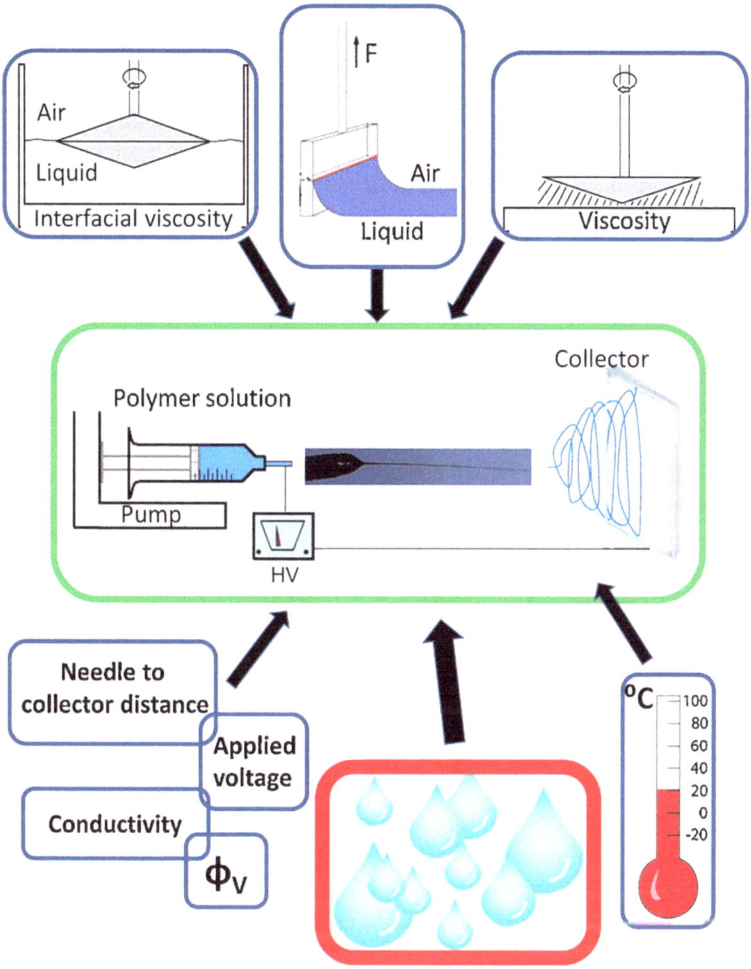

Fig. 5.1 Schematic representation of the electrospinning process with parameters influencing the morphology of the electrospun product. (Reproduced with permission from Pelipenko et al., 2013, © 2013 Elsevier)

characterized in highly ordered structure (Sharma & Maffulli, 2006), thus the aligned electrospun nanofibrous scaffolds mimicking anisotropic morphology shown to have promising results in regeneration. Until now, there has been accountable efforts cemented towards investigating the beneficial effects of electrospun nanofibrous scaffold alignment in tissue regeneration applications. In this chapter, we mainly concentrate to present brief summary on influence of fiber alignment in cell growth and differentiation behavior. Further, examples of various approaches developed for fabricating scaffolds for regeneration application with special focus in neural regeneration are discussed.

5.3.1 Aligned Electrospun Nanofibers

Electrospun nanofibers can be produced either in aligned form or not-aligned form and various materials can be utilized for their production from pure polymeric materials to combinations of various polymeric and biomaterials. Alignment of the fibers has been shown to aid in the alignment of neurites and axons especially when the fibers are functionalized with specific ligands. These nanofibers have immediate potential applications in peripheral nerve tissue regeneration where long-gaps between nerve endings caused by trauma or diseases are already treated with polymeric nerve conduits to enable proper connection between the neurons.

Aligned nanofibers that are fabricated by using different types of polymers have been used for neuroregeneration of neural cell culture purposes. Aligned aminolyzed poly-L-lactide (PLLA) fibers that are coated with graphene oxide have been shown to promote the proliferation of Schwann cells and direct the alignment of their cytoskeleton along the alignment of the fibers themselves. These nanofibers also enhanced the proliferation and differentiation of PC12 cells, which are considered a model neuronal cell line, in the presence of neuronal growth factor (NGF) (K. Zhang, Zheng, Liang, & Gao, 2016). Nebulized solvent patterning of aligned electrospun PLLA enabled formation of different patterns on aligned nanofibers and astrocyte cultures on these substrates showed that the CSPGs and fibronectin produced by the astrocytes on aligned sites exhibited an aligned morphology while the astrocytes that are cultured on other topographical cues did not which resembles the natural anisotropic-to-isotropic transition in neural ECM that restrict neurite outgrowth (Zuidema et al., 2015). In another study, aligned PLLA nanofibers, which were electrically polarized by using a lab-built corona poling device, promoted the retinoic acid-induced differentiation of neuroblastoma cells significantly enhanced neurite outgrowth compared to control groups (Barroca et al., 2018). Aligned PLLA nanofibers which were made electroconductive through coating with polypyrole (PPy) were shown to not only guide the neurites extended by PC12 cells along the fiber alignment but also significantly increase neurite length under electrical stimulation (Y. Zou et al., 2016). In another study, electrically conductive PPy-poly lactic acid (PLA) nanofibers coated with poly-ornithine were shown to facilitate attachment or PC12 cells, and enhance proliferation and differentiation of PC12 cells especially when combined with electrical stimulation (Tian et al., 2016).

Poly(methyl methacrylate) (PMMA) is another widely studied polymer as electrospun materials due to its biocompatible nature. A comparison of aligned and non-aligned electrospun PMMA nanofibers revealed that although both groups supported attachment and viability of astrocytes, these cells formed longer and structurally aligned processes on aligned fibers compared to the cells on non-aligned PMMA fibers. However, this study also showed that astrocytes respond to non-aligned nanofibers and extend processes much earlier than the cells cultured on their aligned counterparts (Xia & Xia, 2018). For neuromuscular junction regeneration, aligned and random electrospun polylactic acid (PLA) nanofibers were tested in vitro, where long-term cocultures of primary embryonic motor neurons and C2C12 cells showed that these nanofibers enables formation and alignment of the myotubes and enhanced their long-term survival in culture (B. Luo et al., 2018). NGF encapsulation into aligned electrospun poly(ε-caprolactone) (PCL) nanofibers enabled their slow release when stabilized with bovine serum albumin (BSA), and enhanced viability and elongated and aligned neurite outgrowth of PC12 cells (Hu, Tian, Prabhakaran, Ding, & Ramakrishna, 2016). In another study, a continuous gradient of stromal cell-derived factor 1α (SDF1α) in combination with a peptide with collagen binding domain was formed on radially aligned electrospun nanofibers, and neural stem cell behavior was observed, where they exhibited directed and enhanced migration along the gradient on aligned fibers (Xiaoran Li, Li, et al., 2016). O_2 plasma treated aligned PCL nanofibers also enhanced differentiation of embryoid bodies, as evidenced by gene expression and microscopical analyses (Abbasi et al., 2016). Neuronal differentiation of other types of stem cells has also been observed on aligned electrospun nanofibers. Human endometrial stem cells were shown to differentiate into motor neuron-like cells on aligned PLGA nanofibers when induced with retinoic acid and sonic hedgehog (Shh) containing culture media (Ebrahimi-Barough et al., 2015).

Other techniques have also been combined with electrospinning to produce ECM mimetic scaffolds with desirable properties for neuroregeneration purposes. When both electrospinning and 3D printing was combined to fabricate a 3D scaffold with aligned nanofibers made of PCL and gelatin, this scaffold increased the average neurite length and enabled directed neurite elongation of primary cortical neurons in culture (Lee, Nowicki, Harris, & Zhang, 2016). In another study, electrospinning was combined with electrospraying forming a PCL scaffold with adjustable bioactive factor embedded poly(D,L-lactide-co-glycolide) core–shell nanospheres. The topography of the modified scaffold provided a favorable environment for PC12 cells and astrocytes and exhibited sustained release of the bioactive factor (Zhu, Masood, O'Brien, & Zhang, 2015). Aligned PCL nanofibers that were functionalized with laminin-derived Gly-Tyr-Ile-Gly-Ser-Arg (GYIGSR) peptides were shown to accelerate the differentiation of mouse embryonic stem cells into neuronal lineage even faster than laminin coated surfaces, which shows that these synthetic substrates can be used as xeno-free materials for stem cell transplantation (Silantyeva et al., 2018). Combination of electrospun nanofibers with self-assembling peptide structures has also resulted in favorable materials for neural regeneration studies. Aligned PLGA nanofibers that were functionalized with RADA16-I-BMHP1 peptides were shown to enhance bipolar extension of Schwann cells in in vitro and promote recovery of sciatic nerve injury in vivo regulating collagen organization and increasing myelination significantly more than control groups (Nune, Subramanian, Krishnan, Kaimal, & Sethuraman, 2017). In another study, aligned PLGA-carbon nanotube composite nanofibers were fabricated through electrospinning and coated with poly-L-lysine. While the use of carbon nanotubes provided electrical conductivity to the nanofibers, poly-L-lysine coating increased its hydrophilicity, creating a better environment for nerve cell attachment. This aligned conductive scaffold not only enabled growth of PC12 cells and dorsal root ganglion neurons along the fiber direction but also enhanced neuronal differentiation of

PC12 cells and viability and myelination of Schwann cells (Jing Wang, Tian, Chen, Ramakrishna, & Mo, 2018).

Aligned electrospun nanofibers have also been investigated in in vivo studies. Aligned electroconductive PPy modified electrospun PLA nanofibers were tested in a rat spinal cord injury model, where these nanofibers were also compared with non-PPy modified fibers in one study. The results showed that implantation of the electrospun conductive fibers enhanced the functional recover after spinal cord injury as evidenced by electrophysiological recordings and BBB scores and reduced apoptosis and autophagy (Shu et al., 2019).

5.3.2 Non-aligned Electrospun Nanofibers

Although alignment of the electrospun nanofibers has been shown to be a desirable factor for many neuronal cultures and neuroregeneration studies, there are also many studies which investigated the effects of non-aligned or randomly aligned electrospun nanofibers for such applications. PCL, PLLA, PLA, and PLGA are again most widely used polymeric materials that are used for these purposes. Human induced pluripotent stem cells (hIPSCs) were shown to completely differentiate into neuronal lineage on bi-electrospun PCL-gelatin nanofibers in neural differentiation medium (KarbalaeiMahdi et al., 2017). In another study, polyallylamine-epidermal growth factor (EGF) functionalized electrospun PLA nanofibers were shown to support viability of neural stem cells for up to 14 days even in the absence of additional growth factors in the culture media (Haddad et al., 2016). Electrospun nanofibers were also used for transplantation of neural stem cells, where gelatin sponge-electrospun PLGA/PEG nanofibers were seeded with neural precursor cells which were differentiated from iPSCs. This system was later transplanted into rats whose spinal cords were transected and motor functional recovery was assessed, which showed improvements in BBB scores compared to PLGA alone and untreated groups (Pang et al., 2016). Mixing materials that are hard to electrospun by themselves with other materials is a frequently applied approach for electrospinning applications. Poly(glycerol sebacate) (PGS) is an example to the materials that is difficult to use for electrospinning. Electrospun nanofibers manufactured from the copolymer of PGS and PMMA mixed with gelatin have also shown to be biocompatible and support the viability and neurite outgrowth of PC12 cells (Hu et al., 2017).

Combination of synthetic polymeric materials with natural biomaterials in electrospinning applications was frequently applied for producing nanomaterials for neural tissue regeneration. Hyaluronic acid is a GAG that is frequently found in the natural neural ECM. ECM mimetic nanofibrous scaffolds made of electrospun PCL and hyaluronic acid nanofibers exhibited controlled porosity and favorable morphological characteristics for cell attachment for neural cells such as SH-SY5Y human neuroblastoma cells (Entekhabi, Haghbin Nazarpak, Moztarzadeh, & Sadeghi, 2016). Another natural biomaterial, lignin, was also combined with PCL in order to fabricate electrospun nanofibers with antioxidant characteristics. These nanofibers not only good antioxidant properties but also enhanced myelin basic protein expression from Schwann cells and enhanced neurite extension from DRG neurons (Jing Wang, Tian, Luo, et al., 2018).

Composites of polymeric materials with other types of nanomaterials have also resulted in improved qualities for electrospun nanofibers in terms of neural tissue engineering applications. In one study where carboxylated carbon nanotubes were used for modifying PLGA nanofibers, the nanofibers were later treated with plasma in order to increase their hydrophilicity and PC12 cells, DRG neurons and Schwann cell were cultured on these nanofibers. Both plasma treatment and carbon nanotube addition were shown to enhance the biocompatibility and differentiation potential of PLGA nanofibers (Jing Wang, Chen, Ramakrishna, Tian, & Mo, 2017). Another composite nanofiber system, where graphene oxide was incorporated into electrospun PLGA nanofibrous mats and insulin-like growth factor (IGF-1) was immobilized on graphene oxide, resulted in enhanced survival of neural stem cells under H_2O_2 pretreatment, which shows its neuroprotective effects. This nanofiber system also increased neuronal differentiation to a certain extent (Qi et al., 2019). In another study, graphene oxide modified PLGA nanofibers were used for controlled release of methylene blue. The slow release of methylene blue from these nanofibers and the favorable environment provided by the scaffold system inhibited apoptosis in neural progenitor cells and diminished tau-phosphorylation (L. Wang et al., 2019). Randomly aligned electrospun PCL/gelatin nanofibers that are modified with graphene oxide were also utilized for slow release of antibacterial drugs (Heidari, Bahrami, Ranjbar-Mohammadi, & Milan, 2019). Non-aligned PLGA nanofibers that are decorated with self-assembling RADA16-I-BMHP1 peptides were shown to enhance Schwann cell proliferation and myelination activity, showing their potential to be used for peripheral nerve regeneration (Nune, Krishnan, & Sethuraman, 2015).

5.4 Natural Biomaterials as Neural ECM Mimics

Natural biomaterials that are obtained from the ECMs of different organisms such as mammals or crustaceans have been of interest for many regenerative medicine applications including neural tissue engineering. Among these natural biomaterials, collagen, fibrin/fibronectin, various GAGs from mammalian ECMs, chitosan and alginate have been the most widely applied materials.

5.4.1 Collagen

Collagen is the most widely found fibrous protein in the mammalian ECM and have been used for various tissue regeneration studies. Collagen is also an ideal material for stem cell culturing. Three-dimensional collagen grafted PCL nanofibers were shown to induce the differentiation of MSCs derived from Wharton's jelly into motor neuron-like cells when cultured in retinoic acid and Shh containing media (Bagher et al., 2016). In another study, collagen containing scaffolds were optimized in order to provide the most suitable scaffold for neural stem cells under simulated spinal cord injury microenvironment. The scaffold that was chosen after optimization studies was later used for neural stem cell transplantation into rat severe spinal cord injury models where a high survival rate and functional differentiation of neural stem cells was observed. This study shows the importance of the use of ECM mimetic materials as training microenvironments prior to stem cell transplantations (Xing Li, Liu, et al., 2016).

Collagen has been used as a functionalization agent for synthetic polymeric systems in order to render these materials bioactivity. These bioactive scaffolds can also be used for understanding the molecular processes behind the activity of certain drugs or proteins. In one example for these studies, human endometrial stem cells were cultured on collagen/PCL nanofibers in the presence of a small molecule inhibitor of PI3K/Akt signaling pathway, LY294002, together with neuronal induction media. This small molecule was shown to specifically induce the differentiation of endometrial stem cells into motor neuronal differentiation, suggesting that it can be used for motor neuronal regeneration (Ebrahimi-Barough et al., 2017). In another study, the effect of microtubule-stabilizing paclitaxel loaded collagen microchannel scaffold was investigated for its effects on neural stem cells both in vitro and in vivo. This scaffold was shown to rescue myelin-inhibited neuronal differentiation of neural stem cells in vitro and provide an instructive microenvironment for the differentiation of neural stem cells after being transplanted to animal models with spinal cord injury in vivo (Xiaoran Li et al., 2018).

Collagen scaffolds have also been used as carriers of biological drugs such as antibodies and growth factors. Collagen scaffolds modified with Cetuximab, which is an EGFR signaling antagonist, were shown to direct neuronal differentiation, maturation, myelination and synapse formation at spinal cord injury sites in dogs and thus enabled neuronal regeneration resulting in significant motor functional recovery (Xing Li et al., 2017). In another study, a collagen-multichannel poly(propylene fumarate) scaffold system was used as a carrier for collagen-binding neurotrophic factor 3 was shown to improve the inhibitory environment of the tissue after spinal cord injury and facilitate axonal and neuronal regeneration in an animal spinal cord injury model improving functional recovery (X. Chen, Zhao, et al., 2018).

The combinations of collagen with other bioactive materials have resulted in increased bioactivity in several studies. Electrospinning was used to combine laminin with collagen in a core Shell nanofiber model in order to provide both the bioactive and topographical cues for the viability and optimal differentiation of neural cells. The aligned nanofibers provided the topographical assistance while the signals on collagen and the slow-released laminin were utilized for bioactivity. The viability and differentiation potential of hippocampal neurons cultured on these nanofibers were elevated compared to controls, as evidenced by enhanced guided neurite outgrowth (Song et al., 2018). In another study, graphene oxide nanosheets were used to fabricate a hybrid scaffold with collagen for providing electroconductivity. When MSCs were cultured on this scaffold system in neuronal induction medium, the gene and protein expression analyses showed an increased level of expression for neuronal markers after 7 days of culture (W. Guo, Wang, et al., 2016).

5.4.2 *Fibrin*

Fibrin is an important ECM component especially for peripheral nervous system regeneration and has been used as a neural ECM mimetic system either by itself or in combination with other materials in several studies. Hierarchically aligned fibrin nanofiber hydrogel was fabricated with self-assembly and electrospinning was implanted into a rat spinal cord injury model to bridge the lesion site where it was shown to accelerate axonal regrowth, vascular reconstruction and recovery of motor function (Yao et al., 2018). In another study, a similarly prepared fibrin hydrogel was utilized for peripheral nerve regeneration, and it was shown that the aligned fibrin nanofiber gel supports cell adhesion and directional migration of Schwann cells and DRG neurons in vitro and Schwann cell cable formation and axonal regrowth in a 10-mm long sciatic nerve gap model in vivo more efficiently than hollow chitosan tubes and non-aligned fibrin nanofiber hydrogel (Du et al., 2017).

Fibrin hydrogels were combined with different materials for neural tissue engineering applications. Salmon fibrin, laminin and hyaluronic acid were combined to fabricate a fibrin-based scaffold for culturing human neural stem/progenitor cells, where salmon fibrin was shown to provide a more favorable environment for these cells compared to human fibrin. This hybrid scaffold was also shown to enhance proliferation and differentiation of neural stem/progenitor cells (Arulmoli et al., 2016). Combination of fibrin with carbon nanotube/polyurethane nanofibers was also tested for neural tissue engineering and human endometrial stem cells cultured on these fibers exhibited higher viability and proliferation compared to fibrin hydrogels (Hasanzadeh et al., 2019). Fibrin hydrogels containing multiple layers of highly oriented electrospun PCL nanofibers were utilized for culturing purified Schwann cells, where it was observed that the topographical cues provided by the nanofibers modulated the branching patterns of the Schwann cells and their proliferation capability (Hodde et al., 2016).

Fibrin hydrogels were also used for culture and differentiation of iPSCs in several studies. The culturing conditions required for the directed differentiation of the iPSCs into neuronal lineage was optimized by using fibrin hydrogels in order to use these cells for spinal cord injury treatment (Montgomery, Wong, Gabers, & Willerth, 2015). The rapid degradation problem of fibrin hydrogels for iPSC culture was overcome by using various cross-linking agents such as genipin, which elongated the degradation rate while still promoting the viability and neurite outgrowth of the neural cells that are differentiated from iPSCs inside these gels (Robinson, Douglas, & Willerth, 2017). When the specific lineages into which iPSCs differentiate in fibrin hydrogels, were investigated, the differentiated cells were found to exhibit dorsal/ventral spinal neuron identities and less number of cells differentiated into astrocytes compared to 2D laminin coated cultures (Edgar, Robinson, & Willerth, 2017). Fibrin hydrogels were also used for 3D printing of neural tissues by using iPSCs where the fibrin hydrogels were used as a bioink (Abelseth et al., 2019).

5.4.3 Laminin

Laminin is a crucial component of basement membrane and has been extensively used for 2D culture of neuronal cells in tissue culture through coating the plate surface with laminin. Thus, various nanomaterials were also coated or functionalized with laminin in order to produce more favorable surfaces for neuronal cell attachment, survival and differentiation. When the surfaces of electrospun silica nanofibers were modified with laminin, it not only made surface properties more favorable for biological applications but also facilitated formation of longer neurites from PC12 cells (W. Chen, Guo, et al., 2018). Laminin modified aligned electrospun poly(3-hydroxybutyrate) nanofibers were also more biocompatible than their unmodified counterparts, and enhanced the proliferation rates of both murine neuroblastoma Neuro2a cells and brain-derived neural stem cells (Sangsanoh, Ekapakul, Israsena, Suwantong, & Supaphol, 2018). Laminin coating also rendered pHEMA-MOETACl hydrogels more favorable for neural cell culture. Transplantation of iPSC-derived neural progenitor cell seeded laminin-coated pHEMA-MOETACl hydrogels in a rat spinal cord injury model resulted in increased survival of the transplanted neural progenitor cells, improved integration with the host system and more recovery in motor functions (Ruzicka et al., 2019).

Since both hyaluronic acid and laminin are found in high amounts in the neural basal membrane ECM, hyaluronic acid-laminin hydrogels provide ideal neural ECM-mimetic environments. Hyaluronic acid-laminin hydrogels that are modified with a SDF-1α gradient were investigated for their effect on neural progenitor cells in culture, where it was found that these hydrogels upregulated the expression of SDF-1α receptor in neural progenitor/stem cells and enhanced their migration in a gradient specific manner (Addington et al., 2015). When this hydrogel was tested for cell transplantation in vivo, it was shown to facilitate the neural progenitor/stem cell response in terms of survival and directed migration (Addington, Dharmawaj, Heffernan, Sirianni, & Stabenfeldt, 2017).

5.4.4 Alginate

Alginate is a biodegradable, biocompatible, and relatively economical material with suitable mechanical characteristics for neural tissue regeneration. Alginate hydrogels were seeded with bone marrow stromal cells which expressed brain-derived neurotrophic factor (BDNF) and they were transplanted into rat spinal cord injury models at the injury site. MSCs were observed to survive at the injury site even after 4 weeks of transplantation and the hydrogel treatment significantly enhanced axonal elongation and regeneration (Günther, Weidner, Müller, & Blesch, 2015). Alginate hydrogels were also used for sustained delivery of drugs and biological factors. Alginate sulfate/alginate hydrogels that were loaded with epidermal growth factor (EGF) and fibroblast growth factor-2 (FGF-2) were seeded with neural progenitor cells and the sustained release of the growth factors and the effect of the hydrogel system on neural progenitor cells were observed. The sustained release of the growth factors and the morphological characteristics of the hydrogel system enabled formation of a significantly higher number of neurospheres and the cells were found to proliferate and differentiate inside the hydrogels (Cizkova et al., 2015). When these dual growth factor carrier scaffolds were used for the treatment of spinal cord injury in vivo, they increased the survival rate of neurons and sensory fibers and facilitated functional recovery (Grulova et al., 2015).

Combinations of alginate with other materials were also investigated in order to modulate their biological and physicochemical characteristics. Alginate scaffolds that are combined with polyvinyl alcohol and graphene nanosheets exhibited superior electrical conductivity and mechanical properties and supported viability and proliferation of PC12 cells (Golafshan, Kharaziha, & Fathi, 2017). Electrical stimulation to these gels further enhanced their supportive characteristics for PC12 cell culture, hence their potential to be used for neuronal regeneration studies (Golafshan et al., 2018).

Alginate hydrogels that are modified with integrin ligands were used for long-term 3D culture of neural progenitor cells and induce their differentiation into oligodendrocyte lineage. Both human and mouse neural progenitor cell containing neurospheroids were successfully encapsulated and cultured in these hydrogels for up to 90 days (Wen et al., 2019).

5.4.5 Chitosan

Chitosan is another example of natural biomaterials that shows good biocompatibility and biodegradability while having an affordable price and tailorable chemical characteristics. Chitosan-based scaffolds have been investigated for many regenerative purposes including neural regeneration. Culture of dental pulp stem cells in chitosan scaffolds enabled neuronal differentiation of these cells in vitro and transplantation of dental pulp stem cell-embedded chitosan scaffolds into spinal cord injury models resulted in motor functional recovery in vivo (J. Zhang, Lu, et al., 2016). When soft chitosan microbeads were used for 3D culture of primary neuronal cells, and analyzed in detail in by using optical and electron microscopy, immunocytochemistry and electrophysiological recordings, the results shows that these microbeads can be used for 3D cultures of neuronal cells (Tedesco et al., 2018).

The mechanical properties of chitosan could be enhanced by using cross-linking agents such as genipin, and genipin treated chitosan nanofibers were shown to facilitate Schwann cell alignment and proliferation and enhance the neurite growth rate of DRG neurons (Lau et al., 2018).

Hybrid materials containing both chitosan and other materials were also developed for neural tissue engineering. Chitosan/silk fibroin-based scaffolds were used to culture bone marrow derived stem cells, and then decellularized to maintain the ECM components that were secreted by these cells. When this acellularized matrix was used for bridging a sciatic nerve gap in dogs, this scaffold facilitated increased axonal growth and Schwann cell proliferation, which was almost comparable to autologous nerve graft, the golden standard for peripheral nerve repair (C. Xue, Ren, et al., 2017). In another study, chitosan/gelatin scaffolds were rendered electroactivity through modification with poly(3,4-ethylenedioxythiophene) (PEDOT) nanoparticles while still maintaining their biocompatibility. These gels also enhanced neurite outgrowth and neuronal marker expression in PC12 cells (S. Wang, Sun, et al., 2017). Chitosan hydrogels that were functionalized with PEDOT on their surface were also electroconductive and exhibited good biocompatibility with neuronal cells (Xu et al., 2017). When PEDOT was used to functionalize chitosan/gelatin scaffolds, these hybrid gels also became electroactive and were used for 3D culturing of neural stem cells. The electroactive hybrid scaffold was shown to facilitate not only adhesion and proliferation of neural stem cells but also induce their differentiation towards neurons and astrocytes which was assessed by protein and gene expression analyses (S. Wang, Guan, Li, et al., 2018).

Hollow polymeric nerve guidance channels are usually used for assisting peripheral nerve repair across critical size defects. In order to enhance their therapeutic potential, electrospun polyurethane nerve guidance conduits were filled with aligned chitosan/gelatin cryogels and evaluated in vitro by culturing DRG neurons. The results showed that these hydrogels were biocompatible and enabled aligned axonal growth and cellular migration for neuronal cells (A. Singh, Shiekh, Das, Seppälä, & Kumar, 2019). Chitosan-intercalated montmorillonite/poly(vinyl alcohol) nanofibers were also shown to be biocompatible and guided differentiation of dental pulp stem cells into neuronlike cells (Ghasemi Hamidabadi et al., 2017).

5.5 Peptide Nanofibers for Neural Regeneration

Many fibrous proteins that make up the structural components of the ECM and the cytoskeleton are self-assembling molecules. This self-assembly strategy of nature has been mimicked by materials scientists in order to fabricate nanomaterials for regenerative medicine purposes through bottom-up approach. Among these self-assembled systems, peptide nanofibers and peptide amphiphile nanofibers are the most widely used materials for neural tissue regeneration and engineering.

5.5.1 Self-Assembling Peptides as Neural ECM Mimics

Self-assembling peptides have been extensively utilized for neural regeneration applications. Among these, peptides that have a base sequence of arginine–alanine–aspartic acid–alanine (RADA) have been first identified by Zhang et al. and were also commercialized as PuraMatrix (S. Zhang et al., 1995). These self-assembling peptides go through spontaneous self-assembly due to ionic self-complementary β-sheet formation and form nanofibrous scaffolds under physiological conditions. They have been used for encapsulation of various types of cells and were shown to support survival and neurite outgrowth of neuronal cells and synapse formation between neurons (Holmes et al., 2002). In traumatic spinal cord injury models, neural progenitor and Schwann cells were transplanted inside these scaffolds to the injury site, and this system was able to facilitate the bridging of the injured spinal cord, migration of host cells, growth of blood vessels, and axonal growth into the scaffolds and promote locomotor function recovery (J. Guo et al., 2007). In traumatic brain injury models, the injection of self-assembling peptide gels enabled fully filling of the gap formed by the lesion due to full integration of the gel with the tissue, and lessened the number of astrocytes and macrophages around the lesion site showing reduced glial action (J. Guo et al., 2009). Realtime imaging of the regeneration process after injection of self-assembling peptides to a chronic injury model by noninvasive manganese-enhanced magnetic resonance imaging (MEMRI) showed how the chronic optic tract lesion was able to heal and axonal regeneration after treatment (Liang et al., 2011). Transplantation of human fatal Schwann cells encapsulated PuraMatrix into a rat spinal cord injury model resulted in reduced astrogliosis and increased infiltration of endogenous S100+ cells into the injury site, which shows the potential of this hydrogel for spinal cord injury treatment (Moradi et al., 2012).

Although the porous, fibrous, ECM-mimetic morphology of the scaffolds that are made of self-assembling peptides are suitable for tissue regeneration studies, addition of specific bioactive ligands has yielded more favorable results for modulating cell behavior. Addition of a neural cell adhesion molecule (NCAM)–derived peptide sequence to this scaffold resulted in a biocompatible ECM-mimetic matrix that promotes adhesion and neurite sprouting of neurons (Z. Zou et al., 2010). When this peptide hydrogel was used for culturing spinal-cord derived neural stem cells, they were shown to have good biocompatibility with these cells and promoted their adhesion, migration, proliferation and differentiation (Jian Wang et al., 2015).

Another bioactive peptide-functionalized $RADA_{16}$ peptide, which was functionalized with laminin-derived IKVAV sequence was used for transplanting neural stem cells into rat brain injury model, and was shown to fill the cavity and bridge the gaps while enhancing the survival of neural stem cells, reducing the formation of glial astrocytes and inducing the differentiation of neural stem cells into neurons (Cheng, Chen, Chang, Huang, & Wang, 2013). When IKVAV containing self-assembling peptides were used for encapsulating embryonic stem cells, the cells formed embryoid bodies by themselves without further induction and these embryoid bodies later differentiated into neuronal lineage with extended neurite growths and high expression of neuronal markers (Q. Li, Chow, & Chau, 2014). When primary microglia were cultured in these peptide nanofiber hydrogels, they were observed to be viable but ramified, while the intracerebral injection of these peptide nanofibers showed that the IKVAV sequence had no effect on microglia activation and astrogliosis (K. M. Koss, Churchward, et al., 2016). When IKVAV containing peptides were mixed with self-assembling peptides containing fibronectin derived-RGD sequence, it was observed that they supported neural progenitor survival and differentiation to a much greater extent compared to non-functionalized RADA peptides. The RGD/IKVAV functionalized peptide nanofiber hydrogels were also employed in three different nerve injury models: sciatic nerve injury, spinal cord injury and intracerebral hemorrhage, all of which showed that these nanofibers provided a more permissive environment for neural regeneration compared to non-functionalized RADA peptides (Y. Sun et al., 2016).

Another ligand that was used for functionalization of RADA peptides was BMP7-derived short peptide RKPS. When RKPS-functionalized peptides were used for culturing nucleus pulposus-derived stem cells in an apoptosis promoting environment induced through tumor necrosis factor-alpha (TNFα) for providing an intravertebral disc degeneration model, the bioactive hydrogels were observed to significantly promote the survival and proliferation of stem cells and decrease the rate of apoptosis (X.-C. Li, Wu, et al., 2016).

RADA peptides were also utilized as controlled drug delivery systems for neural regeneration. In one study, RADA peptides were synthesized together with neurotrophic factors, which were attached to the self-assembling peptide sequence with a cleavage sequences for matrix metalloprotease 2 (MMP2). Two different neurotrophic factor-derived sequences (the brain-derived NTF secretion stimulating peptide MVG and the ciliary NTF analogue DGGL) were attached by using two different MMP2 cleavage sequences (GPQG + IASQ and GPQG + PAGQ). When PC12 cells were cultured on peptide nanofibers formed by different combinations of these functional groups in the presence or absence of varying concentrations of MMP2, signal specific cellular behavior was observed depending on the MMP2 concentration (K. Koss, Tsui, & Unsworth, 2016).

When a longer version of the laminin-derived IKVAV motif, CQAASIKVAV, was used to modify the RADA peptides, it was shown that the longer version of the laminin-derived peptide promoted the viability of human endometrial-derived stromal cells. Its effects were also compared with the effects of bone marrow homing peptide 1 (BMHP-1) modified peptide nanofibers in vivo in spinal cord injury models. The results of this study showed that both longer laminin derived peptide and BMHP-1 peptide had therapeutic effects on spinal cord injury, however, their effects were on different aspects and were complementary (Tavakol et al., 2016). The effects of these two different ligands were also compared for their effects on tyrosine hydroxylase expression in the presence or absence of Noggin protein, since tyrosine hydroxylase is an important protein for the pathogenesis of Alzheimer's or Parkinson's diseases. The results of this study showed that the type of scaffold that is used for therapeutic purposes is fundamental in defining the final fate of cells (Tavakol et al., 2017).

Another self-assembling peptide, $K_2(QL)_6K_2$, was also used for spinal cord injury repair and was shown to reduce inflammation, post-traumatic apoptosis and glial scarring and enhance axonal regeneration and locomotor functional recovery (Y. Liu et al., 2013). Injection of these self-assembling peptides to the injury site prior to cell transplantation was found to significantly improve its therapeutic effects in terms of cell survival, stem cell differentiation and functional recovery (Zweckberger, Ahuja, Liu, Wang, & Fehlings, 2016).

5.5.2 *Peptide Amphiphiles*

Peptide amphiphiles (PAs) are peptide sequences that are attached to alkyl groups and these molecules can self-assemble to form a variety of nanostructures from nanospheres to nanoribbons and nanofibers. They can be tailored to exhibit high-density epitopes on their surface, enabling robust interactions with cells. PA nanofibers are the most widely employed macromolecular structures that are produced by self-assembling peptide nanofibers. The PA molecules that assemble into nanofibers usually have an alkyl tail, which forms the core of the nanofibers in aqueous solutions, a beta sheet forming domain of several mostly hydrophobic amino acids, which enable beta-sheet formation, hence nanofiber formation, after self-assembly, and hydrophilic amino acids that are sometimes followed by bioactive domains which will be presented on the surface of the nanofibers. The self-assembly process is driven by hydrophobic interactions, hydrogen bonding and electrostatic interactions. In one of the early studies, IKVAV functionalized PA nanofibers were shown to induce selective differentiation of neuronal progenitor cells into neurons instead of astrocytes (Silva et al., 2004). These nanofibers were shown to inhibit glial scar formation, reduce cell death, and facilitate axonal elongation in rat spinal cord injury model. This treatment also resulted in significantly enhanced functional recovery (Tysseling-Mattiace et al., 2008). The IKVAV epitope displaying nanofibers were later investigated in two different spinal injury models, contusion and compression, and were shown to promote plasticity of serotonergic fibers and increased the number of serotonergic fibers and the total number of neurons caudal to the lesion site in addition to the axonal regeneration (Tysseling et al., 2010). PA nanofibers displaying IKVAV epitopes were also used for encapsulation and culture of bone marrow mesenchymal stem cells and induced their neuronal differentiation (Ruan et al., 2019).

Combination of the IKVAV epitope with other bioactive signals has resulted in improved response for neural regeneration in several studies. When IKVAV was combined with another laminin-derived epitope, YIGSR, the resulting PA nanofibers significantly enhanced the viability of the granule cells and Purkinje cells and axonal and dendrite growth of Purkinje cells in an epitope density dependent manner. Due to easy adjustment of epitope density and ease of combination of different epitopes, PA molecules can be finely tuned for specific applications mimicking the natural ECM (Sur et al., 2012). In another study, IKVAV epitope containing PAs were mixed with RGDS containing PA molecules and were used for culturing Schwann cells and were observed to support adhesion and proliferation of these cells. When these PA nanofibers were utilized in a critical sized peripheral nerve defect model together with PLGA conduits, the aligned PA gel filled conduits resulted in recovery of both motor and sensory functions which were comparable to autologous nerve transplantation. At histological level, the PA gel filled conduits caused elevated axonal and Schwann cell regeneration, which shows that these gels can be used in combination with polymeric nerve conduits for peripheral nerve regeneration (A. Li, Hokugo, et al., 2014).

The hydrogels that are made of PA nanofibers usually have a comparable stiffness to neural tissue, which is a desirable quality for neural regeneration. In order to analyze whether tuning of the mechanical properties of these gels could change neural behavior, hippocampal neurons were cultured on peptide nanofibers with varying stiffnesses adjusted by the supramolecular interactions. As a result of this study, it was observed that soft nanofibers accelerated neuronal polarity which might be a reason for faster axonal regeneration (Sur, Newcomb, Webber, & Stupp, 2013).

Aligned PA nanofibers were also used for controlled delivery of Sonic hedgehog protein to the cavernous nerve after injuries such as prostatectomy, diabetes or other neuropathies. This treatment facilitated cavernous nerve regeneration, suppressed penile apoptosis and improved erectile function (Angeloni et al., 2011). More detailed analyses revealed that sonic hedgehog treatment by using PA hydrogels facilitates retrograde transport of sonic hedgehog to PG neurons enabling the transport of sonic hedgehog receptors and survival of the neurons. Overall, PA gel-sonic hedgehog treatment protects against neuronal degeneration and protects neuronal, glial and downstream signaling events, enhancing cavernous nerve regeneration (Choe et al., 2017). PA hydrogels were also used for delivery of other biomolecules and drugs. In one example, PA gels were used for encapsulation and delivery of curcumin and other neuroprotective peptides, and were utilized for 2D and 3D culture of neurons. The results of this study showed that these hydrogels provided stability to cytoskeletal elements, enabled slow release of neuroprotective elements and promoted neurite outgrowth (Adak et al., 2017).

GAGs are an important constituent of the neural ECM. Heparan sulfate in particular has been shown to play important roles in many cellular processes, one of which is growth factor binding and localization. Heparan sulfate mimetic PA molecules were shown to bind to several growth factors and increase their expression from cells and enhance their activity (R. Mammadov et al., 2011; Rashad Mammadov, Mammadov, Guler, & Tekinay, 2012b). When laminin mimetic IKVAV-PA molecules were used in combination with heparan sulfate mimetic PAs, the resulting nanofiber system was found to bind to growth factors and significantly promote neurite outgrowth of PC12 cells, even in the presence of chondroitin sulfate proteoglycans, the major inhibitory components of the central nervous system

(B. Mammadov, Mammadov, Guler, & Tekinay, 2012a). This double-bioactive PA system was also used as a filler material in polymeric conduits for the treatment of critical size peripheral nerve defects, and was shown to enhance neuronal regeneration, Schwann cell viability and NGF release, and facilitate functional recovery (B. Mammadov et al., 2016). When heparan sulfate/laminin mimetic PA nanofibers were used in a Parkinson's disease model, they were shown to reduce striatal injury and cleaved-Caspase-3 levels and enhance functional recovery assessed by histological and behavioral analyses. In addition, these PA nanofibers reduced cell loss caused by 6-OHDA treatment in vitro in SH-SY5Y cell cultures (Sever et al., 2016).

Another model for peripheral nerve regeneration is the facial nerve regeneration model, where nerve guidance conduits are used as well. When PA gels were used to fill the conduits in a facial nerve regeneration model, measurements of nerve compound action potentials (nCAPs) and electromyographic responses revealed that PA gel filled conduits exhibited similar results to gold standard autograft transplantations. Immunohistochemical analysis and transmission electron microscopy further confirmed myelinated neural regeneration in PA filled conduits (Greene et al., 2018). In addition to PA gels being used as fillers for nerve conduits, electroactive PA gels have also been proposed for peripheral nerve regeneration. In one study, tetra(aniline) was conjugated to PA nanofibers to render them electroactive. The tetra(aniline)-conjugated PA nanofibers were shown to be biocompatible and induced neural differentiation of PC12 cells (Arioz et al., 2018).

Tenascin C is also an important ECM molecule and functions in crucial neural processes. When Tenascin-C mimetic PA molecules were used for culturing neurosphere-derived cells, they were shown to significantly promote the number and length of neurites and migration of these cells (Berns et al., 2016). Another group developed a Tenascin-C mimetic PA nanofiber system displaying the migratory sequence of this protein. When these nanofibers were injected into the ventral horn of the rostral migratory stream, doublecortin positive cells were observed to migrate inside the PA gels reaching the cortex. These PA nanofibers also did not induce a neuroinflammatory response (Motalleb et al., 2018). Another Tenascin-C mimetic PA nanofiber system was used for 3D culturing of PC12 cells, the length and number of neurites and expression levels of neuronal markers were found to be significantly increased compared to 2D cultures, emphasizing the synergistic effect of 3D conformation and bioactive epitopes for neuronal cultures (Sever, Gunay, Guler, & Tekinay, 2018). Similarly, when 3D culturing system was applied to laminin-mimetic PA nanofibers, 3D culturing was found to be superior to 2D culturing with respect to induction of neuronal differentiation even in the absence of NGF, which is a strong neuronal differentiation inducer (Gunay, Sever, Tekinay, & Guler, 2017). PA nanofibers that display an NGF-β binding epitope, which was identified through phage display, also significantly promoted neurite outgrowth of PC12 cells and primary sensory neurons by enhancing NGF/high-activity NGF receptor (TrkA) interactions and activating MAPK pathway (Okur et al., 2018).

5.6 Carbon-Based Nanomaterials for Neural Tissue Regeneration

Carbon nanotubes, graphene and fullerenes are the most widely employed carbon-based materials for regenerative medicine applications. For neural regeneration and tissue engineering applications, carbon nanotubes and graphene have been especially attractive due to their electroactivity.

5.6.1 Carbon Nanotubes

In addition to neural regeneration applications, carbon nanotubes have also been extensively utilized as neural stimulation probes for various purposes including stimulation of neurons for the treatment of neurodegenerative diseases. In one example, bidirectional soft carbon nanotube fiber microelectrodes were tested bot in vitro and in vivo, and were compared to state-of-the-art metal electrodes. In vitro studies showed that carbon nanotube fiber electrodes can be used for even single neuron recordings, while in vivo stimulation in Parkinson's disease models showed that these electrodes can stimulate 10 times larger areas than metal electrodes (Vitale, Summerson, Aazhang, Kemere, & Pasquali, 2015).

Carbon nanotubes were used for functionalizing various types of nanomaterials to render them electroactivity. In one study, poly(L/D-lactic acid) (PLDLA) nerve conduits were functionalized with carbon nanotubes tethered aligned phosphate glass microfibers to be used for peripheral nerve regeneration. In transected rat sciatic nerve model, these functionalized conduits significantly improved the number of regenerating axons, the cross-sectional area of the re-innervated muscles and the electrophysiological findings (Ahn et al., 2015).

Multiwalled carbon nanotubes were also used for functionalizing and aligning chitosan scaffolds, and these aligned electroactive scaffolds were used for culturing HT-22 hippocampal neurons. These scaffolds not only enhanced the viability of these cells but also directed the alignment of the cells along the alignment of the scaffold (P. Gupta, Sharan, Roy, & Lahiri, 2015). When chitin nanotubes were functionalized with carbon nanotubes and plasma treatment, they enabled good neuronal adhesion and facilitated synaptic function (N. Singh et al., 2016). In another chitin/carbon nanotube preparation, an aqueous solution of urea/NaOH was used for blending carbon nanotubes and chitin, and the resulting composite hydrogels were able to enhance the proliferation of Schwann cells and adhesion, proliferation and neurite outgrowth of neuronal cells (Wu et al., 2017). Carbon nanotube/graphene reinforced chitosan scaffolds were shown to have improved properties compared to chitosan scaffolds containing either one of these carbon-based materials and were biocompatible with hippocampal H22 cells while promoting neurons to spread cellular processes radially (P. Gupta et al., 2019).

Other types of hydrogels were also functionalized with carbon nanotubes for addition of electroconductivity to the scaffold systems. Single-walled carbon nanotube modified collagen I-10% Matrigel hydrogels were observed to significantly enhance neurite outgrowth from DRG neurons compared to control groups, especially when applied together with electrical stimulation (Koppes et al., 2016). A thermally sensitive hydrogel fabricated using copolymerization of *n*-isopropylacrylamide, the oligomeric amphiphilic cross-linker of polyethylene glycol diacrylate-dodecylamine-1-(2-aminoethyl)piperazine (PEGDA–DD–AEP), and single-walled carbon nanotubes was shown to be biocompatible and electrically conductive, and were able to promote significantly more neurite outgrowth from SH-SY5Y cells under electrical stimulation. This hydrogels was also able to facilitate neural regeneration and reduce glial scar formation in a spinal cord injury model (Sang et al., 2016). When carbon nanotubes were used in combination with graphene oxide for functionalizing oligo(poly(ethylene glycol) fumarate) (OPF) hydrogels, the resulting composite hydrogels were electrically conductive, biocompatible and were able to stimulate neurite development of PC12 cells when assisted with NGF addition (X. Liu et al., 2017).

In another study, single-walled carbon nanotubes were combined with PPy in order to fabricate electroactive hyaluronic acid hydrogels, which were shown to promote viability and differentiation of human fetal neural stem cells and human iPSC-derived neural progenitor cells into neuronal lineage as evidenced by electrophysiological recordings and calcium channel measurements in addition to other molecular and imaging techniques (Shin et al., 2017). In another study, 3D scaffolds that were modified with PPy and carbon nanotubes through vapor phase polymerization were observed to have high conductivity values and facilitate viability of C8-D1A astrocytes showing their biocompatibility (Alegret et al., 2018). When molecular mechanisms of how carbon nanotubes induce neural differentiation was investigated, it was found that neural stem cells interact with carbon nanotubes through integrin mediated interactions activating focal adhesion kinase and downstream signaling events regulating neuronal differentiation and synapse formation (Shao et al., 2018).

Another composite system, where carbon nanotubes were combined with gelatin fibrils and PLA, was coated with recombinant human erythropoietin-loaded chitosan nanoparticles for additional bioactivity. This composite system was able to maintain slow release of erythropoietin and support viability of Schwann cells. When Schwann cell seeded conduit was implanted into a 10 mm sciatic nerve defect model in rat, both behavioral and histopathological findings revealed comparable efficacy to autograft treatment (Salehi et al., 2018).

5.6.2 Graphene and Fullerenes

Similar to carbon nanotubes, graphene and fullerenes were also used to mostly alter the physical and electroconductive characteristics of the nanomaterials to be used for regenerative medicine. Graphene oxide has also been used by itself for cell culturing purposes as well. Rolled 15–50 μm thick graphene oxide foams have been used as 3D scaffolds for culturing neural stem cells under electrical stimulation.

The electrical stimulation of the cells cultured on these scaffolds resulted in neuronal differentiation of neural stem cells rather than glial differentiation (Akhavan, Ghaderi, & Shirazian, 2015; Akhavan, Ghaderi, Shirazian, & Rahighi, 2016). Guo et al. also used graphene for culturing neural stem cells and analyzed the membrane potentials of the cells through single-cell electrophysiological recordings. While graphene did not affect basic membrane electrical parameters, it changed resting membrane potentials of the cells and enhanced neural differentiation. In addition, both neural stem cells and their differentiated progeny exhibited elevated firing of action potentials (R. Guo, Zhang, Xiao, et al., 2016).

Graphene oxide nanosheets have also been functionalized with PEDOT and were investigated for culturing neural stem cells. These nanosheets were found to be biocompatible and promoted differentiation of neural stem cells towards neural lineage. When these nanosheets were further covalently attached on their surface with interferon-γ (IFNγ) and platelet-derived growth factor (PDGF) that selectively promote neuronal or oligodendrocyte lineage differentiation, respectively, both bioactive ligands facilitated their respective differentiation lineages (Weaver & Cui, 2015). When a high effective triboelectric nanogenerator was combined with PEDOT-reduced graphene oxide nanofibers for fabricating a self-powered electrical stimulation assisted neural differentiation system, it was found to be biocompatible with mesenchymal stem cells and enhanced their proliferation and neural differentiation. Especially, the electrical pulses generated by the TENG acted as a strong stimulant for differentiation to neuronal lineage (W. Guo, Zhang, Yu, et al., 2016). Graphene/silk fibroin films were fabricated and evaluated for their capability for neuronal differentiation of iPSCs. These electrically conductive films were found to significantly enhance neuronal differentiation at an optimal graphene content of 4% (Niu et al., 2018).

Highly aligned fullerene whiskers were fabricated through a liquid-liquid interfacial precipitation method and were shown to facilitate the orientation of neural stem cells on aligned whiskers and their differentiation towards neural lineage (Hsieh, Shrestha, Ariga, & Hsu, 2017). Self-assembled fullerene nanosheets were used for culturing bacteriorhodopsin-transfected human fibroblasts due to their photoelectric properties and the cells were modulated through optogenetic modulation. The experiments showed that light stimulation promoted reprogramming and neural differentiation of these fibroblasts even in the absence of neural differentiation medium, revealing that fullerene nanosheets may provide an ideal platform for optogenetic modulation (P. W. Luo et al., 2019).

5.7 Conclusion and Future Perspectives

Neuroregeneration has been an inherently more difficult branch of regenerative medicine due to the inhibitory environment in the central nervous system and scarcity of the neural stem cells. The nanomaterials that have been developed for neuroregeneration have focused on several characteristic requirements for nervous system: porous structure, alignment of the fibers, bioactive ligands that favor neural cells, and electroactivity.

The electrospun nanofiber based scaffold fabrication has been wildly recognized in neural tissue regeneration. Alignment of fibers has potential influence in guiding neurites especially for peripheral nerve regeneration. Furthermore, recent developments in fabrication of highly aligned and 3D electrospun nanofibrous scaffolds have paved the way for novel application areas for electrospun fibers for neuroregeneration. However, still more efforts need to explore the designing aspects and evaluate the impacts of electrospun nanofibrous scaffold in neural regeneration using animal models in order to meet clinical significance.

Natural biomaterials have inherent bioactivity; however, the complexity of multiple signals might make getting a desired response from the stem cells more difficult. In addition, natural biomaterials are harder to chemically process and carry the risk of pathogenicity and presence of other immunogenic biomacromolecules that might interfere with the success of tissue regeneration.

Self-assembling peptides and peptide amphiphiles have paved the way of a new type of nanomaterials for neural regenerative medicine and have several advantages including optimal rheological characteristics, biodegradability, alignability of the nanofibrous structure, and tailorability of the biological signals depending on the cell types. Therefore, they have been effectively used in many neural regeneration applications from peripheral nerve regeneration to spinal cord injury or Parkinson's disease treatment. However, the use of these self-assembling systems is limited by the biological signals that have been widely used before and novel signals that can be more potent and specific will make these nanofibers more effective for inducing specific tissue regeneration.

Carbon-based materials have been used either alone or in combination with other materials for fabricating electroactive materials either to be used for regeneration or for probing the neural activity for diagnosis or treatment. However, potential risks that have been associated with the use of some of these materials, especially carbon nanotubes, still limit the clinical use of these materials until further more detailed tests are performed.

Overall, since most of the nervous system related tissue degeneration problems have almost no available cures, nanomaterials provide very promising therapeutic approaches for their treatment. Thus, it is crucial that some of the materials that have been discussed in this chapter to be investigated clinically as soon as possible so that new treatment options would be available to patients with neurodegenerative diseases or severe neural damages.

References

Abbasi, N., Hashemi, S. M., Salehi, M., Jahani, H., Mowla, S. J., Soleimani, M., & Hosseinkhani, H. (2016). Influence of oriented nanofibrous PCL scaffolds on quantitative gene expression during neural differentiation of mouse embryonic stem cells. *Journal of Biomedical Materials Research - Part A, 104*(1), 155–164. https://doi.org/10.1002/jbm.a.35551

Abelseth, E., Abelseth, L., De La Vega, L., Beyer, S. T., Wadsworth, S. J., & Willerth, S. M. (2019). 3D printing of neural tissues derived from human induced pluripotent stem cells using a fibrin-based bioink. *ACS Biomaterials Science & Engineering, 5*(1), 234–243. https://doi.org/10.1021/acsbiomaterials.8b01235
Adak, A., Das, G., Barman, S., Mohapatra, S., Bhunia, D., Jana, B., & Ghosh, S. (2017). Biodegradable neuro-compatible peptide hydrogel promotes neurite outgrowth, shows significant neuroprotection, and delivers anti-alzheimer drug. *ACS Applied Materials and Interfaces, 9*(6), 5067–5076. https://doi.org/10.1021/acsami.6b12114
Addington, C. P., Dharmawaj, S., Heffernan, J. M., Sirianni, R. W., & Stabenfeldt, S. E. (2017). Hyaluronic acid-laminin hydrogels increase neural stem cell transplant retention and migratory response to SDF-1α. *Matrix Biology, 60–61*, 206–216. https://doi.org/10.1016/j.matbio.2016.09.007
Addington, C. P., Heffernan, J. M., Millar-Haskell, C. S., Tucker, E. W., Sirianni, R. W., & Stabenfeldt, S. E. (2015). Enhancing neural stem cell response to SDF-1α gradients through hyaluronic acid-laminin hydrogels. *Biomaterials, 72*, 11–19. https://doi.org/10.1016/j.biomaterials.2015.08.041
Ahn, H. S., Hwang, J. Y., Kim, M. S., Lee, J. Y., Kim, J. W., Kim, H. S., … Hyun, J. K. (2015). Carbon-nanotube-interfaced glass fiber scaffold for regeneration of transected sciatic nerve. *Acta Biomaterialia, 13*, 324–334. https://doi.org/10.1016/j.actbio.2014.11.026
Akhavan, O., Ghaderi, E., & Shirazian, S. A. (2015). Near infrared laser stimulation of human neural stem cells into neurons on graphene nanomesh semiconductors. *Colloids and Surfaces B: Biointerfaces, 126*, 313–321. https://doi.org/10.1016/j.colsurfb.2014.12.027
Akhavan, O., Ghaderi, E., Shirazian, S. A., & Rahighi, R. (2016). Rolled graphene oxide foams as three-dimensional scaffolds for growth of neural fibers using electrical stimulation of stem cells. *Carbon, 97*, 71–77. https://doi.org/10.1016/j.carbon.2015.06.079
Alegret, N., Dominguez-Alfaro, A., González-Domínguez, J. M., Arnaiz, B., Cossío, U., Bosi, S., … Prato, M. (2018). Three-dimensional conductive scaffolds as neural prostheses based on carbon nanotubes and polypyrrole. *ACS Applied Materials and Interfaces, 10*(50), 43904–43914. https://doi.org/10.1021/acsami.8b16462
Angeloni, N. L., Bond, C. W., Tang, Y., Harrington, D. A., Zhang, S., Stupp, S. I., … Podlasek, C. A. (2011). Regeneration of the cavernous nerve by Sonic hedgehog using aligned peptide amphiphile nanofibers. *Biomaterials, 32*(4), 1091–1101. https://doi.org/10.1016/j.biomaterials.2010.10.003
Arioz, I., Erol, O., Bakan, G., Dikecoglu, F. B., Topal, A. E., Urel, M., … Guler, M. O. (2018). Biocompatible electroactive tetra(aniline)-conjugated peptide nanofibers for neural differentiation. *ACS Applied Materials and Interfaces, 10*(1), 308. https://doi.org/10.1021/acsami.7b16509
Arora, P., Sindhu, A., Dilbaghi, N., Chaudhury, A., Rajakumar, G., & Rahuman, A. A. (2012). Nano-regenerative medicine towards clinical outcome of stem cell and tissue engineering in humans. *Journal of Cellular and Molecular Medicine, 16*(9), 1991–2000. https://doi.org/10.1111/j.1582-4934.2012.01534.x
Arulmoli, J., Wright, H. J., Phan, D. T. T., Sheth, U., Que, R. A., Botten, G. A., … Flanagan, L. A. (2016). Combination scaffolds of salmon fibrin, hyaluronic acid, and laminin for human neural stem cell and vascular tissue engineering. *Acta Biomaterialia, 43*, 122–138. https://doi.org/10.1016/j.actbio.2016.07.043
Atala, A. (2012). Regenerative medicine strategies. *Journal of Pediatric Surgery, 47*, 17–28. https://doi.org/10.1016/j.jpedsurg.2011.10.013
Aumailley, M. (2013). The laminin family. *Cell Adhesion & Migration, 7*, 48–55. https://doi.org/10.4161/cam.22826
Baek, S. T., Copeland, B., Yun, E. J., Kwon, S. K., Guemez-Gamboa, A., Schaffer, A. E., … Gleeson, J. G. (2015). An AKT3-FOXG1-reelin network underlies defective migration in human focal malformations of cortical development. *Nature Medicine, 21*(12), 1445–1454. https://doi.org/10.1038/nm.3982

Bagher, Z., Azami, M., Ebrahimi-Barough, S., Mirzadeh, H., Solouk, A., Soleimani, M., … Joghataei, M. T. (2016). Differentiation of Wharton's jelly-derived mesenchymal stem cells into motor neuron-like cells on three-dimensional collagen-grafted nanofibers. *Molecular Neurobiology, 53*(4), 2397–2408. https://doi.org/10.1007/s12035-015-9199-x

Barroca, N., Marote, A., Vieira, S. I., Almeida, A., Fernandes, M. H. V., Vilarinho, P. M., & Da Cruz e Silva, O. A. B. (2018). Electrically polarized PLLA nanofibers as neural tissue engineering scaffolds with improved neuritogenesis. *Colloids and Surfaces B: Biointerfaces, 167*, 93–103. https://doi.org/10.1016/j.colsurfb.2018.03.050

Becker, C. G., Schweitzer, J., Feldner, J., Schachner, M., & Becker, T. (2004). Tenascin-R as a repellent guidance molecule for newly growing and regenerating optic axons in adult zebrafish. *Molecular and Cellular Neuroscience, 26*(3), 376–389. https://doi.org/10.1016/j.mcn.2004.03.003

Berns, E. J., Álvarez, Z., Goldberger, J. E., Boekhoven, J., Kessler, J. A., Kuhn, H. G., & Stupp, S. I. (2016). A tenascin-C mimetic peptide amphiphile nanofiber gel promotes neurite outgrowth and cell migration of neurosphere-derived cells. *Acta Biomaterialia, 37*, 50–58. https://doi.org/10.1016/j.actbio.2016.04.010

Blockus, H., & Chédotal, A. (2016). Slit-Robo signaling. *Development, 143*(17), 3037–3044. https://doi.org/10.1242/dev.132829

Bosworth, L. A., & Downes, S. (2011). *Electrospinning for tissue regeneration*. Cambridge: Woodhead Publishing. https://doi.org/10.1533/9780857092915

Braghirolli, D. I., Steffens, D., & Pranke, P. (2014). Electrospinning for regenerative medicine: A review of the main topics. *Drug Discovery Today, 19*, 743–753. https://doi.org/10.1016/j.drudis.2014.03.024

Brose, K., Bland, K. S., Kuan, H. W., Arnott, D., Henzel, W., Goodman, C. S., … Kidd, T. (1999). Slit proteins bind robo receptors and have an evolutionarily conserved role in repulsive axon guidance. *Cell, 96*(6), 795–806. https://doi.org/10.1016/S0092-8674(00)80590-5

Chan, K. H., Lee, W. H., Zhuo, S., & Ming, N. (2017). Harnessing supramolecular peptide nanotechnology in biomedical applications. *International Journal of Nanomedicine, 12*, 1171–1182. https://doi.org/10.2147/IJN.S126154

Chen, S., Li, R., Li, X., & Xie, J. (2018). Electrospinning: An enabling nanotechnology platform for drug delivery and regenerative medicine. *Advanced Drug Delivery Reviews, 132*, 188–213. https://doi.org/10.1016/j.addr.2018.05.001

Chen, W., Guo, L., Tang, C., Tsai, C., Huang, H., Chin, T., … Chen-Yang, Y. (2018). The effect of laminin surface modification of electrospun silica nanofiber substrate on neuronal tissue engineering. *Nanomaterials, 8*(3), 165. https://doi.org/10.3390/nano8030165

Chen, X., Zhao, Y., Li, X., Xiao, Z., Yao, Y., Chu, Y., … Dai, J. (2018). Functional multichannel poly(propylene fumarate)-collagen scaffold with collagen-binding neurotrophic factor 3 promotes neural regeneration after transected spinal cord injury. *Advanced Healthcare Materials, 7*(14), e1800315. https://doi.org/10.1002/adhm.201800315

Cheng, T. Y., Chen, M. H., Chang, W. H., Huang, M. Y., & Wang, T. W. (2013). Neural stem cells encapsulated in a functionalized self-assembling peptide hydrogel for brain tissue engineering. *Biomaterials, 34*(8), 2005–2016. https://doi.org/10.1016/j.biomaterials.2012.11.043

Choe, S., Bond, C. W., Harrington, D. A., Stupp, S. I., McVary, K. T., & Podlasek, C. A. (2017). Peptide amphiphile nanofiber hydrogel delivery of sonic hedgehog protein to the cavernous nerve to promote regeneration and prevent erectile dysfunction. *Nanomedicine: Nanotechnology, Biology, and Medicine, 13*(1), 95–101. https://doi.org/10.1016/j.nano.2016.08.032

Cizkova, D., Slovinska, L., Grulova, I., Salzet, M., Cikos, S., Kryukov, O., & Cohen, S. (2015). The influence of sustained dual-factor presentation on the expansion and differentiation of neural progenitors in affinity-binding alginate scaffolds. *Journal of Tissue Engineering and Regenerative Medicine, 9*(8), 918–929. https://doi.org/10.1002/term.1797

Colognato, H., Galvin, J., Wang, Z., Relucio, J., Nguyen, T., Harrison, D., … Ffrench-Constant, C. (2007). Identification of dystroglycan as a second laminin receptor in oligodendrocytes, with a role in myelination. *Development, 134*(9), 1723–1736. https://doi.org/10.1242/dev.02819

D'Arcangelo, G., Miao, G. G., Chen, S. C., Scares, H. D., Morgan, J. I., & Curran, T. (1995). A protein related to extracellular matrix proteins deleted in the mouse mutant reeler. *Nature, 374*(6524), 719–723. https://doi.org/10.1038/374719a0

Dominici, C., Moreno-Bravo, J. A., Puiggros, S. R., Rappeneau, Q., Rama, N., Vieugue, P., … Chédotal, A. (2017). Floor-plate-derived netrin-1 is dispensable for commissural axon guidance. *Nature, 545*(7654), 350–354. https://doi.org/10.1038/nature22331

Du, J., Liu, J., Yao, S., Mao, H., Peng, J., Sun, X., … Wang, X. (2017). Prompt peripheral nerve regeneration induced by a hierarchically aligned fibrin nanofiber hydrogel. *Acta Biomaterialia, 55*, 296–309. https://doi.org/10.1016/j.actbio.2017.04.010

Dulabon, L., Olson, E. C., Taglienti, M. G., Eisenhuth, S., McGrath, B., Walsh, C. A., … Anton, E. S. (2000). Reelin binds α3β1 integrin and inhibits neuronal migration. *Neuron, 27*(1), 33–44. https://doi.org/10.1016/S0896-6273(00)00007-6

Durakoglugil, M. S., Chen, Y., White, C. L., Kavalali, E. T., & Herz, J. (2009). Reelin signaling antagonizes β-amyloid at the synapse. *Proceedings of the National Academy of Sciences, 106*(37), 15938–15943. https://doi.org/10.1073/pnas.0908176106

Ebrahimi-Barough, S., Hoveizi, E., Yazdankhah, M., Ai, J., Khakbiz, M., Faghihi, F., … Bayat, N. (2017). Inhibitor of PI3K/Akt signaling pathway small molecule promotes motor neuron differentiation of human endometrial stem cells cultured on electrospun biocomposite polycaprolactone/collagen scaffolds. *Molecular Neurobiology, 54*(4), 2547–2554. https://doi.org/10.1007/s12035-016-9828-z

Ebrahimi-Barough, S., Norouzi Javidan, A., Saberi, H., Joghataei, M. T., Rahbarghazi, R., Mirzaei, E., … Ai, J. (2015). Evaluation of motor neuron-like cell differentiation of hEnSCs on biodegradable PLGA nanofiber scaffolds. *Molecular Neurobiology, 52*(3), 1704–1713. https://doi.org/10.1007/s12035-014-8931-2

Edgar, J. M., Robinson, M., & Willerth, S. M. (2017). Fibrin hydrogels induce mixed dorsal/ventral spinal neuron identities during differentiation of human induced pluripotent stem cells. *Acta Biomaterialia, 51*, 237–245. https://doi.org/10.1016/j.actbio.2017.01.040

Engel, E., Michiardi, A., Navarro, M., Lacroix, D., & Planell, J. A. (2008). Nanotechnology in regenerative medicine: The materials side. *Trends in Biotechnology, 26*, 39–47. https://doi.org/10.1016/j.tibtech.2007.10.005

Entekhabi, E., Haghbin Nazarpak, M., Moztarzadeh, F., & Sadeghi, A. (2016). Design and manufacture of neural tissue engineering scaffolds using hyaluronic acid and polycaprolactone nanofibers with controlled porosity. *Materials Science and Engineering C, 69*, 380–387. https://doi.org/10.1016/j.msec.2016.06.078

Fawcett, J. W., Kwok, J. C. F., Afshari, F., & García-Alías, G. (2008). Proteoglycans in the central nervous system: Plasticity, regeneration and their stimulation with chondroitinase ABC. *Restorative Neurology and Neuroscience, 26*(2-3), 131–145.

Flanagan, L. A., Rebaza, L. M., Derzic, S., Schwartz, P. H., & Monuki, E. S. (2006). Regulation of human neural precursor cells by laminin and integrins. *Journal of Neuroscience Research, 83*(5), 845–856. https://doi.org/10.1002/jnr.20778

Garcion, E., Faissner, A., & Ffrench-Constant, C. (2001). Knockout mice reveal a contribution of the extracellular matrix molecule tenascin-C to neural precursor proliferation and migration. *Development (Cambridge, England), 128*(13), 2485–2496. Retrieved from http://www.ncbi.nlm.nih.gov/pubmed/11493565

Garcion, E., Halilagic, A., Faissner, A., & Ffrench-Constant, C. (2004). Generation of an environmental niche for neural stem cell development bythe extracellular matrix molecule tenascin C. *Development, 131*(14), 3423–3432. https://doi.org/10.1242/dev.01202

Ghasemi Hamidabadi, H., Rezvani, Z., Nazm Bojnordi, M., Shirinzadeh, H., Seifalian, A. M., Joghataei, M. T., … Gholipourmalekabadi, M. (2017). Chitosan-intercalated montmorillonite/poly(vinyl alcohol) nanofibers as a platform to guide neuronlike differentiation of human dental pulp stem cells. *ACS Applied Materials and Interfaces, 9*(13), 11392–11404. https://doi.org/10.1021/acsami.6b14283

Gobaa, S., Hoehnel, S., Roccio, M., Negro, A., Kobel, S., & Lutolf, M. P. (2011). Artificial niche microarrays for probing single stem cell fate in high throughput. *Nature Methods, 8*(11), 949–955. https://doi.org/10.1038/nmeth.1732

Golafshan, N., Kharaziha, M., & Fathi, M. (2017). Tough and conductive hybrid graphene-PVA: Alginate fibrous scaffolds for engineering neural construct. *Carbon, 111*, 752–763. https://doi.org/10.1016/j.carbon.2016.10.042

Golafshan, N., Kharaziha, M., Fathi, M., Larson, B. L., Giatsidis, G., & Masoumi, N. (2018). Anisotropic architecture and electrical stimulation enhance neuron cell behaviour on a tough graphene embedded PVA: Alginate fibrous scaffold. *RSC Advances, 8*(12), 6381–6389. https://doi.org/10.1039/c7ra13136d

Greene, J. J., McClendon, M. T., Stephanopoulos, N., Álvarez, Z., Stupp, S. I., & Richter, C. P. (2018). Electrophysiological assessment of a peptide amphiphile nanofiber nerve graft for facial nerve repair. *Journal of Tissue Engineering and Regenerative Medicine, 12*(6), 1389–1401. https://doi.org/10.1002/term.2669

Grulova, I., Slovinska, L., Blaško, J., Devaux, S., Wisztorski, M., Salzet, M., … Cizkova, D. (2015). Delivery of alginate scaffold releasing two trophic factors for spinal cord injury repair. *Scientific Reports, 5*, 13702. https://doi.org/10.1038/srep13702

Gu, W., Wu, C., Chen, J., & Xiao, Y. (2013). Nanotechnology in the targeted drug delivery for bone diseases and bone regeneration. *International Journal of Nanomedicine, 8*, 2305–2317. https://doi.org/10.2147/IJN.S44393

Guldager Kring Rasmussen, D., & Karsdal, M. A. (2016). Laminins. In *Biochemistry of collagens, laminins and elastin: structure, function and biomarkers* (pp. 163–196). New York, NY: Academic Press. https://doi.org/10.1016/B978-0-12-809847-9.00029-5

Gunay, G., Sever, M., Tekinay, A. B., & Guler, M. O. (2017). Three-dimensional laminin mimetic peptide nanofiber gels for in vitro neural differentiation. *Biotechnology Journal, 12*(12), 1700080. https://doi.org/10.1002/biot.201700080

Günther, M. I., Weidner, N., Müller, R., & Blesch, A. (2015). Cell-seeded alginate hydrogel scaffolds promote directed linear axonal regeneration in the injured rat spinal cord. *Acta Biomaterialia, 27*, 140–150. https://doi.org/10.1016/j.actbio.2015.09.001

Guo, J., Leung, K. K. G., Su, H., Yuan, Q., Wang, L., Chu, T. H., … Wu, W. (2009). Self-assembling peptide nanofiber scaffold promotes the reconstruction of acutely injured brain. *Nanomedicine: Nanotechnology, Biology, and Medicine, 5*(3), 345–351. https://doi.org/10.1016/j.nano.2008.12.001

Guo, J., Su, H., Zeng, Y., Liang, Y. X., Wong, W. M., Ellis-Behnke, R. G., … Wu, W. (2007). Reknitting the injured spinal cord by self-assembling peptide nanofiber scaffold. *Nanomedicine: Nanotechnology, Biology, and Medicine, 3*(4), 311–321. https://doi.org/10.1016/j.nano.2007.09.003

Guo, R., Zhang, S., Xiao, M., Qian, F., He, Z., Li, D., … Tang, M. (2016). Accelerating bioelectric functional development of neural stem cells by graphene coupling: Implications for neural interfacing with conductive materials. *Biomaterials, 106*, 193–204. https://doi.org/10.1016/j.biomaterials.2016.08.019

Guo, W., Wang, S., Yu, X., Qiu, J., Li, J., Tang, W., … Wang, Z. (2016). Construction of a 3D rGO-collagen hybrid scaffold for enhancement of the neural differentiation of mesenchymal stem cells. *Nanoscale, 8*(4), 1897–1904. https://doi.org/10.1039/c5nr06602f

Guo, W., Zhang, X., Yu, X., Wang, S., Qiu, J., Tang, W., … Wang, Z. L. (2016). Self-powered electrical stimulation for enhancing neural differentiation of mesenchymal stem cells on graphene-poly(3,4-ethylenedioxythiophene) hybrid microfibers. *ACS Nano, 10*(5), 5086–5095. https://doi.org/10.1021/acsnano.6b00200

Gupta, A., Avci, P., Sadasivam, M., Chandran, R., Parizotto, N., Vecchio, D., … Hamblin, M. R. (2013). Shining light on nanotechnology to help repair and regeneration. *Biotechnology Advances, 31*, 607–631. https://doi.org/10.1016/j.biotechadv.2012.08.003

Gupta, P., Agrawal, A., Murali, K., Varshney, R., Beniwal, S., Manhas, S., … Lahiri, D. (2019). Differential neural cell adhesion and neurite outgrowth on carbon nanotube and graphene reinforced polymeric scaffolds. *Materials Science and Engineering C, 97*, 539–551. https://doi.org/10.1016/j.msec.2018.12.065

Gupta, P., Sharan, S., Roy, P., & Lahiri, D. (2015). Aligned carbon nanotube reinforced polymeric scaffolds with electrical cues for neural tissue regeneration. *Carbon, 95*, 715–724. https://doi.org/10.1016/j.carbon.2015.08.107

Haddad, T., Noel, S., Liberelle, B., El Ayoubi, R., Ajji, A., & De Crescenzo, G. (2016). Fabrication and surface modification of poly lactic acid (PLA) scaffolds with epidermal growth factor for neural tissue engineering. *Biomatter, 6*(1), e1231276. https://doi.org/10.1080/21592535.2016.1231276

Hall, P. E., Lathia, J. D., Caldwell, M. A., & Ffrench-Constant, C. (2008). Laminin enhances the growth of human neural stem cells in defined culture media. *BMC Neuroscience, 9*, 71. https://doi.org/10.1186/1471-2202-9-71

Hara, Y., Balci-Hayta, B., Yoshida-Moriguchi, T., Kanagawa, M., Beltrán-Valero de Bernabé, D., Gündeşli, H., … Campbell, K. P. (2011). A dystroglycan mutation associated with limb-girdle muscular dystrophy. *New England Journal of Medicine, 364*(10), 939–946. https://doi.org/10.1056/nejmoa1006939

Hasanzadeh, E., Ebrahimi-Barough, S., Mirzaei, E., Azami, M., Tavangar, S. M., Mahmoodi, N., … Ai, J. (2019). Preparation of fibrin gel scaffolds containing MWCNT/PU nanofibers for neural tissue engineering. *Journal of Biomedical Materials Research - Part A, 107*(4), 802–814. https://doi.org/10.1002/jbm.a.36596

Hashimoto-Torii, K., Torii, M., Sarkisian, M. R., Bartley, C. M., Shen, J., Radtke, F., … Rakic, P. (2008). Interaction between reelin and notch signaling regulates neuronal migration in the cerebral cortex. *Neuron, 60*(2), 273–284. https://doi.org/10.1016/j.neuron.2008.09.026

Heidari, M., Bahrami, S. H., Ranjbar-Mohammadi, M., & Milan, P. B. (2019). Smart electrospun nanofibers containing PCL/gelatin/graphene oxide for application in nerve tissue engineering. *Materials Science and Engineering C, 103*, 109768. https://doi.org/10.1016/j.msec.2019.109768

Hiesberger, T., Trommsdorff, M., Howell, B. W., Goffinet, A., Mumby, M. C., Cooper, J. A., & Herz, J. (1999). Direct binding of Reelin to VLDL receptor and ApoE receptor 2 induces tyrosine phosphorylation of Disabled-1 and modulates tau phosphorylation. *Neuron, 24*(2), 481–489. https://doi.org/10.1016/S0896-6273(00)80861-2

Hodde, D., Gerardo-Nava, J., Wöhlk, V., Weinandy, S., Jockenhövel, S., Kriebel, A., … Brook, G. A. (2016). Characterisation of cell-substrate interactions between Schwann cells and three-dimensional fibrin hydrogels containing orientated nanofibre topographical cues. *European Journal of Neuroscience, 43*(3), 376–387. https://doi.org/10.1111/ejn.13026

Holmes, T. C., de Lacalle, S., Su, X., Liu, G., Rich, A., & Zhang, S. (2002). Extensive neurite outgrowth and active synapse formation on self-assembling peptide scaffolds. *Proceedings of the National Academy of Sciences, 97*(12), 6728–6733. https://doi.org/10.1073/pnas.97.12.6728

Hsieh, F. Y., Shrestha, L. K., Ariga, K., & Hsu, S. H. (2017). Neural differentiation on aligned fullerene C60 nanowhiskers. *Chemical Communications, 53*(80), 11024–11027. https://doi.org/10.1039/c7cc06395d

Hu, J., Kai, D., Ye, H., Tian, L., Ding, X., Ramakrishna, S., & Loh, X. J. (2017). Electrospinning of poly(glycerol sebacate)-based nanofibers for nerve tissue engineering. *Materials Science and Engineering C, 70*, 1089–1094. https://doi.org/10.1016/j.msec.2016.03.035

Hu, J., Tian, L., Prabhakaran, M. P., Ding, X., & Ramakrishna, S. (2016). Fabrication of nerve growth factor encapsulated aligned poly(ε-caprolactone) nanofibers and their assessment as a potential neural tissue engineering scaffold. *Polymers, 8*(2), 54. https://doi.org/10.3390/polym8020054

Huang, W., Fan, Q., Wang, W., Zhou, M., Laties, A. M., & Zhang, X. (2013). Collagen: A potential factor involved in the pathogenesis of glaucoma. *Medical Science Monitor Basic Research, 19*, 237–240. https://doi.org/10.12659/msmbr.889061

Huang, W., Zhang, L., Niu, R., & Liao, H. (2009). Tenascin-R distinct domains modulate migration of neural stem/progenitor cells in vitro. *In Vitro Cellular and Developmental Biology - Animal, 45*(1–2), 10–14. https://doi.org/10.1007/s11626-008-9145-6

Ikegami, T., Nakamura, M., Yamane, J., Katoh, H., Okada, S., Iwanami, A., … Okano, H. (2005). Chondroitinase ABC combined with neural stem/progenitor cell transplantation enhances graft cell migration and outgrowth of growth-associated protein-43-positive fibers after rat spinal cord injury. *European Journal of Neuroscience, 22*(12), 3036–3046. https://doi.org/10.1111/j.1460-9568.2005.04492.x

KarbalaeiMahdi, A., Shahrousvand, M., Javadi, H. R., Ghollasi, M., Norouz, F., Kamali, M., & Salimi, A. (2017). Neural differentiation of human induced pluripotent stem cells on polycaprolactone/gelatin bi-electrospun nanofibers. *Materials Science and Engineering C, 78*, 1195–1202. https://doi.org/10.1016/j.msec.2017.04.083

Karumbaiah, L., Enam, S. F., Brown, A. C., Saxena, T., Betancur, M. I., Barker, T. H., & Bellamkonda, R. V. (2015). Chondroitin sulfate glycosaminoglycan hydrogels create endogenous niches for neural stem cells. *Bioconjugate Chemistry, 26*(12), 2336–2349. https://doi.org/10.1021/acs.bioconjchem.5b00397

Khademhosseini, A., & Langer, R. (2016). A decade of progress in tissue engineering. *Nature Protocols, 11*(10), 1775–1781. https://doi.org/10.1038/nprot.2016.123

Kidd, T., Bland, K. S., & Goodman, C. S. (1999). Slit is the midline repellent for the Robo receptor in Drosophila. *Cell, 96*(6), 785–794. https://doi.org/10.1016/S0092-8674(00)80589-9

Koopmans, G., Hasse, B., & Sinis, N. (2009). Chapter 19 The role of collagen in peripheral nerve repair. *International Review of Neurobiology, 87*, 363–379. https://doi.org/10.1016/S0074-7742(09)87019-0

Koppes, A. N., Keating, K. W., McGregor, A. L., Koppes, R. A., Kearns, K. R., Ziemba, A. M., … Thompson, D. M. (2016). Robust neurite extension following exogenous electrical stimulation within single walled carbon nanotube-composite hydrogels. *Acta Biomaterialia, 39*, 34–43. https://doi.org/10.1016/j.actbio.2016.05.014

Koss, K., Tsui, C., & Unsworth, L. D. (2016). Induced neural differentiation of MMP-2 cleaved (RADA)4 drug delivery systems. *Journal of Controlled Release, 243*, 204–213. https://doi.org/10.1016/j.jconrel.2016.09.037

Koss, K. M., Churchward, M. A., Nguyen, A. T., Yager, J. Y., Todd, K. G., & Unsworth, L. D. (2016). Brain biocompatibility and microglia response towards engineered self-assembling (RADA)4 nanoscaffolds. *Acta Biomaterialia, 35*, 127–137. https://doi.org/10.1016/j.actbio.2016.02.001

Kwok, J. C. F., Dick, G., Wang, D., & Fawcett, J. W. (2011). Extracellular matrix and perineuronal nets in CNS repair. *Developmental Neurobiology, 71*(11), 1073–1089. https://doi.org/10.1002/dneu.20974

Lau, Y. T., Kwok, L. F., Tam, K. W., Chan, Y. S., Shum, D. K. Y., & Shea, G. K. H. (2018). Genipin-treated chitosan nanofibers as a novel scaffold for nerve guidance channel design. *Colloids and Surfaces B: Biointerfaces, 162*, 126–134. https://doi.org/10.1016/j.colsurfb.2017.11.061

Lee, S.-J., Nowicki, M., Harris, B., & Zhang, L. G. (2016). Fabrication of a highly aligned neural scaffold via a table top stereolithography 3D printing and electrospinning. *Tissue Engineering Part A, 23*(11–12), 491–502. https://doi.org/10.1089/ten.tea.2016.0353

Li, A., Hokugo, A., Yalom, A., Berns, E. J., Stephanopoulos, N., McClendon, M. T., … Jarrahy, R. (2014). A bioengineered peripheral nerve construct using aligned peptide amphiphile nanofibers. *Biomaterials, 35*(31), 8780–8790. https://doi.org/10.1016/j.biomaterials.2014.06.049

Li, Q., Chow, K. L., & Chau, Y. (2014). Three-dimensional self-assembling peptide matrix enhances the formation of embryoid bodies and their neuronal differentiation. *Journal of Biomedical Materials Research - Part A, 102*(6), 1991–2000. https://doi.org/10.1002/jbm.a.34876

Li, X.-C., Wu, Y.-H., Bai, X.-D., Ji, W., Guo, Z.-M., Wang, C.-F., … Ruan, D. (2016). BMP7-based functionalized self-assembling peptides protect nucleus pulposus-derived stem cells from apoptosis in vitro. *Tissue Engineering Part A, 22*(19–20), 1218–1228. https://doi.org/10.1089/ten.tea.2016.0230

Li, X., Fan, C., Xiao, Z., Zhao, Y., Zhang, H., Sun, J., … Dai, J. (2018). A collagen microchannel scaffold carrying paclitaxel-liposomes induces neuronal differentiation of neural stem cells through Wnt/β-catenin signaling for spinal cord injury repair. *Biomaterials, 183*, 114–127. https://doi.org/10.1016/j.biomaterials.2018.08.037

Li, X., Li, M., Sun, J., Zhuang, Y., Shi, J., Guan, D., … Dai, J. (2016). Radially aligned electrospun fibers with continuous gradient of SDF1α for the guidance of neural stem cells. *Small, 12*(36), 5009–5018. https://doi.org/10.1002/smll.201601285

Li, X., Liu, S., Zhao, Y., Li, J., Ding, W., Han, S., … Dai, J. (2016). Training neural stem cells on functional collagen scaffolds for severe spinal cord injury repair. *Advanced Functional Materials, 26*(32), 5835–5847. https://doi.org/10.1002/adfm.201601521

Li, X., Zhao, Y., Cheng, S., Han, S., Shu, M., Chen, B., … Dai, J. (2017). Cetuximab modified collagen scaffold directs neurogenesis of injury-activated endogenous neural stem cells for acute spinal cord injury repair. *Biomaterials, 137*, 73–86. https://doi.org/10.1016/j.biomaterials.2017.05.027

Liang, Y. X., Cheung, S. W. H., Chan, K. C. W., Wu, E. X., Tay, D. K. C., & Ellis-Behnke, R. G. (2011). CNS regeneration after chronic injury using a self-assembled nanomaterial and MEMRI for real-time in vivo monitoring. *Nanomedicine: Nanotechnology, Biology, and Medicine, 7*(3), 351–359. https://doi.org/10.1016/j.nano.2010.12.001

Lin, L., Rao, Y., & Isacson, O. (2005). Netrin-1 and slit-2 regulate and direct neurite growth of ventral midbrain dopaminergic neurons. *Molecular and Cellular Neuroscience, 28*(3), 547–555. https://doi.org/10.1016/j.mcn.2004.11.009

Little, L., Healy, K. E., & Schaffer, D. (2008). Engineering biomaterials for synthetic neural stem cell microenvironments. *Chemical Reviews, 108*(5), 1787–1796. https://doi.org/10.1021/cr078228t

Liu, W., Thomopoulos, S., & Xia, Y. (2012). Electrospun nanofibers for regenerative medicine. *Advanced Healthcare Materials, 1*(1), 10–25. https://doi.org/10.1002/adhm.201100021

Liu, X., Miller, A. L., Park, S., Waletzki, B. E., Zhou, Z., Terzic, A., & Lu, L. (2017). Functionalized carbon nanotube and graphene oxide embedded electrically conductive hydrogel synergistically stimulates nerve cell differentiation. *ACS Applied Materials and Interfaces, 9*(17), 14677–14690. https://doi.org/10.1021/acsami.7b02072

Liu, Y., Ye, H., Satkunendrarajah, K., Yao, G. S., Bayon, Y., & Fehlings, M. G. (2013). A self-assembling peptide reduces glial scarring, attenuates post-traumatic inflammation and promotes neurological recovery following spinal cord injury. *Acta Biomaterialia, 9*(9), 8075–8088. https://doi.org/10.1016/j.actbio.2013.06.001

Longman, C., Brockington, M., Torelli, S., Jimenez-Mallebrera, C., Kennedy, C., Khalil, N., … Muntoni, F. (2003). Mutations in the human LARGE gene cause MDC1D, a novel form of congenital muscular dystrophy with severe mental retardation and abnormal glycosylation of α-dystroglycan. *Human Molecular Genetics, 12*(21), 2853–2861. https://doi.org/10.1093/hmg/ddg307

Luo, B., Tian, L., Chen, N., Ramakrishna, S., Thakor, N., & Yang, I. H. (2018). Electrospun nanofibers facilitate better alignment, differentiation, and long-term culture in an: In vitro model of the neuromuscular junction (NMJ). *Biomaterials Science, 6*(12), 3262–3272. https://doi.org/10.1039/c8bm00720a

Luo, P. W., Han, H. W., Yang, C. S., Shrestha, L. K., Ariga, K., & Hsu, S. (2019). Optogenetic modulation and reprogramming of bacteriorhodopsin-transfected human fibroblasts on self-assembled fullerene c60 nanosheets. *Advanced Biosystems, 3*(2), 1800254. https://doi.org/10.1002/adbi.201800254

Ma, B., Xie, J., Jiang, J., Shuler, F. D., & Bartlett, D. E. (2013). Rational design of nanofiber scaffolds for orthopedic tissue repair and regeneration. *Nanomedicine, 8*, 1459–1481. https://doi.org/10.2217/nnm.13.132

Ma, W., Tavakoli, T., Derby, E., Serebryakova, Y., Rao, M. S., & Mattson, M. P. (2008). Cell-extracellular matrix interactions regulate neural differentiation of human embryonic stem cells. *BMC Developmental Biology, 8*, 90. https://doi.org/10.1186/1471-213X-8-90

Mammadov, B., Mammadov, R., Guler, M. O., & Tekinay, A. B. (2012a). Cooperative effect of heparan sulfate and laminin mimetic peptide nanofibers on the promotion of neurite outgrowth. *Acta Biomaterialia, 8*(6), 2077–2086. https://doi.org/10.1016/j.actbio.2012.02.006

Mammadov, B., Sever, M., Gecer, M., Zor, F., Ozturk, S., Akgun, H., … Tekinay, A. B. (2016). Sciatic nerve regeneration induced by glycosaminoglycan and laminin mimetic peptide nanofiber gels. *RSC Advances, 6*(112), 110535–110547. https://doi.org/10.1039/C6RA24450E

Mammadov, R., Mammadov, B., Toksoz, S., Aydin, B., Yagci, R., Tekinay, A. B., & Guler, M. O. (2011). Heparin mimetic peptide nanofibers promote angiogenesis. *Biomacromolecules, 12*(10). https://doi.org/10.1021/bm200957s

Mammadov, R., Mammadov, B., Guler, M. O., & Tekinay, A. B. (2012b). Growth factor binding on heparin mimetic peptide nanofibers. *Biomacromolecules, 13*(10), 3311–3319. https://doi.org/10.1021/bm3010897

Mao, A. S., & Mooney, D. J. (2015). Regenerative medicine: Current therapies and future directions. *Proceedings of the National Academy of Sciences, 112*(47), 14452–14459. https://doi.org/10.1073/pnas.1508520112

Maroto, M., Fernández-Morales, J. C., Padín, J. F., González, J. C., Hernández-Guijo, J. M., Montell, E., … García, A. G. (2013). Chondroitin sulfate, a major component of the perineuronal net, elicits inward currents, cell depolarization, and calcium transients by acting on AMPA and kainate receptors of hippocampal neurons. *Journal of Neurochemistry, 125*(2), 205–213. https://doi.org/10.1111/jnc.12159

Matejas, V., Hinkes, B., Alkandari, F., Al-Gazali, L., Annexstad, E., Aytac, M. B., … Zenker, M. (2010). Mutations in the human laminin β2 (LAMB2) gene and the associated phenotypic spectrum. *Human Mutation, 31*(9), 992–1002. https://doi.org/10.1002/humu.21304

Midwood, K. S., & Orend, G. (2009). The role of tenascin-C in tissue injury and tumorigenesis. *Journal of Cell Communication and Signaling, 3*, 287–310. https://doi.org/10.1007/s12079-009-0075-1

Montgomery, A., Wong, A., Gabers, N., & Willerth, S. M. (2015). Engineering personalized neural tissue by combining induced pluripotent stem cells with fibrin scaffolds. *Biomaterials Science, 3*(2), 401–413. https://doi.org/10.1039/c4bm00299g

Moradi, F., Bahktiari, M., Joghataei, M. T., Nobakht, M., Soleimani, M., Hasanzadeh, G., … Maleki, F. (2012). BD PuraMatrix peptide hydrogel as a culture system for human fetal Schwann cells in spinal cord regeneration. *Journal of Neuroscience Research, 90*(12), 2335–2348. https://doi.org/10.1002/jnr.23120

Motalleb, R., Berns, E. J., Patel, P., Gold, J., Stupp, S. I., & Kuhn, H. G. (2018). In vivo migration of endogenous brain progenitor cells guided by an injectable peptide amphiphile biomaterial. *Journal of Tissue Engineering and Regenerative Medicine, 12*(4), e2123–e2133. https://doi.org/10.1002/term.2644

Niu, Y., Chen, X., Yao, D., Peng, G., Liu, H., & Fan, Y. (2018). Enhancing neural differentiation of induced pluripotent stem cells by conductive graphene/silk fibroin films. *Journal of Biomedical Materials Research - Part A, 106*(11), 2973–2983. https://doi.org/10.1002/jbm.a.36486

Nobile, S., & Nobile, L. (2017). Nanotechnology for biomedical applications: Recent advances in neurosciences and bone tissue engineering. *Polymer Engineering and Science, 57*, 644–650. https://doi.org/10.1002/pen.24595

Nune, M., Krishnan, U. M., & Sethuraman, S. (2015). Decoration of PLGA electrospun nanofibers with designer self-assembling peptides: A "nano-on-Nano" concept. *RSC Advances, 5*(108), 88748–88757. https://doi.org/10.1039/c5ra13576a

Nune, M., Subramanian, A., Krishnan, U. M., Kaimal, S. S., & Sethuraman, S. (2017). Self-assembling peptide nanostructures on aligned poly(lactide-co-glycolide) nanofibers for the functional regeneration of sciatic nerve. *Nanomedicine, 12*(3), 219–235. https://doi.org/10.2217/nnm-2016-0323

Okur, Z., Senturk, O. I., Yilmaz, C., Gulseren, G., Mammadov, B., Guler, M. O., & Tekinay, A. B. (2018). Promotion of neurite outgrowth by rationally designed NGF-β binding peptide nanofibers. *Biomaterials Science, 6*(7), 1777–1790. https://doi.org/10.1039/c8bm00311d

Pang, M., Shu, T., Chen, R. Q., Liu, C., He, L., Yang, Y., … Rong, L. M. (2016). Neural precursor cells generated from induced pluripotent stem cells with gelatin sponge-electrospun PLGA/PEG nanofibers for spinal cord injury repair. *International Journal of Clinical and Experimental Medicine, 9*(9), 17985–17994.

Pelaz, B., Alexiou, C., Alvarez-Puebla, R. A., Alves, F., Andrews, A. M., Ashraf, S., … Parak, W. J. (2017). Diverse applications of nanomedicine. *ACS Nano, 11*(3), 2313–2381. https://doi.org/10.1021/acsnano.6b06040

Pelipenko, J., Kristl, J., Janković, B., Baumgartner, S., & Kocbek, P. (2013). The impact of relative humidity during electrospinning on the morphology and mechanical properties of nanofibers. *International Journal of Pharmaceutics, 456*(1), 125–134. https://doi.org/10.1016/j.ijpharm.2013.07.078

Pham, Q. P., Sharma, U., & Mikos, A. G. (2006). Electrospinning of polymeric nanofibers for tissue engineering applications: A review. *Tissue Engineering, 12*(5), 1197. https://doi.org/10.1089/ten.2006.12.ft-65

Pujadas, L., Rossi, D., Andrés, R., Teixeira, C. M., Serra-Vidal, B., Parcerisas, A., … Soriano, E. (2014). Reelin delays amyloid-beta fibril formation and rescues cognitive deficits in a model of Alzheimer's disease. *Nature Communications, 5*, 3443. https://doi.org/10.1038/ncomms4443

Qi, Z., Guo, W., Zheng, S., Fu, C., Ma, Y., Pan, S., … Yang, X. (2019). Enhancement of neural stem cell survival, proliferation and differentiation by IGF-1 delivery in graphene oxide-incorporated PLGA electrospun nanofibrous mats. *RSC Advances, 9*(15), 8315–8325. https://doi.org/10.1039/c8ra10103e

Ramos, A. P., Cruz, M. A. E., Tovani, C. B., & Ciancaglini, P. (2017). Biomedical applications of nanotechnology. *Biophysical Reviews, 9*, 79–89. https://doi.org/10.1007/s12551-016-0246-2

Rasi, K., Hurskainen, M., Kallio, M., Staven, S., Sormunen, R., Heape, A. M., … Pihlajaniemi, T. (2010). Lack of collagen XV impairs peripheral nerve maturation and, when combined with laminin-411 deficiency, leads to basement membrane abnormalities and sensorimotor dysfunction. *Journal of Neuroscience, 30*(43), 14490–14501. https://doi.org/10.1523/jneurosci.2644-10.2010

Robinson, M., Douglas, S., & Willerth, S. M. (2017). Mechanically stable fibrin scaffolds promote viability and induce neurite outgrowth in neural aggregates derived from human induced pluripotent stem cells. *Scientific Reports, 7*(1), 6250. https://doi.org/10.1038/s41598-017-06570-9

Rosenthal, N., & Badylak, S. (2016). Regenerative medicine: Today's discoveries informing the future of medical practice. *NPJ Regenerative Medicine, 1*(1), 16007. https://doi.org/10.1038/npjregenmed.2016.7

Ruan, H., Xiao, R., Jiang, X., Zhao, B., Wu, K., Shao, Z., … Song, Y. (2019). Biofunctionalized self-assembly of peptide amphiphile induces the differentiation of bone marrow mesenchymal stem cells into neural cells. *Molecular and Cellular Biochemistry, 450*(1–2), 199–207. https://doi.org/10.1007/s11010-018-3386-9

Ruzicka, J., Romanyuk, N., Jirakova, K., Hejcl, A., Janouskova, O., Machova, L. U., … Jendelova, P. (2019). The effect of iPS-derived neural progenitors seeded on laminin-coated pHEMA-MOETACl hydrogel with dual porosity in a rat model of chronic spinal cord injury. *Cell Transplantation, 28*, 400. https://doi.org/10.1177/0963689718823705

Salehi, M., Naseri-Nosar, M., Ebrahimi-Barough, S., Nourani, M., Khojasteh, A., Hamidieh, A. A., … Ai, J. (2018). Sciatic nerve regeneration by transplantation of Schwann cells via erythropoietin controlled-releasing polylactic acid/multiwalled carbon nanotubes/gelatin nanofibrils neural guidance conduit. *Journal of Biomedical Materials Research - Part B Applied Biomaterials, 106*(4), 1463–1476. https://doi.org/10.1002/jbm.b.33952

Sang, L., Hua, W., Wang, G., Wang, L., Qiu, X., Liu, Y., … Zhong, W. (2016). Thermally sensitive conductive hydrogel using amphiphilic crosslinker self-assembled carbon nanotube to enhance neurite outgrowth and promote spinal cord regeneration. *RSC Advances, 6*(31), 26341–26351. https://doi.org/10.1039/C5RA20780K

Sangsanoh, P., Ekapakul, N., Israsena, N., Suwantong, O., & Supaphol, P. (2018). Enhancement of biocompatibility on aligned electrospun poly(3-hydroxybutyrate) scaffold immobilized with laminin towards murine neuroblastoma Neuro2a cell line and rat brain-derived neural stem cells (mNSCs). *Polymers for Advanced Technologies, 29*(7), 2050–2063. https://doi.org/10.1002/pat.4313

Sanna, V., Pala, N., & Sechi, M. (2014). Targeted therapy using nanotechnology: Focus on cancer. *International Journal of Nanomedicine, 9*, 467–483. https://doi.org/10.2147/IJN.S36654

Schwarting, G. A., Raitcheva, D., Bless, E. P., Ackerman, S. L., & Tobet, S. (2004). Netrin 1-mediated chemoattraction regulates the migratory pathway of LHRH neurons. *European Journal of Neuroscience, 19*(1), 11–20. https://doi.org/10.1111/j.1460-9568.2004.03094.x

Sekine, K., Kubo, K. I., & Nakajima, K. (2014). How does Reelin control neuronal migration and layer formation in the developing mammalian neocortex? *Neuroscience Research, 86*, 50–58. https://doi.org/10.1016/j.neures.2014.06.004

Senthamizhan, A., Balusamy, B., & Uyar, T. (2016). Glucose sensors based on electrospun nanofibers: A review fiber-based platforms for bioanalytics. *Analytical and Bioanalytical Chemistry, 408*(5), 1285–1306. https://doi.org/10.1007/s00216-015-9152-x

Sever, M., Gunay, G., Guler, M. O., & Tekinay, A. B. (2018). Tenascin-C derived signaling induces neuronal differentiation in a three-dimensional peptide nanofiber gel. *Biomaterials Science, 6*(7), 1859–1868. https://doi.org/10.1039/c7bm00850c

Sever, M., Turkyilmaz, M., Sevinc, C., Cakir, A., Ocalan, B., Cansev, M., … Tekinay, A. B. (2016). Regenerative effects of peptide nanofibers in an experimental model of Parkinson's disease. *Acta Biomaterialia, 46*, 79–90. https://doi.org/10.1016/j.actbio.2016.09.011

Shabafrooz, V., Mozafari, M., Vashaee, D., & Tayebi, L. (2014). Electrospun nanofibers: From filtration membranes to highly specialized tissue engineering scaffolds. *Journal of Nanoscience and Nanotechnology, 14*(1), 522–534. https://doi.org/10.1166/jnn.2014.9195

Shao, H., Li, T., Zhu, R., Xu, X., Yu, J., Chen, S., … He, L. (2018). Carbon nanotube multilayered nanocomposites as multifunctional substrates for actuating neuronal differentiation and functions of neural stem cells. *Biomaterials, 175*, 93–109. https://doi.org/10.1016/j.biomaterials.2018.05.028

Sharma, P., & Maffulli, N. (2006). Biology of tendon injury: Healing, modeling and remodeling. *Journal of Musculoskeletal & Neuronal Interactions, 6*, 181–190. Retrieved from http://www.embase.com/search/results?subaction=viewrecord&from=export&id=L44193219%0A, http://rug.on.worldcat.org/atoztitles/link/?sid=EMBASE&issn=11087161&id=doi:&atitle=Biology+of+tendon+injury%3A+Healing%2C+modeling+and+remodeling&stitle=J.+Musculoskelet

Shen, Y., Tenney, A. P., Busch, S. A., Horn, K. P., Cuascut, F. X., Liu, K., … Flanagan, J. G. (2009). PTPσ is a receptor for chondroitin sulfate proteoglycan, an inhibitor of neural regeneration. *Science, 326*(5952), 592–596. https://doi.org/10.1126/science.1178310

Shin, J., Choi, E. J., Cho, J. H., Cho, A. N., Jin, Y., Yang, K., … Cho, S. W. (2017). Three-dimensional electroconductive hyaluronic acid hydrogels incorporated with carbon nanotubes and polypyrrole by catechol-mediated dispersion enhance neurogenesis of human neural stem cells. *Biomacromolecules, 18*(10), 3060–3072. https://doi.org/10.1021/acs.biomac.7b00568

Shu, B., Sun, X., Liu, R., Jiang, F., Yu, H., Xu, N., & An, Y. (2019). Restoring electrical connection using a conductive biomaterial provides a new therapeutic strategy for rats with spinal cord injury. *Neuroscience Letters, 692*, 33–40. https://doi.org/10.1016/j.neulet.2018.10.031

Silantyeva, E. A., Nasir, W., Carpenter, J., Manahan, O., Becker, M. L., & Willits, R. K. (2018). Accelerated neural differentiation of mouse embryonic stem cells on aligned GYIGSR-functionalized nanofibers. *Acta Biomaterialia, 75*, 129–139. https://doi.org/10.1016/j.actbio.2018.05.052

Silva, G. A., Czeisler, C., Niece, K. L., Beniash, E., Harrington, D. A., Kessler, J. A., & Stupp, S. I. (2004). Selective differentiation of neural progenitor cells by high-epitope density nanofibers. *Science, 303*(5662), 1352–1355. https://doi.org/10.1126/science.1093783

Singh, A., Shiekh, P. A., Das, M., Seppälä, J., & Kumar, A. (2019). Aligned chitosan-gelatin cryogel-filled polyurethane nerve guidance channel for neural tissue engineering: fabrication, characterization, and in vitro evaluation. *Biomacromolecules, 20*(2), 662–673. https://doi.org/10.1021/acs.biomac.8b01308

Singh, N., Chen, J., Koziol, K. K., Hallam, K. R., Janas, D., Patil, A. J., … Rahatekar, S. S. (2016). Chitin and carbon nanotube composites as biocompatible scaffolds for neuron growth. *Nanoscale, 8*(15), 8288–8299. https://doi.org/10.1039/c5nr06595j

Singhal, N., & Martin, P. T. (2011). Role of extracellular matrix proteins and their receptors in the development of the vertebrate neuromuscular junction. *Developmental Neurobiology, 71*(11), 982–1005. https://doi.org/10.1002/dneu.20953

Sirko, S., von Holst, A., Wizenmann, A., Gotz, M., & Faissner, A. (2007). Chondroitin sulfate glycosaminoglycans control proliferation, radial glia cell differentiation and neurogenesis in neural stem/progenitor cells. *Development, 134*(15), 2727–2738. https://doi.org/10.1242/dev.02871

Solanki, A., Kim, J. D., & Lee, K. B. (2008). Nanotechnology for regenerative medicine: Nanomaterials for stem cell imaging. *Nanomedicine, 3*, 567–578. https://doi.org/10.2217/17435889.3.4.567

Song, S. J., Shin, Y. C., Kim, S. E., Kwon, I. K., Lee, J. H., Hyon, S. H., … Kim, B. (2018). Aligned laminin core-polydioxanone/collagen shell fiber matrices effective for neuritogenesis. *Scientific Reports, 8*(1), 5570. https://doi.org/10.1038/s41598-018-23958-3

Stein, E., & Tessier-Lavigne, M. (2001). Hierarchical organization of guidance receptors: Silencing of netrin attraction by slit through a Robo/DCC receptor complex. *Science, 291*(5510), 1928–1938. https://doi.org/10.1126/science.1058445

Sun, K. L. W., Correia, J. P., & Kennedy, T. E. (2011). Netrins: Versatile extracellular cues with diverse functions. *Development, 138*(11), 2153–2169. https://doi.org/10.1242/dev.044529

Sun, Y., Li, W., Wu, X., Zhang, N., Zhang, Y., Ouyang, S., … Wu, W. (2016). Functional self-assembling peptide nanofiber hydrogels designed for nerve degeneration. *ACS Applied Materials and Interfaces, 8*(3), 2348–2359. https://doi.org/10.1021/acsami.5b11473

Sur, S., Newcomb, C. J., Webber, M. J., & Stupp, S. I. (2013). Tuning supramolecular mechanics to guide neuron development. *Biomaterials, 34*(20), 4749–4757. https://doi.org/10.1016/j.biomaterials.2013.03.025

Sur, S., Pashuck, E. T., Guler, M. O., Ito, M., Stupp, S. I., & Launey, T. (2012). A hybrid nanofiber matrix to control the survival and maturation of brain neurons. *Biomaterials, 33*(2), 545–555. https://doi.org/10.1016/j.biomaterials.2011.09.093

Suzuki, H., Ahuja, C. S., Salewski, R. P., Li, L., Satkunendrarajah, K., Nagoshi, N., … Fehlings, M. G. (2017). Neural stem cell mediated recovery is enhanced by chondroitinase ABC pretreatment in chronic cervical spinal cord injury. *PLoS One, 12*(8), e0182339. https://doi.org/10.1371/journal.pone.0182339

Tan, A., Chawla, R., Natasha, G., Mahdibeiraghdar, S., Jeyaraj, R., Rajadas, J., … Seifalian, A. M. (2016). Nanotechnology and regenerative therapeutics in plastic surgery: The next frontier. *Journal of Plastic, Reconstructive & Aesthetic Surgery, 69*, 1–13. https://doi.org/10.1016/j.bjps.2015.08.028

Tang, X., Davies, J. E., & Davies, S. J. A. (2003). Changes in distribution, cell associations, and protein expression levels of NG2, neurocan, phosphacan, brevican, versican v2, and tenascin-C during acute to chronic maturation of spinal cord scar tissue. *Journal of Neuroscience Research, 71*(3), 427–444. https://doi.org/10.1002/jnr.10523

Tavakol, S., Musavi, S. M. M., Tavakol, B., Hoveizi, E., Ai, J., & Rezayat, S. M. (2017). Noggin along with a self-assembling peptide nanofiber containing long motif of laminin induces tyrosine hydroxylase gene expression. *Molecular Neurobiology, 54*(6), 4609–4616. https://doi.org/10.1007/s12035-016-0006-0

Tavakol, S., Saber, R., Hoveizi, E., Tavakol, B., Aligholi, H., Ai, J., & Rezayat, S. M. (2016). Self-assembling peptide nanofiber containing long motif of laminin induces neural differentiation, tubulin polymerization, and neurogenesis: In vitro, ex vivo, and in vivo studies. *Molecular Neurobiology, 53*(8), 5288–5299. https://doi.org/10.1007/s12035-015-9448-z

Tedesco, M. T., Di Lisa, D., Massobrio, P., Colistra, N., Pesce, M., Catelani, T., … Pastorino, L. (2018). Soft chitosan microbeads scaffold for 3D functional neuronal networks. *Biomaterials, 156*, 159–171. https://doi.org/10.1016/j.biomaterials.2017.11.043

Telese, F., Ma, Q., Perez, P. M., Notani, D., Oh, S., Li, W., … Rosenfeld, M. G. (2015). LRP8-Reelin-regulated neuronal enhancer signature underlying learning and memory formation. *Neuron, 86*(3), 696–710. https://doi.org/10.1016/j.neuron.2015.03.033

Tian, L., Prabhakaran, M. P., Hu, J., Chen, M., Besenbacher, F., & Ramakrishna, S. (2016). Synergistic effect of topography, surface chemistry and conductivity of the electrospun nanofibrous scaffold on cellular response of PC12 cells. *Colloids and Surfaces B: Biointerfaces, 145*, 420–429. https://doi.org/10.1016/j.colsurfb.2016.05.032

Tissir, F., & Goffinet, A. M. (2003). Reelin and brain development. *Nature Reviews Neuroscience, 4*(6), 496–505. https://doi.org/10.1038/nrn1113

Tucker, R. P., & Chiquet-Ehrismann, R. (2015). Tenascin-C: Its functions as an integrin ligand. *International Journal of Biochemistry and Cell Biology, 65*, 165–168. https://doi.org/10.1016/j.biocel.2015.06.003

Turney, S. G., & Bridgman, P. C. (2005). Laminin stimulates and guides axonal outgrowth via growth cone myosin II activity. *Nature Neuroscience, 8*(6), 717–719. https://doi.org/10.1038/nn1466

Tysseling, V. M., Sahni, V., Pashuck, E. T., Birch, D., Hebert, A., Czeisler, C., … Kessler, J. A. (2010). Self-assembling peptide amphiphile promotes plasticity of serotonergic fibers following spinal cord injury. *Journal of Neuroscience Research, 88*(14), 3161–3170. https://doi.org/10.1002/jnr.22472

Tysseling-Mattiace, V. M., Sahni, V., Niece, K. L., Birch, D., Czeisler, C., Fehlings, M. G., … Kessler, J. A. (2008). Self-assembling nanofibers inhibit glial scar formation and promote axon elongation after spinal cord injury. *Journal of Neuroscience, 28*(14), 3814–3823. https://doi.org/10.1523/jneurosci.0143-08.2008

Uyar, T., & Kny, E. (2017). *Electrospun materials for tissue engineering and biomedical applications*. Cambridge: Woodhead Publishing. https://doi.org/10.1016/c2015-0-05802-6

Verma, S., Domb, A. J., & Kumar, N. (2011). Nanomaterials for regenerative medicine. *Nanomedicine, 6*, 157–181. https://doi.org/10.2217/nnm.10.146

Vitale, F., Summerson, S. R., Aazhang, B., Kemere, C., & Pasquali, M. (2015). Neural stimulation and recording with bidirectional, soft carbon nanotube fiber microelectrodes. *ACS Nano, 9*(4), 4465–4474. https://doi.org/10.1021/acsnano.5b01060

Wang, A. Z., & Tepper, J. E. (2014). Nanotechnology in radiation oncology. *Journal of Clinical Oncology, 32*, 2879–2885. https://doi.org/10.1200/JCO.2014.55.0699

Wang, J., Zheng, J., Zheng, Q., Wu, Y., Wu, B., Huang, S., … Guo, X. (2015). FGL-functionalized self-assembling nanofiber hydrogel as a scaffold for spinal cord-derived neural stem cells. *Materials Science and Engineering C, 46*, 140–147. https://doi.org/10.1016/j.msec.2014.10.019

Wang, J., Chen, N., Ramakrishna, S., Tian, L., & Mo, X. (2017). The effect of plasma treated PLGA/MWCNTs-COOH composite nanofibers on nerve cell behavior. *Polymers, 9*(12), 713. https://doi.org/10.3390/polym9120713

Wang, J., Tian, L., Chen, N., Ramakrishna, S., & Mo, X. (2018). The cellular response of nerve cells on poly-L-lysine coated PLGA-MWCNTs aligned nanofibers under electrical stimulation. *Materials Science and Engineering C, 91*, 715–726. https://doi.org/10.1016/j.msec.2018.06.025

Wang, J., Tian, L., Luo, B., Ramakrishna, S., Kai, D., Loh, X. J., … Mo, X. (2018). Engineering PCL/lignin nanofibers as an antioxidant scaffold for the growth of neuron and Schwann cell. *Colloids and Surfaces B: Biointerfaces, 169*, 356–365. https://doi.org/10.1016/j.colsurfb.2018.05.021

Wang, L., Liu, X., Fu, J., Ning, X., Zhang, M., Jiang, Z., … Zhang, Z. (2019). Release of methylene blue from graphene oxide-coated electrospun nanofibrous scaffolds to modulate functions of neural progenitor cells. *Acta Biomaterialia, 88*, 346–356. https://doi.org/10.1016/j.actbio.2019.02.036

Wang, S., Guan, S., Li, W., Ge, D., Xu, J., Sun, C., … Ma, X. (2018). 3D culture of neural stem cells within conductive PEDOT layer-assembled chitosan/gelatin scaffolds for neural tissue engineering. *Materials Science and Engineering C, 93*, 890–901. https://doi.org/10.1016/j.msec.2018.08.054

Wang, S., Sun, C., Guan, S., Li, W., Xu, J., Ge, D., … Ma, X. (2017). Chitosan/gelatin porous scaffolds assembled with conductive poly(3,4-ethylenedioxythiophene) nanoparticles for neural tissue engineering. *Journal of Materials Chemistry B, 5*(24), 4774–4788. https://doi.org/10.1039/c7tb00608j

Weaver, C. L., & Cui, X. T. (2015). Directed neural stem cell differentiation with a functionalized graphene oxide nanocomposite. *Advanced Healthcare Materials, 4*(9), 1408–1416. https://doi.org/10.1002/adhm.201500056

Weeber, E. J., Beffert, U., Jones, C., Christian, J. M., Förster, E., David Sweatt, J., & Herz, J. (2002). Reelin and apoE receptors cooperate to enhance hippocampal synaptic plasticity and learning. *Journal of Biological Chemistry, 277*(42), 39944–39952. https://doi.org/10.1074/jbc.M205147200

Wen, H., Xiao, W., Biswas, S., Cong, Z. Q., Liu, X. M., Lam, K. S., … Deng, W. (2019). Alginate hydrogel modified with a ligand interacting with α3β1 integrin receptor promotes the differentiation of 3D neural spheroids toward oligodendrocytes in vitro. *ACS Applied Materials and Interfaces, 11*(6), 5821–5833. https://doi.org/10.1021/acsami.8b19438

Wu, S., Duan, B., Lu, A., Wang, Y., Ye, Q., & Zhang, L. (2017). Biocompatible chitin/carbon nanotubes composite hydrogels as neuronal growth substrates. *Carbohydrate Polymers, 174*, 830–840. https://doi.org/10.1016/j.carbpol.2017.06.101

Xia, H., & Xia, Y. (2018). An in vitro study of non-aligned or aligned electrospun poly(methyl methacrylate) nanofibers as primary rat astrocytes-loading scaffold. *Materials Science and Engineering C, 91*, 228–235. https://doi.org/10.1016/j.msec.2018.05.050

Xie, J., MacEwan, M. R., Schwartz, A. G., & Xia, Y. (2010). Electrospun nanofibers for neural tissue engineering. *Nanoscale, 2*(1), 35–44. https://doi.org/10.1039/b9nr00243j

Xu, C., Guan, S., Wang, S., Gong, W., Liu, T., Ma, X., & Sun, C. (2017). Biodegradable and electroconductive poly(3,4-ethylenedioxythiophene)/carboxymethyl chitosan hydrogels for neural tissue engineering. *Materials Science and Engineering C, 84*, 32–43. https://doi.org/10.1016/j.msec.2017.11.032

Xue, C., Ren, H., Zhu, H., Gu, X., Guo, Q., Zhou, Y., … Gu, X. (2017). Bone marrow mesenchymal stem cell-derived acellular matrix-coated chitosan/silk scaffolds for neural tissue regeneration. *Journal of Materials Chemistry B, 5*(6), 1246–1257. https://doi.org/10.1039/C6TB02959K

Xue, J., Xie, J., Liu, W., & Xia, Y. (2017). Electrospun nanofibers: New concepts, materials, and applications. *Accounts of Chemical Research, 50*(8), 1976–1987. https://doi.org/10.1021/acs.accounts.7b00218

Yang, S., Kwok, J. C. F., & Fawcett, J. W. (2014). Neural ECM in regeneration and rehabilitation. *Progress in Brain Research, 214*, 179–192. https://doi.org/10.1016/B978-0-444-63486-3.00008-6

Yao, S., Yu, S., Cao, Z., Yang, Y., Yu, X., Mao, H. Q., … Wang, X. (2018). Hierarchically aligned fibrin nanofiber hydrogel accelerated axonal regrowth and locomotor function recovery in rat spinal cord injury. *International Journal of Nanomedicine, 13*, 2883–2895. https://doi.org/10.2147/IJN.S159356

Yebra, M., Montgomery, A. M. P., Diaferia, G. R., Kaido, T., Silletti, S., Perez, B., … Cirulli, V. (2003). Recognition of the neural chemoattractant netrin-1 by integrins α6β4 and α3β1 regulates epithelial cell adhesion and migration. *Developmental Cell, 5*(5), 695–707. https://doi.org/10.1016/S1534-5807(03)00330-7

Zhang, J., Lu, X., Feng, G., Gu, Z., Sun, Y., Bao, G., … Cui, Z. (2016). Chitosan scaffolds induce human dental pulp stem cells to neural differentiation: Potential roles for spinal cord injury therapy. *Cell and Tissue Research, 366*(1), 129–142. https://doi.org/10.1007/s00441-016-2402-1

Zhang, K., Zheng, H., Liang, S., & Gao, C. (2016). Aligned PLLA nanofibrous scaffolds coated with graphene oxide for promoting neural cell growth. *Acta Biomaterialia, 37*, 131–142. https://doi.org/10.1016/j.actbio.2016.04.008

Zhang, S., Holmes, T. C., DiPersio, C. M., Hynes, R. O., Su, X., & Rich, A. (1995). Self-complementary oligopeptide matrices support mammalian cell attachment. *Biomaterials, 16*(18), 1385–1393. https://doi.org/10.1016/0142-9612(95)96874-Y

Zhu, W., Masood, F., O'Brien, J., & Zhang, L. G. (2015). Highly aligned nanocomposite scaffolds by electrospinning and electrospraying for neural tissue regeneration. *Nanomedicine: Nanotechnology, Biology, and Medicine, 11*(3), 693–704. https://doi.org/10.1016/j.nano.2014.12.001

Zou, Y., Qin, J., Huang, Z., Yin, G., Pu, X., & He, D. (2016). Fabrication of aligned conducting PPy-PLLA fiber films and their electrically controlled guidance and orientation for neurites. *ACS Applied Materials and Interfaces, 8*(20), 12576–12582. https://doi.org/10.1021/acsami.6b00957

Zou, Z., Zheng, Q., Wu, Y., Guo, X., Yang, S., Li, J., & Pan, H. (2010). Biocompatibility and bioactivity of designer self-assembling nanofiber scaffold containing FGL motif for rat dorsal root ganglion neurons. *Journal of Biomedical Materials Research - Part A, 95*(4), 1125–1131. https://doi.org/10.1002/jbm.a.32910

Zuidema, J. M., Desmond, G. P., Rivet, C. J., Kearns, K. R., Thompson, D. M., & Gilbert, R. J. (2015). Nebulized solvent ablation of aligned PLLA fibers for the study of neurite response to anisotropic-to-isotropic fiber/film transition (AFFT) boundaries in astrocyte-neuron co-cultures. *Biomaterials, 46*, 82–94. https://doi.org/10.1016/j.biomaterials.2014.12.046

Zweckberger, K., Ahuja, C. S., Liu, Y., Wang, J., & Fehlings, M. G. (2016). Self-assembling peptides optimize the post-traumatic milieu and synergistically enhance the effects of neural stem cell therapy after cervical spinal cord injury. *Acta Biomaterialia, 42*, 77–89. https://doi.org/10.1016/j.actbio.2016.06.016

Chapter 6
Nanomaterials for Cartilage Regeneration

Huseyin Ozkan and Orhan Yanmis

Abstract Repair and regeneration of cartilage tissue has always posed difficulties owing to its avascular, aneural structure and its sparsely distributed cellular arrangement within a dense extracellular matrix. This is why damage to cartilage tissue such as acute trauma, repetitive trauma, inflammatory disease, or wear due to aging eventually results in osteoarthritis. In addition to being a challenging disease for both patients and physicians, osteoarthritis is also a significant public health issue that requires the attention of healthcare planners. In the osteochondral tissue damage, it is important to provide a tissue scaffold and support biosignaling molecules due to its dense extracellular structure as well as cell-based treatments. Therefore, the importance of nanomaterials in tissue regeneration studies is gradually increasing. Nanomaterials are defined as structural elements smaller than 100 nm in at least one dimension, and they offer us the ability to control various properties of materials by assembling them at nanometric proportions. With their excellent biomimetic and physicochemical properties, nanomaterials open up new possibilities and horizons as integration, interaction and signaling in structural and cellular dimensions. This chapter will discuss the opportunities provided by nanomaterials in cartilage regeneration.

6.1 Introduction

In the musculoskeletal system, cartilage is a light blue or white bright tissue that covers the end-bone of all synovial joints. It allows movement of the joints due to its low-friction gliding surface, increased compressive strength, and bio-lubrication. Any damage to the chondral tissue might harm not just the cartilage but also the subchondral osseous structures. Because of this, chondral pathologies may be generally described as "osteochondral injuries." These injuries are commonly encoun-

H. Ozkan (✉)
University of North Carolina, Chapel Hill, NC, USA

O. Yanmis
Department of Orthopaedics Surgery and Rehabilitation, Wake Forest University School of Medicine, Winston-Salem, NC, USA

A. B. Tekinay (ed.), *Nanomaterials for Regenerative Medicine*, Stem Cell Biology and Regenerative Medicine, https://doi.org/10.1007/978-3-030-31202-2_6

tered conditions in orthopedic practice; and they cause pain, swelling, and loss of range-of-motion in the affected joint. Joint instability may also be evident, due to accompanying ligamentous or capsular injuries. Such processes, influenced by numerous other etiological factors, usually lead to osteoarthritis (OA). While OA prevalence increases dramatically with age, it can be encountered in all age groups. It is more prevalent in groups such as soldiers, professional athletes, and females. Current approach to cartilage injury includes conservative and surgical treatment options; mainly utilizing bone marrow stimulation techniques (microfracture), autograft and allograft transplantations (mosaicplasty), transplantations of autologous chondrocytes, and application of bioactive agents (Simon & Jackson, 2018). All of these treatments, however, are still far from achieving the desired curative outcomes, and their usage is subject to significant limitations and obstacles.

The goal of treatment in joint cartilage injury is to ensure the formation of new cartilage tissue that is similar to healthy hyaline cartilage. This requires the restoration of the extracellular matrix, existence of healthy chondrocytes, and presence of appropriate morphogenetic signals. Unfortunately, our treatment options are still very limited in terms of ensuring these factors. Regenerative medicine is a branch of medical science that is still in development, and cartilage regeneration currently receives a great deal of attention from researchers. The main goal of nanomaterial-focused tissue engineering is to discover new treatment options using biocompatible, biodegradable, and bioactive materials in order to restore or regenerate one or more of these three components that ensure cartilage viability. Nanomaterials are versatile components that are able to mimic the surface characteristics of extracellular matrix elements, thereby providing a wide array of possibilities for tissue engineering efforts. The high level of interest in this field of research accurately reflects the immense potential for improvement. This chapter will specifically focus on the role of nanomaterials-based tissue regeneration research in cartilage injury, with a discussion of the possibilities and future goals.

6.1.1 Cartilage Tissue

During the embryonic period, in the 5th week of gestation, some mesenchymal cells form the blastema. These cells begin producing cartilage matrix and then reside within this specialized matrix and are called respectively chondroblasts and chondrocytes. Eventually, the mesenchymal tissue encircling the blastema forms the perichondrial membrane (Bhosale & Richardson, 2008).

Cartilage structures form the temporary skeleton until the onset of ossification, and hypertrophic chondrocytes eventually form the bone tissue. During this process, cartilage also differentiates into various types according to their matrix structures. The four types of cartilage in the human body are elastic, fibro-cartilage, fibro-elastic, and hyaline cartilage. Hyaline cartilage is the type that covers the bone surfaces in synovial joints, connects the ribs and the sternum, and supports the trachea. It is the most abundant type of cartilage in the human body, and we shall call the type that covers the joint surfaces "articular cartilage" (Bhosale & Richardson, 2008; Simon & Jackson, 2018).

6.1.2 Articular Cartilage

Articular Cartilage is an extremely specialized, white-blue colored, smooth, tough, multilayered tissue. It has a water content of over 70% and its main organic components are collagen II and aggrecan. Collagen fibrillary structure provides tensile strength, while proteoglycan aggrecans and hydrophilic glycosaminoglycans allow the water to change compartments when bearing weight, thereby ensuring pressure resilience. The joint surface also has a very low friction coefficient due to biolubricants produced by the cartilage and synovial cells, such as lubricin and hyaluronic acid (S. R. Goldring & Goldring, 2016). Articular cartilage does not have a direct blood, lymphatic, or neural supply; instead, it receives nutrients from surrounding tissues through diffusion. Apart from providing unique biomechanical properties, this complex and multilayered architecture of the articular cartilage also causes great difficulties when it comes to repair and regeneration. There are differences between these layers not only limited to matrix structure but also between their chondrocyte phenotypes and the functional properties of these components.

6.1.2.1 Zonal Structure

Articular cartilage has a highly organized structure composed of four distinct zones, which are identified as the superficial (tangential) zone, middle (transitional) zone, deep (radial) zone, and calcified zone.

The superficial (tangential) zone contains flattened chondrocytes that are elongated and lie parallel to the surface. Also parallel to the surface are the densely deposited type II collagen fibers, and the type I collagen content is minimal. This zone constrains resistance to shear stress and swelling pressures imposed by the negatively charged glycosaminoglycans (GAGs). This allows the cartilage to retain its shape under pressure (Simon & Jackson, 2018). The surface of the superficial zone does not contain any cells and a distinct layer that is several hundred nanometers thick, called "lamina splendens," forms the joint surfaces. While it is currently assumed that this layer provides a low friction surface for the cartilage, its actual role is still poorly understood (Camarero-Espinosa, Rothen-Rutishauser, Foster, & Weder, 2016).

The middle (transitional) zone is where shear forces from the superficial layer transform into compressional forces. Cell density in this region is low and the chondrocytes, which now take a spherical shape, express large amounts of collagen II and proteoglycans. The collagen fibers are arranged in random orientation, and this zone has the richest proteoglycan content (S. R. Goldring & Goldring, 2016; Simon & Jackson, 2018).

In the deep (radial) zone, the chondrocytes begin to assume oval shapes. Collagen fibers are thicker and perpendicular to the subchondral bone, and are distributed to resist load compression. In this zone, the cell density is decreased, while the proteoglycan concentration is increased (Camarero-Espinosa et al., 2016; Simon & Jackson, 2018).

The calcified zone is a thin layer of tissue between the subchondral bone and the cartilage. Some of its chondrocytes are completely encapsulated in calcified lacunae and their metabolic activity is minimal. Type X collagen is also present in the calcified zone. This zone is far more dense and mineralized compared to the adjacent subchondral bone, and it also contains the tidemark that separates the calcified and non-calcified cartilage. The calcified tidemark and subchondral zone cut off the cartilage completely from bone marrow blood supply. During joint movement and loading, the calcified zone transforms shear stress into compressive and tensile stresses (S. R. Goldring & Goldring, 2016).

Underneath the calcified cartilage, there is the subchondral bone that is mostly similar to cortical bone. This zone gives way to cancellous bone that is more porous and metabolically active.

6.1.2.2 Chondrogenesis

Chondrocytes, the main cell type in cartilage, are distributed within a dense extracellular matrix. Near the joint surface, the ratio of chondrocytes to matrix is 1.65%, while in deeper zones this can reach to 2.6% (Hunziker, Quinn, & Häuselmann, 2002). Chondrocytes are located in small spaces called lacunae, similar to osteocytes. However, the specialized cytoplasmic structures seen in osteocytes that allow communication are not present in chondrocytes. Mature chondrocytes lose their migration, proliferation and repair abilities as they settle in their matrix. At the same time, the potential for matrix production is limited by the synthesis and stimulation of certain types of proteoglycans and the growth factor response is reduced (Simon & Jackson, 2018).

Compared to other cell types, chondrocytes reside in a lower oxygen tension environment. Oxygen tension in areas close to the articular surface is around 10%, while near the deep zone it is less than 1%. Intracellular survival factors such as hypoxia-inducible factor 1α (HIF1α) ensure the cell viability of these cells in such hypoxic environments (S. R. Goldring & Goldring, 2016).

Chondrogenesis starts with the production of collagen I, III and V by the mesenchymal cells. Chondroprogenitor cell differentiation, however, requires the expression of the cartilage specific collagens II, IX, and XI. Proliferative chondrocytes express collagen VI and matrilin 1. Matrix restructuring involves the activities of matrix metalloproteinases (MMP) 9, 13, and 14. One of the earliest signals triggering chondrogenesis is the expression of transforming growth factor-β (TGF-β). Also involved in chondrogenesis in complex ways are fibroblast growth factor (FGF), bone morphogenetic protein (BMP) and Wnt signaling pathways that serve to regulate the development of the skeletal system (M. B. Goldring, 2012). For this reason, TGF-β is important for both natural cartilage development and cartilage tissue engineering approaches. TGF-β is produced by chondrocytes as part of a larger molecular complex, and then stored in the extracellular matrix. Afterwards it undergoes an activation process and is released from storage as activated-TGF-β. The active form induces the chondrocytes to produce extracellular matrix components such as type II collagen and proteoglycans (especially aggrecan) (M. J. Chen et al., 2019). Spagnoli et al. have demonstrated that TGF-β signaling is essential for joint morphogenesis (Spagnoli et al., 2007). Because of its effects upon progenitor cells and its critical importance in joint and growth plate development, TGF-β has the potential to become one of the target components for the treatment of osteoarthritis (T. Li et al., 2012; T. Li, Chubinskaya, et al., 2019; Longobardi et al., 2012).

6.1.3 Articular Cartilage Injury

Degeneration of articular cartilage (arthritis) is a commonly encountered clinical scenario. Its treatment involves difficulties for both patients and physicians, as cartilage has no blood supply and low regeneration potential. Increasing life expectancies, prevalence of senile osteoarthritis, and numerous diseases that affect the joints in every age group lead us to conclude that arthritis is a public health issue. Osteoarthritis affects 10–12% of all humans, and it is the most common musculoskeletal system disease in the world. This ratio increases to 49.7% among people over the age of 65. In the United States, over one million total joint arthroplasties are performed annually, and this number is expected to reach four million by 2030 thanks to an aging population and increased obesity prevalence (Etkin & Springer, 2017; Medvedeva et al., 2018).

Cartilage insufficiency can be described as the deterioration of the balance between destructive forces on the joint and synthesis of the extracellular matrix, in favor of destructive forces (Armiento, Stoddart, Alini, & Eglin, 2018). The exact mechanisms of this decay are still not fully understood.

We can classify cartilage injury into three groups in terms of the depth of the damage: (1) Group, only affecting the superficial or middle zones; (2) Group, injury reaching all the way down to the subchondral bone but not penetrating into the bone marrow; and (3) Group, injuries that have reached the bone marrow (Simon & Jackson, 2018). Each of these categories has different clinical symptoms, findings, treatment options, and treatment response characteristics.

At the present, progressive degeneration of the joint cartilage as a result of trauma and degenerative diseases is most commonly named as "osteoarthritis" (OA). OA can be divided into primary and secondary forms: Primary OA is observed when there is no underlying abnormality, while secondary OA, by definition, involves a primary cause that damages the extracellular matrix of the cartilage. Various risk factors for OA have been described, including trauma, obesity, high-impact physical activity, joint malalignment, age, gender, hypermobility syndromes (Ehlers–Danlos), degenerative joint diseases (Perthes), and metabolic diseases (diabetes mellitus). Until recently, OA was considered to be a disease that was limited to the wear and tear of joint cartilage. However, new molecular pathophysiological knowledge indicates that it is a disease that involves all components of the joint (Loeser, Goldring, Scanzello, & Goldring, 2012). In response to cartilage injury, paracrine and autocrine mechanisms act to disrupt the regular chondrocyte functions.

Catabolic enzymes such as matrix metalloproteinases (MMPs) 1, 3, 13 and a-disintegrin-and-metalloproteinase-with-thrombospondin-motifs (ADAMTS) 4, 5 play a part in furthering the degeneration. While these processes initially serve to activate the chondrocytes to speed up their metabolism, they end up contributing to a catabolic process with decreased proteoglycan and type II collagen synthesis. This, in turn, decreases the capability of cartilage to hold water and lowers its compression resistance, resulting in further cartilage damage (Armiento et al., 2018; Martel-Pelletier et al., 2016).

6.1.4 *Treatment Options*

The main goal of current medical and surgical therapies in chondral injuries is to relieve the symptoms. Anti-inflammatory analgesic medication and rest comprises the early treatment options, while later options include physical exercise and activity modifications. In the next stage, an attempt to stimulate cartilage regeneration is made through intra-articular injections and arthroscopic surgery.

Microfracture method is the most common of these surgeries. It can be applied in defects where subchondral bone tissue is intact and the surrounded cartilage is healthy. This minimally invasive surgery arthroscopically opens holes in the defect area until marrow material enters the joint (Fig. 6.1). Fibrocartilaginous cartilage tissue is formed after a long period of weight-bearing restriction postoperatively. However the symptoms return quickly due to the biomechanical weakness of fibrocartilage tissue.

Another method is osteochondral autograft or allograft transplantation (mosaicplasty). This method involves the transfer of full-thickness cartilage and subchondral bone tissue to the defect area (Fig. 6.2). Allograft transplantation offers a wider range of possibilities in terms of size and shape since the graft is removed from a cadaver, while immune response and disease transmission are the main drawbacks. On the other hand, autografts have problems such as limited donor tissue and donor site morbidity (hematomas, inflammation, deterioration of articular function).

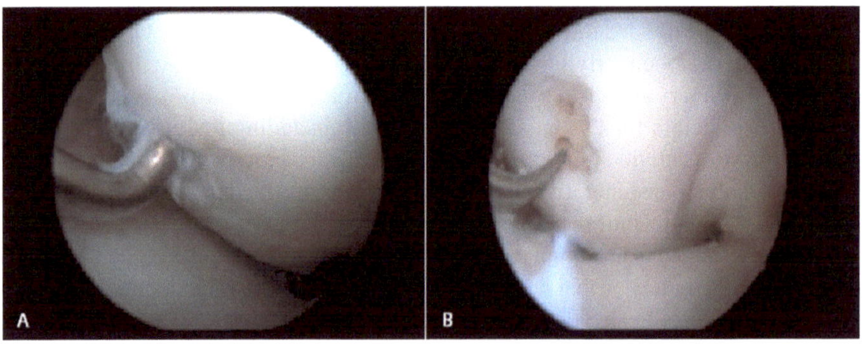

Fig. 6.1 Arthroscopic intra-articular view. (**a**) Chondral injury, (**b**) Microfracture

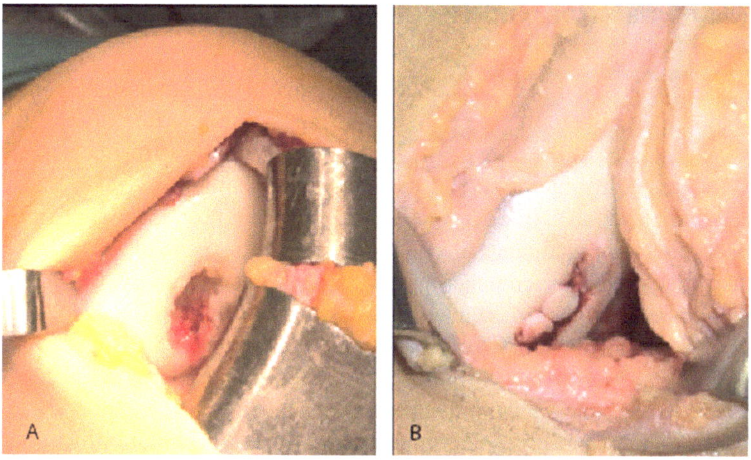

Fig. 6.2 Intraoperative view. (**a**) Osteochondral injury, (**b**) Autograft transplantation (mosaicplasty)

Autologous Chondrocyte Implantation (ACI) and Matrix-induced Autologous Chondrocyte Implantation (MACI) are two-stage surgical procedures that are used for the treatment of OA. First surgery collects healthy chondral tissue and these cells are induced to proliferate in vitro. In ACI, cultured cells are directly applied to the defect area and periosteal or synthetic collagen patches are then used to cover them. In MACI, the cells are cultured on a scaffold and inserted into the defect site with fibrin glue. These methods are partially successful, but the excessive proliferation of cells from the periosteal flap (leading to osseous overgrowth of the defect cavity) and the time-consuming and expensive nature of the two-stage procedure remain the most significant problems associated with these treatment methods (Kwon et al., 2019; Zylińska, Silmanowicz, Sobczyńska-Rak, Jarosz, & Szponder, 2018).

All these regenerative methods require donor material, are invasive, and require long-term postoperative treatment. If these methods fail, total joint arthroplasties are widely used for pain relief and restoration of function. In arthroplasty surgeries, joint surfaces are cut and internal prostheses consisting of metal and polyethylene components are applied to bone surfaces. While complications due to infection and poor surgical technique are seen in the early period, the biggest problem in the late period is the need for revision surgery because of aseptic loosening.

As is evident, none of these treatments can achieve true joint regeneration, and major surgical procedures often pave the way for later invasive interventions. As all these treatments are costly and do not result in full improvement, treatment always continues through various modalities. Regenerative therapies, a new alternative, aim to restructure cartilage with a different approach in order to obtain healthy cartilage tissue. Although there is still much to be done on this subject, research efforts continue to make progress. In the following section, the current state and goals of nanomaterial-based approaches in cartilage regenerative treatment will be discussed.

6.2 Nanomaterials and Cartilage Tissue Engineering

The main methods used in the treatment of damaged cartilage tissue aim to relieve the symptoms by alleviating pain and restoring function, while not focusing on the tissue structures at the core of the problem. Regenerative treatments, however, are focused on the formation of new cartilage tissue that is identical to the original tissue (Armiento et al., 2018). Tissue engineering is a promising approach in the field of regenerative medicine. It requires interdisciplinary work of various scientific disciplines such as engineering, material science, biology, and chemistry. Tissue engineering today is founded upon cells, scaffolds, and signals (the tissue engineering triad) (Zhang, Hu, & Athanasiou, 2012). However, tissues have a complex hierarchical arrangement from the nano level to the micro and macro levels. This affects tissue biology, the transition between layers, and the regulation of tissue interactions. The macro level determines aspects like biocompatibility, biodegradability and mechanical properties. Micro level manages tissue architecture, surface chemistry, surface stiffness, cell migration, nutrient delivery, and vascularization. On the nano scale, the functions of bioactive factors, cell adhesion, mineralization and gene expression are regulated (Santo, Gomes, Mano, & Reis, 2012). Since natural tissues achieve homeostasis through these nano- and micro-level interactions, attaining this complex structure is also crucial for tissue engineering. Therefore, nanomaterials, with their excellent physicochemical structures and biomimetic properties, have attracted great interest in improving cell growth and function and facilitating and directing tissue regeneration (Eslahi, Abdorahim, & Simchi, 2016).

In conventional tissue engineering, cells planted in tissue scaffolds are first statically cultured. They are then transferred to a bioreactor that cultures them under loads similar to the tissue. This immature tissue is expected to undergo remodeling after implantation into the body. The current approach, however, aims to avoid the cell seeding and maturation stages of these methods. By implanting only the scaffold and performing microfractures in order to recruit mesenchymal stem cells from the bone marrow, it is expected that adequate mesenchymal stem cells differentiation can be stimulated (Camarero-Espinosa et al., 2016). Nanomaterials also offer new possibilities for the production, transport, release and timed-activation of bioactive substances. These bring a wealth of approaches for cartilage regeneration, allowing researchers to try new combinations.

6.2.1 Biomaterials

Biomaterials are composed of mainly ceramics, metals, and polymers that can be obtained naturally or synthetically. The biomaterials that are designed to be used for cartilage tissue engineering should be biocompatible, allowing cell retention, allowing the passage of bioactive substances, and able to support cell viability, proliferation and secretory activities (Vinatier & Guicheux, 2016). Since osteochondral tissues are connective tissues that are regularly subjected to loads, the scaffolds used must be made of materials that can withstand mechanical loading. As a result, rigid polymers have been used for three-dimensional structural support, while hydrogel scaffolds are more suitable as cell carriers. Composite biomaterials are still being developed in order to mimic the multilayered hierarchical architecture of the cartilage tissue, which includes a special osteochondral interface with bone tissue (Manoukian et al., 2018).

Natural polymers used as cartilage scaffolds are physiological and nontoxic materials which usually have bioadhesive surfaces for cells. However, they are mechanically weaker due to their rapid degradation profiles. Some of the natural polymers being studied extensively are: polysaccharide structures (chitosan, hyaluronan, alginate, agarose, chondroitin, methylcellulose); and protein structures (collagen, gelatin, fibrin, silk, keratin). Collagen and hyaluronan adapt and degrade more easily since they are a part of natural cartilage tissue (Eslahi et al., 2016; Vinatier & Guicheux, 2016).

Synthetic polymers offer stronger mechanical structures, ease of processing, and can be sterilized. Unfortunately they are less biologically active, degrade poorly, and might cause a stronger inflammatory response. Some synthetic polymers that have been used as cartilage scaffolds are poly(ethylene glycol) (PEG), polylactide (PLA), poly(L-lactide) (PLLA), poly(lactic-co-glycolic acid) (PLGA), poly(ε-caprolactone) (PCL), poly(vinyl alcohol) (PVA), polyurethanes (PUR), polyglycolide (PGA), and PGA/PLA copolymers (Eslahi et al., 2016; Manoukian et al., 2018).

6.2.2 Nanomaterials and Scaffolds

Nanomaterials are emerging as versatile components that can control conventional tissue engineering approaches structurally, mechanically and chemically at the nano level through the use of nanofabrication technologies. They can exhibit a high level of cellular compatibility and bioactivity by enabling molecular interactions with the cell, while providing unique mechanical, optical, electrical, and magnetic properties to better control cell functions. Nanomaterials can also mimic extracellular matrix surface properties such as energy and topography. All these features have led researchers to study nanomaterials to be used as tissue scaffolds, biomolecular carriers or biosensor nanovehicles (H. Chen et al., 2013).

Cartilage tissue scaffolds can be built as hydrogels, fibrous meshes or foam (sponges) depending on their preparation techniques. Hydrogels are water-swollen polymers or protein structures that are physically, chemically or hybrid cross-linked. While chemically cross-linked hydrogels have covalent bonds, physically cross-linked hydrogels are bound by non-covalent (hydrogen bonding, hydrophobic, and electrostatic) weak molecular interactions. Hydrogels are currently the most popular cartilage scaffolds and are being studied extensively. This is because they are easy to apply and they can be injected or applied through minimally invasive surgery such as arthroscopy. The hydrogels can easily fill lesion areas within the joint and quickly adapt to the surrounding healthy tissue. They have a high water content allowing the diffusion of nutrients and waste. Cells within these gel structures can become embedded, as opposed to simple adhesion, and they can differentiate in the direction of a chondrogenic phenotype. Commonly used polymers are PEG, agarose, alginate, hyaluronan, collagen, and chitosan, which can provide high biocompatibility and potential biodegradability. However, hydrogels are mechanically weak, especially for the cartilage surfaces of the load-bearing joints; thus, efforts are still being made to increase the mechanical strength of hydrogels (Camarero-Espinosa et al., 2016; Eslahi et al., 2016; J. Li, Chen, et al., 2019).

Reducing the gel to nano scales is one of the key strategies to increase the response of hydrogel dynamics to stimuli. The large surface-to-volume ratios obtained by the addition of nanoparticles to the structure strengthen the mechanical properties as well as improving function (J. Li, Chen, et al., 2019). However, the number of nanostructuring techniques that can be used for this purpose are limited due to the high water content of hydrogels. But self-assembly of hydrophobic segments of polymers in aqueous environments can still take place. Nanoparticle hydrogels (nanogels), nanofiber hydrogels and hybrid hydrogels constitute promising materials that are being researched extensively for this purpose.

Among the techniques that are used for the production of nanofiber hydrogels, molecular self-assembly and electrospinning techniques are tested extensively by the researchers. The extracellular matrix mimicking biomolecular structures of self-assembled supramolecular materials enable them to support cell adhesion, proliferation, and migration through their bioactive interactions. Among self-assembling materials, peptide nanofiber gels are emerging as an alternative treatment for cartilage tissue. Yaylacı et al. have shown that glycosaminoglycan-mimetic peptide nanofibers can support chondrogenic differentiation in mesenchymal stem cells, and hyaluronic acid mimetic self-assembled peptide nanofiber gels may enhance cartilage regeneration (Yaylaci et al., 2016). Likewise, Arslan et al. investigated the therapeutic effects of a hybrid peptide nanofiber-hyaluronic acid membrane they developed in an in vivo rat OA model. They demonstrated that this hybrid nanofiber membrane was more effective than the commercially available gels they compared it to, and that the hybrid peptide nanofiber-hyaluronic acid membrane could be a suitable alternative for the treatment of OA (Arslan et al., 2018).

Scaffolds produced by electrospinning technique are obtained from polymer solutions through the use of high-voltage collectors, thereby forming a non-woven fabric made of fibers. These simple yet robust structures also enable the discovery of new material combinations, since they also allow the co-spinning of multiple synthetics or biological materials. The surfaces of the scaffolds produced through electrospinning can also be tailored to have surface modifications, chemical modifications, and add biologically active materials. Combined materials can be made with growth factors and biological signal proteins. Electrospinning is a carefully studied topic for 3D scaffold production, which is particularly important for joint cartilage (Zhou, Chyu, & Zumwalt, 2018). Mahboudi et al. fabricated a nanofiber-based polyethersulfone scaffold via electrospinning, and they were able to demonstrate osteochondral differentiation of human bone marrow mesenchymal stem cells on this scaffold (Mahboudi et al., 2018). In another study, Erisken et al. loaded two bioactive substances on a poly(ε-caprolactone) scaffold they produced with the electrospinning technique. Added to the scaffold were insulin (for chondrogenic differentiation) and β-glycerophosphate (for mineralization). They showed that human adipose-derived stromal cells cultured in this scaffold for 8 weeks were heavily directed towards chondrogenic differentiation (Erisken, Kalyon, Wang, Örnek-Ballanco, & Xu, 2011). Yu et al. applied a bioactive resveratrol—PLA—gelatin porous nano-scaffold that they built with the electrospinning technique to a rat cartilage defect. They showed that the scaffold was able to heal the defect and demonstrated how this structure affects the repair through the PI3K/AKT signaling pathway (Yu et al., 2018).

Nano-level controllable composite tissue scaffolds are being studied extensively in cartilage tissue engineering. These composite structures offer new horizons and possibilities not only for cartilage, but also in the formation of a cartilage-bone hierarchical interface. Nanomaterials are also used in tissue engineering to produce and distribute tissue stimulating agents such as growth factors and peptides. Chahine et al. evaluated the biocompatibility of single-wall carbon nanotubes (SWNTs) in articular cartilage tissue engineering in 2D and 3D composite tissue scaffolds. They showed that SWNTs increased GAG content in composite scaffolds, and that it biomechanically increased pressure resistance and tensile modulus. They demonstrated that SWNTs may be able to provide functionalization with bioactive molecules and biomechanical strengthening (Chahine, Collette, Thomas, Genetos, & Loots, 2014). In another study, Ribeiro et al. developed biofunctional hierarchical scaffolds consisted of a horseradish peroxidase (HRP)-cross-linked silk fibroin (SF) cartilage-like layer (HRP-SF layer) fully integrated into a HRP-SF/ZnSr-doped β-tricalcium phosphate (β-TCP) subchondral bone-like layer (HRP-SF/dTCP layer) aimed at regenerating chondral and subchondral tissue. Human osteoblasts (hOBs) and human articular chondrocytes (hACs) planted on this bilayer hierarchical structure demonstrated sufficient integration, proliferation, and appropriate ECM production to the respective interfaces. With this study, they showed that osteochondral-like tissue formation can be promoted with appropriate stimuli in a culture system (Ribeiro et al., 2019). Karami et al. designed a composite scaffold that would firmly adhere to tissues such as cartilage and meniscus. This hydrogel system consisted of poly(ethylene glycol) dimethacrylate, alginate, and nanofibrillated cellulose. They showed that their composition provided a significant increase in cartilage adhesion compared to the existing commercial tissue adhesives (Karami et al., 2018).

6.3 Summary and Future Outlook

Studies on regenerative medicine are advancing with great momentum, challenging our current perceptions of medicine. Our focus is now shifting from alleviation of symptoms to the full regeneration of tissue. We are also witnessing the most promising advances for cartilage, one of the most challenging tissues for regeneration.

The scaffold necessary for the regeneration of cartilage tissue is slowly taking shape; involving new composite structures with bioactive extracellular matrix mimicking materials, and chondrocytes differentiated according to the multilayered architecture required by the natural tissue. As we have seen in the recent studies above, developments in this area have reached an exciting level. However, there are still issues ahead, such as the hierarchical relationship of cartilage tissue with bone and joint fluid, the complex unforeseen interactions that may be observed in vivo, and the unique biomechanical and lubricative properties of the cartilage tissue. These still pose difficulties to be overcome by researchers of this field, and the biomechanical shortcomings of current scaffolds is another problem encountered in clinical practice.

Cartilage tissue is very difficult to reach with medications administered systemically. This is why intra-articular drug administrations are an important subject, and this also poses various hardships. Nanomaterials can also be used to overcome the difficulties in distributing intra-articular medications. Moreover, the ability of nanostructures to overcome extracellular matrix and cell barriers provides great advantages in drug distribution and diffusion (Brown, Kumar, & Sharma, 2019; Wang et al., 2018). Currently, the effective delivery of drugs is limited to indirect pharmacological measures. However, imaging agents added to nanoemulsions will soon be able to monitor the distribution, release, and efficacy of drugs in vivo. This area called *Theranostics* enables delivery, treatment, and imaging with the same molecule using nanomaterials and bioimaging technologies (Herneisey et al., 2016; Patel, Beaino, Anderson, & Janjic, 2015).

All these issues are related and interacting with each other. Future research goals range from regeneration of damaged tissue, to the manufacture of organs and extremities. This subject stands before all humanity as a common goal to be striven towards in order to promote human welfare.

References

Armiento, A. R., Stoddart, M. J., Alini, M., & Eglin, D. (2018). Biomaterials for articular cartilage tissue engineering: Learning from biology. *Acta Biomaterialia, 65*, 1–20. https://doi.org/10.1016/j.actbio.2017.11.021

Arslan, E., Sardan Ekiz, M., Eren Cimenci, C., Can, N., Gemci, M. H., Ozkan, H., … Tekinay, A. B. (2018). Protective therapeutic effects of peptide nanofiber and hyaluronic acid hybrid membrane in in vivo osteoarthritis model. *Acta Biomaterialia, 73*, 263–274. https://doi.org/10.1016/j.actbio.2018.04.015

Bhosale, A. M., & Richardson, J. B. (2008). Articular cartilage: Structure, injuries and review of management. *British Medical Bulletin, 87*, 77–95. https://doi.org/10.1093/bmb/ldn025

Brown, S., Kumar, S., & Sharma, B. (2019). Intra-articular targeting of nanomaterials for the treatment of osteoarthritis. *Acta Biomaterialia, 93*, 239–257. https://doi.org/10.1016/j.actbio.2019.03.010

Camarero-Espinosa, S., Rothen-Rutishauser, B., Foster, E. J., & Weder, C. (2016). Articular cartilage: From formation to tissue engineering. *Biomaterials Science, 4*, 734–767. https://doi.org/10.1039/c6bm00068a

Chahine, N. O., Collette, N. M., Thomas, C. B., Genetos, D. C., & Loots, G. G. (2014). Nanocomposite scaffold for chondrocyte growth and cartilage tissue engineering: Effects of carbon nanotube surface functionalization. *Tissue Engineering Part A, 20*(17–18), 2305–2315. https://doi.org/10.1089/ten.tea.2013.0328

Chen, H., Zeng, Y., Liu, W., Zhao, S., Wu, J., & Du, Y. (2013). Multifaceted applications of nanomaterials in cell engineering and therapy. *Biotechnology Advances, 31*, 638. https://doi.org/10.1016/j.biotechadv.2012.08.002

Chen, M. J., Whiteley, J. P., Please, C. P., Ehlicke, F., Waters, S. L., & Byrne, H. M. (2019). Identifying chondrogenesis strategies for tissue engineering of articular cartilage. *Journal of Tissue Engineering, 10*, 2041731419842431. https://doi.org/10.1177/2041731419842431

Erisken, C., Kalyon, D. M., Wang, H., Örnek-Ballanco, C., & Xu, J. (2011). Osteochondral tissue formation through adipose-derived stromal cell differentiation on biomimetic polycaprolactone nanofibrous scaffolds with graded insulin and beta-glycerophosphate concentrations. *Tissue Engineering Part A, 17*(9–10), 1239–1252. https://doi.org/10.1089/ten.tea.2009.0693

Eslahi, N., Abdorahim, M., & Simchi, A. (2016). Smart polymeric hydrogels for cartilage tissue engineering: A review on the chemistry and biological functions. *Biomacromolecules, 17*, 3441–3463. https://doi.org/10.1021/acs.biomac.6b01235

Etkin, C. D., & Springer, B. D. (2017). The American Joint Replacement Registry—The first 5 years. *Arthroplasty Today, 3*(2), 67–69. https://doi.org/10.1016/j.artd.2017.02.002

Goldring, M. B. (2012). Chondrogenesis, chondrocyte differentiation, and articular cartilage metabolism in health and osteoarthritis. *Therapeutic Advances in Musculoskeletal Disease, 4*(4), 269–285. https://doi.org/10.1177/1759720X12448454

Goldring, S. R., & Goldring, M. B. (2016). Changes in the osteochondral unit during osteoarthritis: Structure, function and cartilage bone crosstalk. *Nature Reviews Rheumatology, 12*, 632–644. https://doi.org/10.1038/nrrheum.2016.148

Herneisey, M., Williams, J., Mirtic, J., Liu, L., Potdar, S., Bagia, C., … Janjic, J. M. (2016). Development and characterization of resveratrol nanoemulsions carrying dual-imaging agents. *Therapeutic Delivery, 7*(12), 795–808. https://doi.org/10.4155/tde-2016-0050

Hunziker, E. B., Quinn, T. M., & Häuselmann, H. J. (2002). Quantitative structural organization of normal adult human articular cartilage. *Osteoarthritis and Cartilage, 10*(7), 564–572. https://doi.org/10.1053/joca.2002.0814

Karami, P., Wyss, C. S., Khoushabi, A., Schmocker, A., Broome, M., Moser, C., … Pioletti, D. P. (2018). Composite double-network hydrogels to improve adhesion on biological surfaces. *ACS Applied Materials and Interfaces, 10*(45), 38692–38699. https://doi.org/10.1021/acsami.8b10735

Kwon, H., Brown, W. E., Lee, C. A., Wang, D., Paschos, N., Hu, J. C., & Athanasiou, K. A. (2019). Surgical and tissue engineering strategies for articular cartilage and meniscus repair. *Nature Reviews Rheumatology, 15*, 550. https://doi.org/10.1038/s41584-019-0255-1

Li, J., Chen, G., Xu, X., Abdou, P., Jiang, Q., Shi, D., & Gu, Z. (2019). Advances of injectable hydrogel-based scaffolds for cartilage regeneration. *Regenerative Biomaterials, 6*(3), 129–140. https://doi.org/10.1093/rb/rbz022

Li, T., Chubinskaya, S., Esposito, A., Jin, X., Tagliafierro, L., Loeser, R., … Spagnoli, A. (2019). TGF-β type 2 receptor-mediated modulation of the IL-36 family can be therapeutically targeted in osteoarthritis. *Science Translational Medicine, 11*(491), eaan2585. https://doi.org/10.1126/scitranslmed.aan2585

Li, T., Longobardi, L., Myers, T. J., Temple, J. D., Chandler, R. L., Ozkan, H., … Spagnoli, A. (2012). Joint TGF-β type II receptor-expressing cells: Ontogeny and characterization as joint progenitors. *Stem Cells and Development, 22*(9), 1342–1359. https://doi.org/10.1089/scd.2012.0207

Loeser, R. F., Goldring, S. R., Scanzello, C. R., & Goldring, M. B. (2012). Osteoarthritis: A disease of the joint as an organ. *Arthritis and Rheumatism, 64*, 1697–1707. https://doi.org/10.1002/art.34453

Longobardi, L., Li, T., Myers, T. J., O'Rear, L., Ozkan, H., Li, Y., … Spagnoli, A. (2012). TGF-β type II receptor/MCP-5 axis: At the crossroad between joint and growth plate development. *Developmental Cell, 23*(1), 71–81. https://doi.org/10.1016/j.devcel.2012.05.004

Mahboudi, H., Kazemi, B., Soleimani, M., Hanaee-Ahvaz, H., Ghanbarian, H., Bandehpour, M., … Barati, G. (2018). Enhanced chondrogenesis of human bone marrow mesenchymal Stem Cell (BMSC) on nanofiber-based polyethersulfone (PES) scaffold. *Gene, 643*, 98–106. https://doi.org/10.1016/j.gene.2017.11.073

Manoukian, O. S., Dieck, C., Milne, T., Dealy, C. N., Rudraiah, S., & Kumbar, S. G. (2018). Nanomaterials/nanocomposites for osteochondral tissue. *Advances in Experimental Medicine and Biology, 1058*, 79–95. https://doi.org/10.1007/978-3-319-76711-6_4

Martel-Pelletier, J., Barr, A. J., Cicuttini, F. M., Conaghan, P. G., Cooper, C., Goldring, M. B., … Pelletier, J.-P. (2016). Osteoarthritis. *Nature Reviews. Disease Primers, 13*(2), 16702.

Medvedeva, E. V., Grebenik, E. A., Gornostaeva, S. N., Telpuhov, V. I., Lychagin, A. V., Timashev, P. S., & Chagin, A. S. (2018). Repair of damaged articular cartilage: Current approaches and future directions. *International Journal of Molecular Sciences, 19*, E2366. https://doi.org/10.3390/ijms19082366

Patel, S. K., Beaino, W., Anderson, C. J., & Janjic, J. M. (2015). Theranostic nanoemulsions for macrophage COX-2 inhibition in a murine inflammation model. *Clinical Immunology, 160*(1), 59–70. https://doi.org/10.1016/j.clim.2015.04.019

Ribeiro, V. P., Pina, S., Costa, J. B., Cengiz, I. F., García-Fernández, L., Fernández-Gutiérrez, M. D. M., … Reis, R. L. (2019). Enzymatically cross-linked silk fibroin-based hierarchical scaffolds for osteochondral regeneration. *ACS Applied Materials and Interfaces, 11*(4), 3781–3799. https://doi.org/10.1021/acsami.8b21259

Santo, V. E., Gomes, M. E., Mano, J. F., & Reis, R. L. (2012). From nano-to macro-scale: Nanotechnology approaches for spatially controlled delivery of bioactive factors for bone and cartilage engineering. *Nanomedicine, 7*, 1045–1066. https://doi.org/10.2217/nnm.12.78

Simon, T. M., & Jackson, D. W. (2018). Articular cartilage: Injury pathways and treatment options. *Sports Medicine and Arthroscopy Review, 26*, 146–154. https://doi.org/10.1097/JSA.0000000000000182

Spagnoli, A., O'Rear, L., Chandler, R. L., Granero-Molto, F., Mortlock, D. P., Gorska, A. E., … Moses, H. L. (2007). TGF-β signaling is essential for joint morphogenesis. *Journal of Cell Biology, 177*(6), 1105–1117. https://doi.org/10.1083/jcb.200611031

Vinatier, C., & Guicheux, J. (2016). Cartilage tissue engineering: From biomaterials and stem cells to osteoarthritis treatments. *Annals of Physical and Rehabilitation Medicine, 59*, 139–144. https://doi.org/10.1016/j.rehab.2016.03.002

Wang, S., Wei, X., Sun, X., Chen, C., Zhou, J., Zhang, G., … Wei, L. (2018). A novel therapeutic strategy for cartilage diseases based on lipid nanoparticle-RNAi delivery system. *International Journal of Nanomedicine, 13*, 617–631. https://doi.org/10.2147/IJN.S142797

Yaylaci, S. U., Sen, M., Bulut, O., Arslan, E., Guler, M. O., & Tekinay, A. B. (2016). Chondrogenic differentiation of mesenchymal stem cells on glycosaminoglycan-mimetic peptide nanofibers. *ACS Biomaterials Science & Engineering, 2*(5), 871–878. https://doi.org/10.1021/acsbiomaterials.6b00099

Yu, F., Li, M., Yuan, Z., Rao, F., Fang, X., Jiang, B., … Zhang, P. (2018). Mechanism research on a bioactive resveratrol–PLA–gelatin porous nano-scaffold in promoting the repair of cartilage defect. *International Journal of Nanomedicine, 13*, 7845–7858. https://doi.org/10.2147/IJN.S181855

Zhang, L., Hu, J., & Athanasiou, K. A. (2012). The role of tissue engineering in articular cartilage repair and regeneration. *Critical Reviews in Biomedical Engineering, 37*(1–2), 1–57. https://doi.org/10.1615/critrevbiomedeng.v37.i1-2.10

Zhou, Y., Chyu, J., & Zumwalt, M. (2018). Recent progress of fabrication of cell scaffold by electrospinning technique for articular cartilage tissue engineering. *International Journal of Biomaterials, 2018*, 1953636. https://doi.org/10.1155/2018/1953636

Zylińska, B., Silmanowicz, P., Sobczyńska-Rak, A., Jarosz, Ł., & Szponder, T. (2018). Treatment of articular cartilage defects: Focus on tissue engineering. *In Vivo, 32*, 1289–1300. https://doi.org/10.21873/invivo.11379

Chapter 7
Nanotechnology in Dentistry: Past, Present, and Future

Metin Calisir

Abstract Nanotechnology is the production of functional materials and structures ranging from 0.1 to 100 nm using various chemical and physical methods. The fields of application of nanotechnology cover a wide range including machinery production, defense industry, space and aircraft technologies, information and communication systems, energy systems, chemistry, environment, molecular biology, gene engineering, medicine, and dentistry. Nano dentistry may be described as the science and technology of diagnosing, treating and preventing oral and dental diseases, reducing/eliminating pain and improving oral health, by using nanosized materials, tissue engineering and dental nanorobotics. Although many diseases in dentistry can be treated by conventional methods, the new era of nanotechnology in dentistry will bring revolutionary approaches in diagnosis and treatment of dental diseases. Current research in nanodentistry includes preventive, diagnostic, reconstructive, regenerative, restorative, and rehabilitative fields.

7.1 Introduction

Search for better, more efficient, advanced technologies and materials have led to the emergence of "Nanotechnology," which deals with structures between 0.1 and 100 nm in size. The term nanotechnology comes from "nanos" which means "dwarf" in Greek. The physicist Richard P. Feynman, who won the Nobel Prize for his pioneering work in nanotechnology field first used the term in his speech to the American Physical Society in 1959 (Feynman, 1960). Nanotechnology (Taniguchi, 1974), was defined by Norio Taniguchi in 1974 to describe semiconductor processes that involved control at nanometer scale. Later, a book by Eric Drexler titled "Engines of Creation: The Coming Era of Nanotechnology" helped publicize the concept of nanotechnology (Drexler, 1986). The first applications in the field of

M. Calisir (✉)
Department of Periodontology, Faculty of Dentistry, Adiyaman University, Adiyaman, Turkey
e-mail: metincalisir@adiyaman.edu.tr

A. B. Tekinay (ed.), *Nanomaterials for Regenerative Medicine*, Stem Cell Biology and Regenerative Medicine, https://doi.org/10.1007/978-3-030-31202-2_7

nanotechnology began with the discovery of scanning tunneling microscopes, carbon nanotubes, and fullerenes (Iijima, 1991; Krätschmer, Lamb, Fostiropoulos, & Huffman, 1990; Kroto, Heath, O'Brien, Curl, & Smalley, 1985).

Ongoing research in the field of nanotechnology utilizes the unique properties of nanoparticles. Nanoparticles are more reactive than micro and macro particles due to their larger surface area compared to their mass (Binns, 2010; Li et al., 2008). Therefore, nanotechnology has brought a whole new dimension in fields from medicine to engineering. One of these areas is the dental medicine. The use of nanotechnology in dentistry is known as nanodentistry. The goal of nanodentistry is to provide a perfect oral health and the regeneration of missing oral tissues using nanomaterials, bioengineering methods and nanorobotics (Farr, 1997; Shi, Tsai, Garrison, Ferrari, & Ratner, 1999; Slavkin, 1999; West & Halas, 2000). In addition, the mechanical and physical properties of the dental materials are improved through the use of nanomaterials and new diagnostic and treatment methods are developed to enable the use of nano-delivery systems (Kanaparthy & Kanaparthy, 2011).

7.2 Diagnostic Dentistry

The dental biofilm layer is considered to be the main cause of many diseases related to the teeth and the periodontium. Specific pathogenic microorganisms within the biofilm layer have been associated with the development of periodontal diseases induced by the dental plaque. During treatment of dental problems, identification and destruction of this biofilm layer is very important (Carpio, Santos, Wei, & Rodrigues, 2012; Chalmers et al., 2007; Hu et al., 2010; Kulshrestha, Khan, Meena, Singh, & Khan, 2014; Son, Hong, & Lee, 2016). For this purpose, different applications have been developed in the field of nanodentistry.

7.2.1 Nanosized Quantum Dots

Quantum dots provide enhanced stability and flexibility for immunofluorescence-based analyses that provide single cell resolution for in vivo and in vitro identification of specific periodontal pathogens. Thus, use of such tools is suggested to aid in the removal of these pathogens (Chalmers et al., 2007).

7.2.2 Graphene

The fact that electrons move very rapidly in graphene allows them to destroy a variety of bacterial biofilm layers and form an antibacterial surface (Carpio et al., 2012; Hu et al., 2010). It was shown that graphene/zinc oxide nanocomposites (GZNC) inhibited biofilm formation on the dental implant surface (Kulshrestha et al., 2014).

7.2.3 Atomic Force Microscopy (AFM)

AFM is the leading nanotechnology technique used in the analysis of cells and biofilm surfaces. In dentistry, AFM was used to investigate nanometer scale topographic changes in the oral biofilm caused by *Streptococcus mutans* after various mouthwashes (Sharma et al., 2010).

7.2.4 Other Nano Agents for Diagnosis

For the early diagnosis and treatment of benign, malignant and autoimmune diseases of the oral cavity, it is important that antibodies, dysplastic cells and salivary markers are detected very precisely. There is ongoing research to measure the saliva markers at lower costs (Gau & Wong, 2007; Hasanzadeh & Shadjou, 2016; Son et al., 2016). Carbon nanotubes, for example, have been shown to specifically bind to cancer cells and were proposed to be an effective and useful diagnostic tool (Son et al., 2016).

The Oral Fluid Nano Sensor Test (OFNASET, The Wong Lab, University of California, Los Angeles) is a highly sensitive, specific, portable, and automated nanoelectromechanical system that facilitates the identification of proteomic biomarkers and nucleic acids that are specific for oral cancer species (Gau & Wong, 2007).

7.3 Preventive Dentistry

The higher restorative capacity of nanomaterials compared to the same material on the micro or macro scale may be attributed to the size of the inorganic building blocks in the enamel which are 20–40 nm in size (Tao, Pan, Zeng, Xu, & Tang, 2007). Small size and higher active surface area are important features to be considered when developing new materials.

7.3.1 Dentifrices

Since nanosized calcium carbonate particles and hydroxyapatite crystals resemble the crystal structure and morphology of the enamel, it has been reported that the presence of these substances in toothpastes may help remineralization of tooth enamels (Nakashima, Yoshie, Sano, & Bahar, 2009; Vandiver, Dean, Patel, Bonfield, & Ortiz, 2005). Similarly, the *nano*-carbonate apatite (*n*-CAP) particles also resemble the inorganic parts of the teeth and may exert a short-term desensitization effect when added into toothpastes (Lee, Kwon, & Kim, 2008). Inclusion of 3% nanosized sodium trimetaphosphate in toothpastes has been reported to result in a higher increase in the remineralization in early stages of caries formation compared to the traditional toothpastes (Danelon, Pessan, Neto, de Camargo, & Delbem, 2015).

7.3.2 *Toothbrush Development*

A nano-toothbrush concept has been developed by adding colloidal particles embedding nanogold or nanosilver to toothbrush bristles. It was suggested that microbial plaque or the biofilm layer could be removed more easily by the silver particles present in these structures (Raval, Vyas, Gandhi, Patel, & Patel, 2016).

7.3.3 *Mouthwashes*

Mouthwashes containing triclosan and silver nanoparticles have been shown to be effective in plaque control. Colloidal suspensions of triclosan nanoparticles form a controlled release system due to their bio-adhesive properties. Therefore, these nanoparticles can be incorporated into gels, toothpastes and mouthwashes in the treatment and prevention of periodontal diseases (Pragati, Ashok, & Kuldeep, 2009). In addition, mouthwashes containing biomimetic carbonate-hydroxyapatite nanocrystals have also been shown to reduce plaque build-up and help prevent inflammation in peri implant tissues (Lelli et al., 2013). In other another study, nano-calcium fluoride has been included in mouthwash products to reduce caries, reduce dentin permeability and increase the concentration of fluoride concentration in oral fluids (Sun & Chow, 2008).

7.3.4 *Other Nanomaterials Used in Preventive Dentistry*

Nanosized hydroxyapatite particles may easily integrate into dentin tubules, and protect the them from external stimuli. The use of these nanoparticles has been reported to be effective in reducing hypersensitivity in teeth (Ebadifar, Nomani, & Fatemi, 2017; Kulal, Jayanti, Sambashivaiah, & Bilchodmath, 2016). In addition to the desensitizing effect, they hold firmly on the tooth enamel and protect the teeth against erosion formed by acidic food and beverages (Li et al., 2008).

Modified silica nanoparticles have also been used to reduce tooth sensitivity. These nanoparticles show their desensitizing effects by reducing the hypersensitivity of dentin tubules by blocking them (Petrou et al., 2009).

7.4 Restorative Dentistry

Restorative dentistry is the branch of dentistry that aims to replace damaged or missing teeth by repairing the damaged teeth by fillings or crowns and replacing the missing teeth with bridges and implants. Nanotechnology research has provided invaluable insight on efforts to generate biofriendly and more durable materials. Below are a few examples of how nanotechnology has been utilized for generating novel treatment options for the benefit of restorative dentistry.

7.4.1 Nanofillers

With the addition of nanofiller particles to composite resins, a new class of materials has emerged with improved properties compared to micro and macro-filled composites. These nanoparticles reduce shrinkage caused by polymerization and thermal expansion and increase the polishing ability, stiffness and wear resistance of composites. One of them, silica-based nanoparticles, are used as filling materials in dentistry (Chen, 2010). In addition, using silica-based nanoparticles as polishing material has also been shown to protect the enamel surface and prevent caries caused by cariogenic bacteria (Roulet & Roulet-Mehrens, 1982).

7.4.2 Nanocomposites

Nanocomposites offer many advantages over conventional microfilled and hybrid resin based composite systems. Nano composites containing alumina silicate powder are more rigid, have superior bending strength, better elastic modulus, semi-transparency, and superior handling properties and result in a 50% reduction in filling shrinkage (de Andrade et al., 2011). In another study, surface modifications by organosilanes of TiO_2 nanoparticles in a resin matrix have been shown to increase the microhardness and flexural strength of dental resin-based composites (Xia, Zhang, Xie, & Gu, 2008). Adhesives formed by Radiopaque Ta_2O_5/SiO_2 filler nanoparticles dispersed in the methacrylic matrix have also been found to have higher adhesion strength and higher radiopacity than enamel and dentin (Schulz, Schimmoeller, Pratsinis, Salz, & Bock, 2008).

A novel restorative material has been developed by combining photopolymerized dendritic copolymers and particulate composite filler materials (Viljanen, Skrifvars, & Vallittu, 2007). Furthermore, by incorporating polymerizable 2,3-dihydroxybenzyl ether dendrimers or dendrons into dental composite resins, new dental materials with improved physical properties have also been produced (Paul, Bader, Schricker, & Parquette, 2006).

Chen, Clarkson, Sun, and Mansfield (2005) have prepared self-assembling hydroxyapatite nanorods with a similar morphology to the crystal structure of the enamel. These nanorods may provide a model of the natural enamel of the tooth. Similarly, Sadat-Shojai, Atai, Nodehi, and Khanlar (2010) reported that the hydroxyapatite nanorods could be an alternative to fillers such as silicates that are currently used in dental adhesives.

Alves, Pilla, Murgo, and Munin (2010) reported that dental resins soaked with different concentrations of CdSe/ZnS core–shell quantum dots showed fluorescence characteristics similar to the human teeth. In another study, carbon nanotubes have been reported to be promising candidates for use as tooth fillers due to their excellent electrical and mechanical properties (Ajayan & Zhou, 2001).

Nano light-curing glass ionomer is a system that combines nanotechnology with fluoro alumino silicate technology, and this system shows better aesthetics, advanced wear resistance and superior polishing properties (Korkmaz, Ozel, Attar, & Bicer, 2010). Addition of nanosized filler materials to resin-modified glass ionomer cement (RMGIC) was also shown to increase polishing and aesthetic properties with the addition of nanosized fillers (Vaikuntam, 1997). In addition to abovementioned materials, solutions containing nanomaterials are already used in next generation bonding agents to ensure homogeneity and a perfect blend every time (Sadat-Shojai et al., 2010).

Silver nanoparticles (AgNPs) have strong antibacterial properties, in addition to their high biocompatibility, low toxicity, low bacterial resistance, and long-term antibacterial activity; thus, they have been used in dental prosthesis and implants as restorative materials in endodontics (Damm, Münstedt, & Rösch, 2007; García-Contreras et al., 2011; Jia, Hou, Wei, Xu, & Liu, 2008; Nam, 2011; Percival, Bowler, & Russell, 2005; Samiei et al., 2013; Sheikh et al., 2010; Slenters, Hauser-Gerspach, Daniels, & Fromm, 2008). A nanocomposite coating consisting of silver nanoparticles and lactose-modified chitosan (Chitlac) has been reported to significantly reduce biofilm formation on the restoration surface 48 h after administration compared to conventional composites (Ionescu et al., 2015).

Dental resin nanocomposites containing calcium fluoride (CaF_2) nanoparticles were shown to release fluoride ions, and they were suggested to be a superior restorative material by reducing the formation of secondary caries and fracture formation in restorations (Xu, Moreau, Sun, & Chow, 2008). On the other hand, composite resins containing nano-amorphous calcium phosphate (nACP) have been shown to improve the remineralization process and control the release of calcium and phosphorus through charge-and-release. It has been reported that the nACP also prevents the formation of secondary caries (Wu et al., 2015; Xie et al., 2016). Similar results were also observed by adding nano-amorphous calcium phosphate to the bonding agents (Zhang, Weir, Hack, Fouad, & Xu, 2015).

Cross-linked quaternized polyethyleneimine (QPEI) nanoparticles incorporated into resin composites have been reported to exhibit antibacterial activity against various oral pathogens such as *Enterococcus faecalis*, *Streptococcus mutans*, *Actinomyces viscosus*, and *Lactobacilus casei*. It was reported that this long-lasting antibacterial effect could prevent secondary caries (Shvero, Zatlsman, Hazan, Weiss, & Beyth, 2015).

Use of calcium peroxide nanoparticles as tooth whitening resulted in a deeper penetration into the tissues, while minimizing the detrimental effects of tooth whitening. In one study, nano-lipobelle H-EQ10 was added to the bleaching agents and their efficacy was investigated. Nano-lipobelle H-EQ10 is liposomes containing 10% vitamin E and 5% coenzyme. By using these nano-liposome carriers, the chemical and mechanical properties of the bleached enamel were significantly improved. This effect may be related to the restorative effects of vitamin E and coenzymes in the preservation of tooth structure and stimulation of cell renewal (AlKahtani, 2018).

7.5 Regenerative Dentistry

7.5.1 Bone Regeneration

Recently, there have been an increasing number of studies on nanoparticle composites and nanofiber scaffolds that support cell growth and differentiation by increasing the mechanical support. In addition, nanodelivery systems that can enhance the production of osteogenic growth factors are also being studied (Kim & Fisher, 2007).

Different alloplastic bone grafts have been developed by using nanoparticles. The most widely investigated one among these grafts are the *nano*-HAP (*n*-HAP) bone grafts available in crystalline, chitosan-associated, and titanium-reinforced forms. Hydroxyapatite nanoparticles are frequently used for bone regeneration and are highly biocompatible (Chesnutt et al., 2009; Chesnutt, Yuan, Buddington, Haggard, & Bumgardner, 2009; Kailasanathan, Selvakumar, & Naidu, 2012; Reves, Jennings, Bumgardner, & Haggard, 2012; Singh, Nayak, Uppoor, & Shah, 2012). It was reported that nanocrystalline and hydroxyapatite mixture bone grafts have low cytotoxicity and high biocompatibility (Chitsazi, Shirmohammadi, Faramarzie, Pourabbas, & Rostamzadeh, 2011; Liu et al., 2011). These properties have been reported to cause better clinical results in the treatment of intra-bone defects (Kasaj, Röhrig, Zafiropoulos, & Willershausen, 2008).

Although hydroxyapatite has been the most widely studied material for regenerative purpose in dentistry, other nanomaterials have also been investigated. For example, "nano-bioactive glass" developed for bone regeneration has been found to be biocompatible with gingival fibroblasts (Tavakoli et al., 2012). Calcium sulfate ($CaSO_4$) bone grafts with nanosized crystals were also developed for regenerative dentistry. In addition to being a biodegradable and osteoconductive bone-like structure, while calcium sulfate is dissolved, calcium phosphate is formed, which assists the binding of osteoblasts and formation of new bone. The nanotization of the particles makes the bone graft more resistant to degradation and resulting graft materials that endure longer (12–14 weeks) than conventional $CaSO_4$ (4–6 weeks). This slower biodegradation rate matches the duration of bone development in intra-bone defects, thus providing better treatment (Kelly et al., 2001; Pandit et al., 2015).

A graft material that is cured with light has been reported as a bone grafting agent. This material is produced by combining the methacrylate resin matrix with nACP and is injectable and has the ability to strongly adhere to the wet bone and crystallize nACP to hydroxyapatite within a few minutes (Schneider et al., 2010).

Recently, an antibacterial nanoceramic composite material has been developed by impregnation of calcium phosphate and zinc oxide (ZnO) nanoparticles surrounded by carbon nanotubes into an alginate polymer matrix. While the ZnO nanoparticles provide antibacterial effect, the carbon nanotubes act as a strong, flexible and inert skeleton in which new bone cells can gather and proliferate. This system also increases HAP formation in bone defects (Beherei, El-Magharby, & Abdel-Aal, 2011).

7.5.2 *Soft Tissue Regeneration*

Different types of nanomaterials have been developed for tissue regeneration in dental tissues. The health of the soft tissues around the teeth are especially important for oral health, thus most of the tissue regeneration efforts have focused on these tissues. Membranes that are used for guided tissue regeneration are among these materials. These membranes consist of a double layer structure with a porous surface on one side (for cellular ingrowth) and a smooth surface on the other side (for cellular occlusion). A new three-layer guided tissue regeneration membrane system has been developed to further improve tissue regeneration. This new system has an inner layer made of a porous membrane composed of 8% nanocarbonate hydroxyapatite/collagen/polylactic-coglycolic acid (nCHAC/PLGA), a middle layer composed of layer comprising 4% nCHAC/PLGA, and a nonporous PLGA outer layer. This composite membrane is highly flexible, biocompatible, osteoconductive, and biodegradable and induces higher activity by osteoblasts than a pure PLGA membrane (Liao et al., 2005).

A platform that combines tissue engineering and local gene delivery systems, the gene-activated matrix, acts as a structural template for therapeutic gene expression, eliminating problems in cell adhesion and proliferation as well as the synthesis of the extracellular matrix. The latest development in this field is a gene-activated matrix consisting of a chitosan/collagen scaffold that acts as a three-dimensional carrier which is loaded with chitosan/plasmid nanoparticles containing the cDNA that codes for the platelet-derived growth factor. An increase in periodontal ligament proliferation was observed with this system (Peng et al., 2009).

Nanoparticles may be used to reconstruct damaged nerves by self-assembled nanofibers composed of amphiphilic monomers. This application may have a great potential in the field of dentistry, such as reconstruction of the nerve after inferior alveolar nerve damage which may occur during oral surgery (Ellis-Behnke et al., 2006).

In addition to growth factors, hormones have also been utilized for tissue regeneration purposes. An example is the α-melanocyte-stimulating hormone (α-MSH) which is a hormone with anti-inflammatory properties. It has been suggested that α-MSH-containing nanofilms can help revitalize teeth, but more studies are needed (Fioretti et al., 2010).

7.6 Rehabilitative Dentistry

7.6.1 *Prosthetic Dentistry*

Nanomaterials can be used for various applications in prosthetic dentistry. One of these applications is to develop better impression materials. An example to better impression materials are vinylpolysiloxanes with nanofillers that are more fluid, more hydrophilic, and have more precise measurement properties compared to traditional materials (Jhaveri & Balaji, 2005).

Another application is to develop better materials for prosthetic tooth production. Nano-composite prosthetic teeth which consist of polymethylmethacrylate (PMMA) and homogeneously dispersed nano-sized filler particles, have well-polished surfaces and are resistant to discoloration and mechanical abrasion (Totu, Nechifor, Nechifor, Aboul-Enein, & Cristache, 2017). Nanomodified zirconium oxide particles, on the other hand, have been shown to provide additional transverse strength not only for the mobile prosthesis but also for all prosthetic procedures (Gad, ArRejaie, Abdel-Halim, & Rahoma, 2016).

Prosthetic cements constitute another application area for nanomaterials in prosthetic dentistry. Nanoparticle impregnated luting cements have been shown to have higher elastic modulus and lower polymerization shrinkage than traditional cements and therefore bind to dentin more strongly (Sadat-Shojai et al., 2010). Zinc-polycarboxylate cement produced using ZnO and MgO nanoparticles has been shown to have enhanced physical and mechanical strength compared to traditional zinc-poly carboxylate cements (Karimi et al., 2011). Similarly, the addition of nano-hydroxyapatite/fluoroapatite particles into the glass ionomer cements has been reported to improve the physical properties of the cement compared to the conventional glass ionomer cements (Moshaverinia et al., 2008). Lava resin nano ceramic blocks have also been shown to have better aesthetics, durability and higher fracture resistance (Chen, Trindade, de Jager, Kleverlaan, & Feilzer, 2014).

7.6.2 *Endodontics*

Endodontics is the branch of dentistry that specializes on the dental pulp and the surrounding tissues which constitute the major living tissues of the teeth. Endodontic treatment is usually known as "root canal" treatment and aims to treat the infections and other problems within the dental pulp. The addition of bio-ceramic nanoparticles such as bioglass, zirconia, and glass ceramics into the endodontic sealers has been shown to result in improvement in material characteristics, such as faster curing time, lower dissolution in body fluids, and better adaptation to dental tissue compared to traditional endodontic sealers (Utneja, Nawal, Talwar, & Verma, 2015). These nanoparticles also accelerate the remineralization process and increase the adhesion strength to dentin (Wang et al., 2017).

It was also shown that adding silver nanoparticles to the root canal fillers provides an effective antibacterial protection against *Enterococcus faecalis*, especially during early stages after treatment, compared to canal fillers containing only calcium hydroxide or containing calcium hydroxide mixed with chlorhexidine (Afkhami, Pourhashemi, Sadegh, Salehi, & Fard, 2015).

In a study comparing the gutta percha containing nano-diamond particles with the traditional gutta percha, the gutta percha containing nano-diamond showed superior chemical and mechanical properties, higher biocompatibility, better adaptation to root canal walls, and a lower gap in canal filling occurrence compared to the traditional gutta percha (Lee et al., 2015).

7.6.3 Dental Implants

Dental implants have become popular recently in rehabilitative dental treatments. In order to have better treatment efficiency, various implant surface modifications have been tested to improve osseointegration. Surface properties are one of the most important factors that determine the biocompatibility and bio-integration of implants by regulating their surface energies, composition, roughness, and topography. The surfaces significantly affect the chemical and physical surface properties of the nanomaterials (Nayar, Bhuminathan, & Muthuvignesh, 2011; Simon & Watson, 2002).

Nanoporous anodic alumina, porous silica, and titanium nanotubes have been shown to play significant roles in the development of drug-releasing implants (Lee, Alhoshan, & Smyrl, 2006; Paulose et al., 2006). In vitro titanium oxide nanotubes have been demonstrated to accelerate the kinetics of hydroxyapatite formation, thus helping the maturation of bone cells around dental implants (Oh, Finones, Daraio, Chen, & Jin, 2005).

The surface morphology of the implants has also been shown to significantly alter the efficacy of implant treatment. For example, it has been reported that the presence of 50 nm nanodots on the surface of the implants increase the number of osteoblast cells accumulated on these surfaces (Pan et al., 2012). Nanostructured hydroxyapatite coatings on the implants have also been shown to enhance osteointegration of the implants. Biocompatibility studies showed that hydroxyapatite has very high binding ability to bone. It can be used to increase osteointegration of titanium implants to the alveolar bone due to their chemical bond forming properties. In addition, such implants have been shown to increase osteoblast adhesion, proliferation and mineralization (Ma, Wong, Kong, & Peng, 2003; Pepla, Besharat, Palaia, Tenore, & Migliau, 2014). Nanosized carbon fibers have also been shown to cause a significant increase in osteoblast adhesion required for successful orthopedic/dental implant applications because of their high surface area (Price, Ellison, Haberstroh, & Webster, 2004). In addition, modification of the implant surface with mechanical nano-properties such as nano-grooves or nano-pillars allows the fibrin clots acting as a bridge for osteogenic cells to adhere to the surface of the implant and thus provide a good osseointegration between the implant and bone (Tomsia et al., 2011).

The infections that are associated with implants have been another major problem for the utilization of the implant treatment. Due to the highly humid and bacteria friendly environment of the oral cavity, bacterial infections constitute a major problem for most dental treatment options. The bacterial biofilm layer formed on dental implants is difficult to remove with antibiotics. The protective layer of the exopolymers protects the microorganisms against antibiotics and immune response cells. This leads to inflammation in the peri-implant tissues. The use of iron oxide nanoparticles was shown to be effective for the elimination of this biofilm layer (Sathyanarayanan, Balachandranath, Genji Srinivasulu, Kannaiyan, & Subbiahdoss, 2013). Another modification agent, zirconia is a water-insoluble material; and it can reduce bacterial attachment. Zirconia has low cytotoxicity and is a biologically inert material, which makes ZrO_2 a suitable biomaterial for dental implants. In addition, it has high fracture resistance and is osseo-conductive, that is, it stimulates bone formation. Implants coated with zirconia nanoparticles on the surface are reported to be highly efficient in osseointegration (Akagawa, Hosokawa, Sato, & Kamayama, 1998; Josset et al., 1999; Lughi & Sergo, 2010). Several studies have shown that silver nitrate charged nanotitania surfaces and silver nanoparticle-modified titanium (Ti-nAg) surfaces also have antimicrobial properties. Silver nanoparticles have a broad antibacterial spectrum and low microbial resistance. Although the antibacterial mechanism of silver nitrate nanoparticles is not clear, clinical use has been shown to reduce the risk of peri-implantitis (Feng, Cao, Li, Liu, & Dong, 2008; Juan, Zhimin, Anchun, Lei, & Jingchao, 2010).

7.7 Periodontology

Periodontology is the branch of dentistry that specializes in prevention and treatment of periodontal diseases and oral inflammation and implant treatment. Since the use of nanotechnology for dental implants has been explained in previous sections, here, other periodontal applications of nanomaterials and nanotechnology will be discussed. Laser plasma application is a technique that utilizes lasers at different wavelengths and can be used for the treatment of periodontal problems. When nanosized TiO_2 particles are applied to human skin in a gel-like form and irradiated with laser, they cause micro-wear of hard tissues and stimulate collagen production. They are currently clinically used in periodontal treatment, depigmentation and soft tissue incision without anesthesia (Freitas Jr., 2005). Zirconium oxide nanoparticles also attract much attention for dental applications. Due to their anti-biofilm activity against some bacteria, zirconium oxide nanoparticles have been suggested to be used as a polishing agent in dentistry (Guerreiro-Tanomaru et al., 2014).

Another area where nanotechnology has been utilized for medical applications is the development of lab-on-a-chip (LOC) systems which enable the use of smaller sizes of biological samples to gather information by using different biomarkers. In periodontology, LOC systems are used to determine the salivary levels of biomarkers such as interleukin-1beta (IL-1β), C-reactive protein (CRP), and Matrix Metallo-Proteinase-8 (MMP-8), which play a role in the pathogenesis of periodontal disease (Christodoulides et al., 2007).

Nanomaterials can provide efficient solutions for specific drug delivery systems with controlled release and biodegradation. These systems are based on slow release of certain drugs loaded into a biologically degradable matrix to have a longer-lasting effect in the desired region (Kong, Peng, Li, & Bartold, 2006). The disadvantage of conventional drug delivery systems is that drugs can stay in the oral cavity for a short time due to saliva, irregular swallowing, and food and drink intake and soft tissue movements. The use of liposomes as a dental drug delivery system is envisaged as a new approach to overcome this problem (Nguyen, Hiorth, Rykke, & Smistad, 2011). In addition, triclosan-laden nanoparticles have been shown to be effective in inhibiting periodontal inflammation (Piñón-Segundo, Ganem-Quintanar, Alonso-Pérez, & Quintanar-Guerrero, 2005). It has also been shown that tetracyclines added to the microspheres may help reduce inflammation when administered in combination with regular methods during treatment of periodontal disease (Kong et al., 2006). For this purpose, niosomes and fullerenes were used. Niosomes are chemically stable, non-ionic vesicles, and fullerenes are hollow carbon molecules of different forms (spheres, tubes, and ellipsoids) (Kazi et al., 2010).

The researchers at South West Research Institute developed nanocapsules containing new vaccines, antibiotics and drug delivery systems with reduced side effects and targeted release. In the future, specialized nanoparticles may be designed to target oral tissues, including cells derived from periodontium (Kanaparthy & Kanaparthy, 2011).

Especially after surgical treatments, wound healing in the oral cavity can be problematic due to the bacteria friendly humid and warm environment and the wear-and-tear that is endured by the oral tissues during eating and drinking. Thus, enhanced wound healing rate and protection against bacterial infection during healing are desirable qualities in materials that are developed for oral wound healing. In order to provide hemostasis, antimicrobial wound dressings including biodegradable nanofibers and nano-crystalline silver particles were developed for wound healing applications in the oral cavity (Saravana & Vijayalakshmi, 2006). In addition, with the development of suture needles that contain nano-sized stainless-steel crystals, micro-surgical procedures have significantly gained speed (Saravana & Vijayalakshmi, 2006). A fine surgical procedure can be performed on a live cell by using a nano-sized needle. With these nanosized stainless steel needles called nanoneedles, higher resolution surgery in the root canals may be possible in the near future (Dunphy Guzman, Taylor, & Banfield, 2006).

7.8 Nanorobotics

As for many other medical applications, nanorobotics systems have been studied for dental medicine as well. One vision is that if the first micro-size dental nanorobots can be produced, they can be controlled by a built-in nanocomputer that executes preprogrammed instructions in response to local sensor stimuli in detecting and treating problems in dental tissues. In addition, dentists will be able to deliver the desired applications directly to in vivo nanorobots by acoustic signals or other means (Freitas Jr., 2005). Although it does not seem likely in the near future, dentists also envision nanorobotic toothpaste called dentifrobots, which are carried by mouthwash or toothpaste at least once a day, may detect and destroy pathogenic bacteria in the plaque and oral cavity and also allow the development of harmless or beneficial oral microflora species. This will help eliminate bad breath (Gambhir et al., 2013; Satyanarayana & Rai, 2011). Dental nanorobots using indigenous biological materials can selectively clog dentine tubules, which cause dentin sensitivity, selectively and precisely, and provide patients with a fast, relaxing and lasting treatment (Cummins, 2009). Another future envision for dental treatments is the use of orthodontic nanorobots that can affect the periodontal tissues, allowing them to quickly and painlessly correct, rotate and vertically reposition teeth in minutes or hours (Kohli & Martin, 2003).

In near future, however, nanorobotic analgesics could be used in the dental offices and cause reduced anxiety because of lack of use of dental injectors, greater selectivity and controllability of the analgesic effect (effective anesthesia only in the tooth/teeth to be treated), rapid and fully reversible effect and the ability to avoid many side effects and complications (Estafan, 1998; Freitas Jr., 2005; Jhaveri & Balaji, 2005).

7.9 Conclusion and Future Perspectives

Nanoparticles are widely studied due to their superior properties and have become the focus of attention for researchers in recent years due to their advantages. Nanotechnology will enable many new developments and novel approaches in the field of dentistry, as well as the advancement of oral diagnosis and treatment methods. However, despite the considerable increase in the number of studies in this area, there are difficulties that prevent the clinical applications of these newly developed nanomaterials. Long-term antimicrobial, toxic, physical, chemical and clinical effects of the nanoparticles on already clinically used dental biomaterials and living tissues are important issues to be investigated in further studies. Since materials may acquire new physical and chemical properties at the 1–100 nm size, we have to be cautious when dealing with nanoparticles. Recent evidence suggests that some of the nanoparticles are able to pass through the blood–brain barrier, which might cause several side effects for the use of nanomaterials in dental tissues in case the nanomaterials pass to the blood stream. Therefore, future investigations will probably be directed to engineer nanomaterials to maximize their positive effects while minimizing nanomaterial associated toxicity. Although the use of nanotechnology in dentistry today is limited to a small number of products available on the market, new studies show that nanotechnology will become an indispensable part of oral health in the future.

References

Afkhami, F., Pourhashemi, S. J., Sadegh, M., Salehi, Y., & Fard, M. J. (2015). Antibiofilm efficacy of silver nanoparticles as a vehicle for calcium hydroxide medicament against Enterococcus faecalis. *Journal of Dentistry, 43*(12), 1573–1579.

Ajayan, P. M., & Zhou, O. Z. (2001). Applications of carbon nanotubes. In M. S. Dresselhaus, G. Dresselhaus, & P. Avouris (Eds.), *Carbon nanotubes. Topics in applied physics* (Vol. 80, pp. 391–425). Berlin: Springer.

Akagawa, Y., Hosokawa, R., Sato, Y., & Kamayama, K. (1998). Comparison between freestanding and tooth-connected partially stabilized zirconia implants after two years' function in monkeys: A clinical and histologic study. *The Journal of Prosthetic Dentistry, 80*(5), 551–558.

AlKahtani, R. N. (2018). The implications and applications of nanotechnology in dentistry: A review. *The Saudi Dental Journal, 30*(2), 107–116.

Alves, L. P., Pilla, V., Murgo, D. O., & Munin, E. (2010). Core-shell quantum dots tailor the fluorescence of dental resin composites. *Journal of Dentistry, 38*(2), 149–152.

Beherei, H. H., El-Magharby, A., & Abdel-Aal, M. S. (2011). Preparation and characterization of novel antibacterial nanoceramic-composites for bone grafting. *Der Pharma Chemica, 3*(6), 10–27.

Binns, C. (2010). *Introduction to nanoscience and nanotechnology*. New York, NY: Wiley & Sons.

Carpio, I. E. M., Santos, C. M., Wei, X., & Rodrigues, D. F. (2012). Toxicity of a polymer–graphene oxide composite against bacterial planktonic cells, biofilms, and mammalian cells. *Nanoscale, 4*(15), 4746–4756.

Chalmers, N. I., Palmer, R. J., Jr., Du-Thumm, L., Sullivan, R., Shi, W., & Kolenbrander, P. E. (2007). Use of quantum dot luminescent probes to achieve single-cell resolution of human oral bacteria in biofilms. *Applied and Environmental Microbiology, 73*(2), 630–636.

Chen, C., Trindade, F. Z., de Jager, N., Kleverlaan, C. J., & Feilzer, A. J. (2014). The fracture resistance of a CAD/CAM Resin Nano Ceramic (RNC) and a CAD ceramic at different thicknesses. *Dental Materials, 30*(9), 954–962.

Chen, H., Clarkson, B. H., Sun, K., & Mansfield, J. F. (2005). Self-assembly of synthetic hydroxyapatite nanorods into an enamel prism-like structure. *Journal of Colloid and Interface Science, 288*(1), 97–103.

Chen, M. H. (2010). Update on dental nanocomposites. *Journal of Dental Research, 89*(6), 549–560.

Chesnutt, B. M., Viano, A. M., Yuan, Y., Yang, Y., Guda, T., Appleford, M. R., … Bumgardner, J. D. (2009). Design and characterization of a novel chitosan/nanocrystalline calcium phosphate composite scaffold for bone regeneration. *Journal of Biomedical Materials Research. Part A, 88*(2), 491–502.

Chesnutt, B. M., Yuan, Y., Buddington, K., Haggard, W. O., & Bumgardner, J. D. (2009). Composite chitosan/nano-hydroxyapatite scaffolds induce osteocalcin production by osteoblasts in vitro and support bone formation in vivo. *Tissue Engineering. Part A, 15*(9), 2571–2579.

Chitsazi, M. T., Shirmohammadi, A., Faramarzie, M., Pourabbas, R., & Rostamzadeh, A. (2011). A clinical comparison of nano-crystalline hydroxyapatite (Ostim) and autogenous bone graft in the treatment of periodontal intrabony defects. *Medicina Oral, Patología Oral y Cirugía Bucal, 16*(3), 448–453.

Christodoulides, N., Floriano, P. N., Miller, C. S., Ebersole, J. L., Mohanty, S., Dharshan, P., ... McDevitt, J. T. (2007). Lab-on-a-chip methods for point-of-care measurements of salivary biomarkers of periodontitis. *Annals of the New York Academy of Sciences, 1098*, 411–428.

Cummins, D. (2009). Dentin hypersensitivity: From diagnosis to a breakthrough for everyday sensitivity relief. *The Journal of Clinical Dentistry, 20*(1), 1–9.

Damm, C., Münstedt, H., & Rösch, A. (2007). Long-term antimicrobial polyamide 6/silver-nanocomposites. *Journal of Materials Science, 42*(15), 6067–6073.

Danelon, M., Pessan, J. P., Neto, F. N., de Camargo, E. R., & Delbem, A. C. (2015). Effect of toothpaste with nano-sized trimetaphosphate on dental caries: In situ study. *Journal of Dentistry, 43*(7), 806–813.

de Andrade, A. K., Duarte, R. M., Medeiros e Silva, F. D., Batista, A. U., Lima, K. C., Pontual, M. L., & Montes, M. A. (2011). 30-Month randomised clinical trial to evaluate the clinical performance of a nanofill and a nanohybrid composite. *Journal of Dentistry, 39*(1), 8–15.

Drexler, K. E. (1986). *Engines of creation: The coming era of nanotechnology* (pp. 99–129). New York, NY: Anchor Press.

Dunphy Guzman, K. A., Taylor, M. R., & Banfield, J. F. (2006). Environmental risks of nanotechnology: National nanotechnology initiative funding, 2000-2004. *Environmental Science & Technology, 40*(5), 1401–1407.

Ebadifar, A., Nomani, M., & Fatemi, S. A. (2017). Effect of nano-hydroxyapatite toothpaste on microhardness ofartificial carious lesions created on extracted teeth. *Journal of Dental Research, Dental Clinics, Dental Prospects, 11*(1), 14–17.

Ellis-Behnke, R. G., Liang, Y. X., You, S. W., Tay, D. K., Zhang, S., So, K. F., & Schneider, G. E. (2006). Nano neuro knitting: Peptide nanofiber scaffold for brain repair and axon regeneration with functional return of vision. *Proceedings of the National Academy of Sciences of the United States of America, 103*(13), 5054–5059.

Estafan, D. J. (1998). Invasive and non-invasive dental analgesia techniques. *General Dentistry, 46*(6), 600–603.

Farr, C. (1997). Biotech in periodontics: Molecular engineering yields new therapies. *Dentistry Today, 16*(10), 92,94–92,97.

Feng, Y., Cao, C., Li, B. E., Liu, X. Y., & Dong, Y. Q. (2008). [Primary study on the antibacterial property of silver-loaded nanotitania coatings]. *Zhonghua Yi Xue Za Zhi, 88*(29), 2077–2080.

Feynman, R. P. (1960). There's plenty of room at the bottom. *Engineering and Science, 23*, 22–36.

Fioretti, F., Mendoza-Palomares, C., Helms, M., Al Alam, D., Richert, L., Arntz, Y., ... Benkirane-Jessel, N. (2010). Nanostructured assemblies for dental application. *ACS Nano, 4*(6), 3277–3287.

Freitas, R. A., Jr. (2005). Nanotechnology, nanomedicine andnanosurgery. *International Journal of Surgery, 3*(4), 243–246.

Gad, M., ArRejaie, A. S., Abdel-Halim, M. S., & Rahoma, A. (2016). The reinforcement effect of nano-zirconia on the transverse strength of repaired acrylic denture base. *International Journal of Dentistry, 2016*, 7094056.

Gambhir, R. S., Sogi, G. M., Nirola, A., Brar, R., Sekhon, T., & Kakar, H. (2013). Nanotechnology in dentistry: Current achievements and prospects. *Journal of Orofacial Sciences, 5*(1), 9–14.

García-Contreras, R., Argueta-Figueroa, L., Mejía-Rubalcava, C., Jiménez-Martínez, R., Cuevas-Guajardo, S., Sánchez-Reyna, P. A., & Mendieta-Zeron, H. (2011). Perspectives for the use of silver nanoparticles in dental practice. *International Dental Journal, 61*(6), 297–301.

Gau, V., & Wong, D. (2007). Oral fluid nanosensor test (OFNASET) with advanced electrochemical-based molecular analysis platform. *Annals of the New York Academy of Sciences, 1098*, 401–410.

Guerreiro-Tanomaru, J. M., Trindade-Junior, A., Costa, B. C., da Silva, G. F., Drullis Cifali, L., Basso Bernardi, M. I., & Tanomaru-Filho, M. (2014). Effect of zirconium oxide and zinc oxide nanoparticles on physicochemical properties and antibiofilm activity of a calcium silicate-based material. *Scientific World Journal, 2014*, 975213.

Hasanzadeh, M., & Shadjou, N. (2016). Electrochemical nanobiosensing in whole blood: Recent advances. *TrAC Trends in Analytical Chemistry, 80*, 167–176.

Hu, W., Peng, C., Luo, W., Lv, M., Li, X., Li, D., … Fan, C. (2010). Graphene-based antibacterial paper. *ACS Nano, 4*(7), 4317–4323.
Iijima, S. (1991). Helical microtubules of graphite carbon. *Nature, 354*, 56–58.
Ionescu, A., Brambilla, E., Travan, A., Marsich, E., Donati, I., Gobbi, P., … Breschi, L. (2015). Silver–polysaccharide antimicrobial nanocomposite coating for methacrylic surfaces reduces Streptococcus mutans biofilm formation in vitro. *Journal of Dentistry, 43*(12), 1483–1490.
Jhaveri, H. M., & Balaji, P. R. (2005). Nanotechnology: The future of dentistry. *The Journal of Indian Prosthodontic Society, 5*(1), 15–17.
Jia, H., Hou, W., Wei, L., Xu, B., & Liu, X. (2008). The structures and antibacterial properties of nano-SiO 2 supported silver/zinc–silver materials. *Dental Materials, 24*(2), 244–249.
Josset, Y., Oum'Hamed, Z., Zarrinpour, A., Lorenzato, M., Adnet, J. J., & Laurent-Maquin, D. (1999). In vitro reactions of human osteoblasts in culture with zirconia and alumina ceramics. *Journal of Biomedical Materials Research, 47*(4), 481–493.
Juan, L., Zhimin, Z., Anchun, M., Lei, L., & Jingchao, Z. (2010). Deposition of silver nanoparticles on titanium surface for antibacterial effect. *International Journal of Nanomedicine, 5*, 261–267.
Kailasanathan, C., Selvakumar, N., & Naidu, V. (2012). Structure and properties of titania reinforced nano-hydroxyapatite/gelatin bio-composites for bone graft materials. *Ceramics International, 38*(1), 571–579.
Kanaparthy, R., & Kanaparthy, A. (2011). The changing face of dentistry: Nanotechnology. *International Journal of Nanomedicine, 6*, 2799–2804.
Karimi, M. A., Rouzbahani, H. S., Asadinia, R., Hatefi-Mehrjerdi, A., Mashhadizadeh, M. H., Ardakani, B. R., … Zebarjad, S. M. (2011). Synthesis and characterization of nanoparticles and nanocomposite of ZnO and MgO by sonochemical method and their application for zinc polycarboxylate dental cement preparation. *International Nano Letters, 1*(1), 43–51.
Kasaj, A., Röhrig, B., Zafiropoulos, G. G., & Willershausen, B. (2008). Clinical evaluation of nanocrystalline hydroxyapatite paste in the treatment of human periodontal bony defects—A randomized controlled clinical trial: 6-month results. *Journal of Periodontology, 79*(3), 394–400.
Kazi, K. M., Mandal, A. S., Biswas, N., Guha, A., Chatterjee, S., Behera, M., & Kuotsu, K. (2010). Niosome: A future of targeted drug delivery systems. *Journal of Advanced Pharmaceutical Technology & Research, 1*(4), 374–380.
Kelly, C. M., Wilkins, R. M., Gitelis, S., Hartjen, C., Watson, J. T., & Kim, P. T. (2001). The use of a surgical grade calcium sulfate as a bone graft substitute: Results of a multicenter trial. *Clinical Orthopaedics and Related Research, 382*, 42–50.
Kim, K., & Fisher, J. P. (2007). Nanoparticle technology in bone tissue engineering. *Journal of Drug Targeting, 15*(4), 241–252.
Kohli, P., & Martin, C. R. (2003). Smart nanotubes for biomedical and biotechnological applications. *Drug News & Perspectives, 16*(9), 566–573.
Kong, L. X., Peng, Z., Li, S. D., & Bartold, P. M. (2006). Nanotechnology and its role in the management of periodontal diseases. *Periodontology 2000, 40*, 184–196.
Korkmaz, Y., Ozel, E., Attar, N., & Bicer, C. O. (2010). Influence of different conditioning methods on the shear bond strength of novel light-curing nano-ionomer restorative to enamel and dentin. *Lasers in Medical Science, 25*(6), 861–866.
Krätschmer, W., Lamb, L. D., Fostiropoulos, K., & Huffman, D. R. (1990). Solid C60: A new form of carbon. *Nature, 347*, 354–358.
Kroto, H. W., Heath, J. R., O'Brien, S. C., Curl, R. F., & Smalley, R. E. (1985). C60: Buckminsterfullerene. *Nature, 318*, 162–163.
Kulal, R., Jayanti, I., Sambashivaiah, S., & Bilchodmath, S. (2016). An in-vitro comparison of nano hydroxyapatite, novamin and proargin desensitizing toothpastes - A SEM study. *Journal of Clinical and Diagnostic Research, 10*(10), ZC51–ZC54.
Kulshrestha, S., Khan, S., Meena, R., Singh, B. R., & Khan, A. U. (2014). A graphene/zinc oxide nanocomposite film protects dental implant surfaces against cariogenic Streptococcus mutans. *Biofouling, 30*(10), 1281–1294.

Lee, D. K., Kim, S. V., Limansubroto, A. N., Yen, A., Soundia, A., Wang, C. Y., ... Ho, D. (2015). Nanodiamond–gutta percha composite biomaterials for root canal therapy. *ACS Nano, 9*(11), 11490–11501.
Lee, S. Y., Kwon, H. K., & Kim, B. I. (2008). Effect of dentinal tubule occlusion by dentifrice containing nano-carbonate apatite. *Journal of Oral Rehabilitation, 35*(11), 847–853.
Lee, W. J., Alhoshan, M., & Smyrl, W. H. (2006). Titanium dioxide nanotube arrays fabricated by anodizing processes electrochemical properties. *Journal of the Electrochemical Society, 153*(11), B499–B505.
Lelli, M., Marchisio, O., Foltran, I., Genovesi, A., Montebugnoli, G., Marcaccio, M., ... Roveri, N. (2013). Different corrosive effects on hydroxyapatite nanocrystals and amine flüoride-based mouthwashes on dental titanium brackets: A comparative in vitro study. *International Journal of Nanomedicine, 8*, 307–314.
Li, L., Pan, H., Tao, J., Xu, X., Mao, C., Gu, X., & Tang, R. (2008). Repair of enamel by using hydroxyapatite nanoparticles as the building blocks. *Journal of Materials Chemistry, 18*(34), 4079–4084.
Liao, S., Wang, W., Uo, M., Ohkawa, S., Akasaka, T., Tamura, K., ... Watari, F. (2005). A three-layered nano-carbonated hydroxyapatite/collagen/PLGA composite membrane for guided tissue regeneration. *Biomaterials, 26*(36), 7564–7571.
Liu, Q., Douglas, T., Zamponi, C., Becker, S. T., Sherry, E., Sivananthan, S., ... Warnke, P. H. (2011). Comparison of in vitro biocompatibility of NanoBone and BioOss for human osteoblasts. *Clinical Oral Implants Research, 22*(11), 1259–1264.
Lughi, V., & Sergo, V. (2010). Low temperature degradation-aging-of zirconia: A critical review of the relevant aspects in dentistry. *Dental Materials, 26*(8), 807–820.
Ma, J., Wong, H., Kong, L. B., & Peng, K. W. (2003). Biomimetic processing of nanocrystallite bioactive apatite coating on titanium. *Nanotechnology, 14*(6), 619.
Moshaverinia, A., Ansari, S., Movasaghi, Z., Billington, R. W., Darr, J. A., & Rehman, I. U. (2008). Modification of conventional glass-ionomer cements with N-vinylpyrrolidone containing polyacids, nano-hydroxy and fluoroapatite to improve mechanical properties. *Dental Materials, 24*(10), 1381–1390.
Nakashima, S., Yoshie, M., Sano, H., & Bahar, A. (2009). Effect of a test dentifrice containing nano-sized calcium carbonate on remineralization of enamel lesions in vitro. *Journal of Oral Science, 51*(1), 69–77.
Nam, K.-Y. (2011). *In vitro* antimicrobial effect of the tissue conditioner containing silver nanoparticles. *The Journal of Advanced Prosthodontics, 3*(1), 20–24.
Nayar, S., Bhuminathan, S., & Muthuvignesh, J. (2011). Upsurge of nanotechnology in dentistry and dental implants. *Indian Journal of Multidisciplinary Dentistry, 1*(5), 264–268.
Nguyen, S., Hiorth, M., Rykke, M., & Smistad, G. (2011). The potential of liposomes as dental drug delivery systems. *European Journal of Pharmaceutics and Biopharmaceutics, 77*(1), 75–83.
Oh, S. H., Finones, R. R., Daraio, C., Chen, L. H., & Jin, S. (2005). Growth of nanoscale hydroxyapatite using chemically treated titanium oxide nanotubes. *Biomaterials, 26*(24), 4938–4943.
Pan, H. A., Hung, Y. C., Chiou, J. C., Tai, S. M., Chen, H. H., & Huang, G. S. (2012). Nanosurface design of dental implants for improved cell growth and function. *Nanotechnology, 23*(33), 335703.
Pandit, N., Sharma, A., Jain, A., Bali, D., Malik, R., & Gugnani, S. (2015). The use of nanocrystalline and two other forms of calcium sulfate in the treatment of infrabony defects: A clinical and radiographic study. *Journal of Indian Society of Periodontology, 19*(5), 545–553.
Paul, N. M., Bader, S. J., Schricker, S. R., & Parquette, J. R. (2006). 2,3-Branching benzyl ether dendrimers for the enhancement of dental composites. *Reactive and Functional Polymers, 66*(12), 1684–1695.
Paulose, M., Shankar, K., Yoriya, S., Prakasam, H. E., Varghese, O. K., Mor, G. K., ... Grimes, C. A. (2006). Anodic growth of highly ordered TiO2 nanotube arrays to 134 microm in length. *The Journal of Physical Chemistry. B, 110*(33), 16179–16184.

Peng, L., Cheng, X., Zhuo, R., Lan, J., Wang, Y., Shi, B., & Li, S. (2009). Novel gene-activated matrix with embedded chitosan/plasmid DNA nanoparticles encoding PDGF for periodontal tissue engineering. *Journal of Biomedical Materials Research, 90*(2), 564–576.

Pepla, E., Besharat, L. K., Palaia, G., Tenore, G., & Migliau, G. (2014). Nano-hydroxyapatite and its applications in preventive, restorative and regenerative dentistry: A review of literature. *Annali di Stomatologia, 5*(3), 108–114.

Percival, S. L., Bowler, P. G., & Russell, D. (2005). Bacterial resistance to silver in wound care. *The Journal of Hospital Infection, 60*(1), 1–7.

Petrou, I., Heu, R., Stranick, M., Lavender, S., Zaidel, L., Cummins, D., … Gimzewski, J. K. (2009). A breakthrough therapy for dentin hypersensitivity: How dental products containing 8% arginine and calcium carbonate work to deliver effective relief of sensitive teeth. *The Journal of Clinical Dentistry, 20*(1), 23–31.

Piñón-Segundo, E., Ganem-Quintanar, A., Alonso-Pérez, V., & Quintanar-Guerrero, D. (2005). Preparation and characterization of triclosan nanoparticles for periodontal treatment. *International Journal of Pharmaceutics, 294*(1-2), 217–232.

Pragati, S., Ashok, S., & Kuldeep, S. (2009). Recent advances in periodontal drug delivery systems. *International Journal of Drug Delivery, 1*, 1–14.

Price, R. L., Ellison, K., Haberstroh, K. M., & Webster, T. J. (2004). Nanometer surface roughness increases select osteoblasts adhesion on carbon nanofiber compacts. *Journal of Biomedical Materials Research. Part A, 70*(1), 129–138.

Raval, C., Vyas, K., Gandhi, U., Patel, B., & Patel, P. (2016). Nanotechnology in dentistry: A review. *Journal of Advanced Medical and Dental Sciences Research, 4*(3), 51–53.

Reves, B. T., Jennings, J. A., Bumgardner, J. D., & Haggard, W. O. (2012). Preparation and functional assessment of composite chitosan-nano-hydroxyapatite scaffolds for bone regeneration. *Journal of Functional Biomaterials, 3*(1), 114–130.

Roulet, J. F., & Roulet-Mehrens, T. K. (1982). The surface roughness of restorative materials and dental tissues after polishing with prophylaxis and polishing pastes. *Journal of Periodontology, 53*(4), 257–266.

Sadat-Shojai, M., Atai, M., Nodehi, A., & Khanlar, L. N. (2010). Hydroxyapatite nanorods as novel fillers for improving the properties of dental adhesives: Synthesis and application. *Dental Materials, 26*(5), 471–482.

Samiei, M., Aghazadeh, M., Lotfi, M., Shakoei, S., Aghazadeh, Z., & Pakdel, S. M. V. (2013). Antimicrobial efficacy of mineral trioxide aggregate with and without silver nanoparticles. *The Iranian Endodontic Journal, 8*(4), 166–170.

Saravana, K. R., & Vijayalakshmi, R. (2006). Nanotechnology in dentistry. *Indian Journal of Dental Research, 17*(2), 62–65.

Sathyanarayanan, M. B., Balachandranath, R., Genji Srinivasulu, Y., Kannaiyan, S. K., & Subbiahdoss, G. (2013). The effect of gold and iron-oxide nanoparticles on biofilm-forming pathogens. *ISRN Microbiology, 2013*, 272086.

Satyanarayana, T. S. V., & Rai, R. (2011). Nanotechnology: The future. *Journal of Interdisciplinary Dentistry, 1*(2), 93–100.

Schneider, O. D., Stepuk, A., Mohn, D., Luechinger, N. A., Feldman, K., & Stark, W. J. (2010). Light-curable polymer/calcium phosphate nanocomposite glue for bone defect treatment. *Acta Biomaterialia, 6*(7), 2704–2710.

Schulz, H., Schimmoeller, B., Pratsinis, S. E., Salz, U., & Bock, T. (2008). Radiopaque dental adhesives: Dispersion of flame-made Ta2O5/SiO2 nanoparticles in methacrylic matrices. *Journal of Dentistry, 36*(8), 579–587.

Sharma, S., Cross, S. E., Hsueh, C., Wali, R. P., Stieg, A. Z., & Gimzewski, J. K. (2010). Nanocharacterization in dentistry. *International Journal of Molecular Sciences, 11*(6), 2523–2545.

Sheikh, F. A., Barakat, N. A., Kanjwal, M. A., Nirmala, R., Lee, J. H., Kim, H., & Kim, H. Y. (2010). Electrospun titanium dioxide nanofibers containing hydroxyapatite and silver nanoparticles as future implant materials. *Journal of Materials Science. Materials in Medicine, 21*(9), 2551–2559.

Shi, H., Tsai, W. B., Garrison, M. D., Ferrari, S., & Ratner, B. D. (1999). Template-imprinted nanostructured surfaces for protein recognition. *Nature, 398*(6728), 593–597.

Shvero, D. K., Zatlsman, N., Hazan, R., Weiss, E. I., & Beyth, N. (2015). Characterisation of the antibacterial effect of polyethyleneimine nanoparticles in relation to particle distribution in resin composite. *Journal of Dentistry, 43*(2), 287–294.

Simon, Z., & Watson, P. A. (2002). Biomimetic dental implants—New ways to enhance osseointegration. *Journal of the Canadian Dental Association, 68*(5), 286–288.

Singh, V. P., Nayak, D. G., Uppoor, A. S., & Shah, D. (2012). Clinical and radiographic evaluation of nano-crystalline hydroxyapatite bone graft (Sybograf) in combination with bioresorbable collagen membrane (Periocol) in periodontal intrabony defects. *Dental Research Journal, 9*(1), 60–67.

Slavkin, H. C. (1999). Entering the era of molecular dentistry. *Journal of the American Dental Association (1939), 130*(3), 413–417.

Slenters, T. V., Hauser-Gerspach, I., Daniels, A. U., & Fromm, K. M. (2008). Silver coordination compounds as light-stable, nano-structured and anti-bacterial coatings for dental implant and restorative materials. *Journal of Materials Chemistry, 18*, 5359–5362.

Son, K. H., Hong, J. H., & Lee, J. W. (2016). Carbon nanotubes as cancer therapeutic carriers and mediators. *International Journal of Nanomedicine, 11*, 5163–5185.

Sun, L., & Chow, L. C. (2008). Preparation and properties of nano-sized calcium fluoride for dental applications. *Dental Materials, 24*(1), 111–116.

Taniguchi, N. (1974). On the basic concept of nanotechnology. In *Proceedings of the International Conference on Precision Engineering (ICPE). Tokyo, Japan* (pp. 18–23).

Tao, J., Pan, H., Zeng, Y., Xu, X., & Tang, R. (2007). Roles of amorphous calcium phosphate and biological additives in the assembly of hydroxyapatite nanoparticles. *The Journal of Physical Chemistry. B, 111*(47), 13410–13418.

Tavakoli, M., Bateni, E., Rismanchian, M., Fathi, M., Doostmohammadi, A., Rabiei, A., … Mirian, M. (2012). Genotoxicity effects of nano bioactive glass and Novabone bioglass on gingival fibroblasts using single cell gel electrophoresis (comet assay): An in vitro study. *Dental Research Journal, 9*(3), 314–320.

Tomsia, A. P., Launey, M. E., Lee, J. S., Mankani, M. H., Wegst, U. G. K., & Saiz, E. (2011). Nanotechnology approaches for better dental implants. *The International Journal of Oral & Maxillofacial Implants, 26*, 25–49.

Totu, E. E., Nechifor, A. C., Nechifor, G., Aboul-Enein, H. Y., & Cristache, C. M. (2017). Poly (methyl methacrylate) with TiO_2 nanoparticles inclusion for stereolithographic complete denture manufacturing - The future in dental care for elderly edentulous patients? *Journal of Dentistry, 59*, 68–77.

Utneja, S., Nawal, R. R., Talwar, S., & Verma, M. (2015). Current perspectives of bio-ceramic technology in endodontics: Calcium enriched mixture cement - Review of its composition, properties and applications. *Restorative Dentistry & Endodontics, 40*(1), 1–13.

Vaikuntam, J. (1997). Resin-modified glass ionomer cements (RM GICs) implications for use in pediatric dentistry. *ASDC Journal of Dentistry for Children, 64*(2), 131–134.

Vandiver, J., Dean, D., Patel, N., Bonfield, W., & Ortiz, C. (2005). Nanoscale variation in surface charge of synthetic hydroxyapatite detected by chemically and spatially specific high-resolution force spectroscopy. *Biomaterials, 26*(3), 271–283.

Viljanen, E. K., Skrifvars, M., & Vallittu, P. K. (2007). Dendritic copolymers and particulate filler composites for dental applications: Degree of conversion and thermal properties. *Dental Materials, 23*(11), 1420–1427.

Wang, L., Xie, X., Li, C., Liu, H., Zhang, K., Zhou, Y., … Xu, H. H. K. (2017). Novel bioactive root canal sealer to inhibit endodontic multispecies biofilms with remineralizing calcium phosphate ions. *Journal of Dentistry, 60*, 25–35.

West, J. L., & Halas, N. J. (2000). Applications of nanotechnology to biotechnology: Commentary. *Current Opinion in Biotechnology, 11*(2), 215–217.

Wu, J., Zhou, H., Weir, M. D., Melo, M. A. S., Levine, E. D., & Xu, H. H. K. (2015). Effect of dimethylaminohexadecyl methacrylate mass fraction on fracture toughness and antibacterial properties of CaP nanocomposite. *Journal of Dentistry, 43*(12), 1539–1546.

Xia, Y., Zhang, F., Xie, H., & Gu, N. (2008). Nanoparticle-reinforced resinbased dental composites. *Journal of Dentistry, 36*(6), 450–455.
Xie, X., Wang, L., Xing, D., Arola, D. D., Weir, M. D., Bai, Y., & Xu, H. H. (2016). Protein-repellent and antibacterial functions of a calcium phosphate rechargeable nanocomposite. *Journal of Dentistry, 52*, 15–22.
Xu, H. H., Moreau, J. L., Sun, L., & Chow, L. C. (2008). Strength and fluoride release characteristics of a calcium fluoride based dental nanocomposite. *Biomaterials, 29*(32), 4261–4267.
Zhang, L., Weir, M. D., Hack, G., Fouad, A. F., & Xu, H. H. (2015). Rechargeable dental adhesive with calcium phosphate nanoparticles for long-term ion release. *Journal of Dentistry, 43*(12), 1587–1595.

Index

A

Acute wound healing
 hemostasis, 82, 83
 inflammation, 83, 84
 proliferation
  granulation tissue formation, 85–87
  re-epithelialization, 85
 scar tissue formation, 82
Acute wounds, 82
Adaptive immunity, 120
Adipocytes, 7
Adipogenic differentiation, 8
Adipose derived mesenchymal stem cells (ADMSCs), 8, 9, 50, 70
Alginate hydrogels, 156, 157
Alginates, 3, 15, 19–22, 97, 156, 157
Aligned electrospun nanofibers
 ECM mimetic scaffolds, 151
 in vivo, 152
 neurites and axons, 150
 neuronal differentiation, 151
 NGF encapsulation, 151
 PCL, 151
 PLGA-carbon nanotube composite nanofibers, 151
 PLLA, 150
 PMMA, 150
 polymers, 150
Aligned micro fibers (AMF), 57
Aligned nanofibers (ANF), 57
Alkaline phosphatase (ALP), 8
Allograft transplantation, 188
α-melanocyte-stimulating hormone (α-MSH), 205
Amorphous silica nanoparticles, 124
Angiogenesis, 85–87
Angled U-shaped patterns, 65
Anionic polymers, 21
Antibacterial activity, 15
Antitumor immune response, 124
Articular cartilage, 185, 187
 calcified cartilage, 186
 calcified zone, 186
 chondrocytes, 186
 chondrogenesis, 186
 collagen II and proteoglycans, 185
 zonal structure, 185–186
Atomic force microscopy (AFM), 199
Autograft transplantation, 189
Autoimmune diseases, 129
Autoimmune disorders, 124
Autologous chondrocyte implantation (ACI), 189

B

Basic fibroblast growth factors (bFGF), 51
Battery-operated e-spinning apparatus (BOEA), 54, 55
Biocompatibility studies, 206
Bioimaging, 12, 14, 26
Bioinert, 19
Biological nanomaterials
 amino acids, 19
 DNA and RNA, 25–26
 living cells/dead organisms, 19
 nanoscale characteristics, 19
 natural, 19–22
 self-assembling synthetic, 22–25
 structural units, 19
 synthetics, 19

A. B. Tekinay (ed.), *Nanomaterials for Regenerative Medicine*, Stem Cell Biology and Regenerative Medicine, https://doi.org/10.1007/978-3-030-31202-2

Biomedical applications, 13
Biosensing, 14
Biosensors, 12
Blastocysts, 9
BMP-mimetic peptides, 8
Bone cells, 10
Bone marrow derived MSCs, 7, 8
Bone marrow homing peptide 1 (BMHP-1), 160
Bone morphogenetic protein 2 (BMP-2), 69
Bone morphogenetic proteins (BMPs), 7–8, 186
Bone regeneration, 68–69, 203–204
4-Butanediol diglycidyl ether (BDDGE), 56

C
Cancer immunotherapy, 120
Cancer therapy, 12
Cancer vaccines, 123
Carbohydrates, 5, 19, 24
Carbon-based nanomaterials
  biomedical application, 12
  carbon nanotubes, 13, 164, 165
  electrical and mechanical characteristics, 12
  fullerenes, 12
  functionalization, 12
  graphene, 14
  graphene and fullerenes, 165, 166
  implantable-sensors, 12
  neurological applications, 12
  one-dimensional carbon nanotubes, 12
  regenerative medicine applications, 164
  two-dimensional graphene, 12
  zero-dimensional fullerenes, 12
Carbon nanoparticles, 128
Carbon nanotubes, 13, 130, 164, 165, 199
Cardiomyocytes, 9, 11
Cartilage insufficiency, 187
Cartilage regeneration, 71, 73, 188, 190, 192
Cartilage tissue, 191
  bioactive interactions, 192
  biomaterials, 190
  electrospinning technique, 192
  nanomaterials, 190
  PI3K/AKT signaling pathway, 193
  structures, 184
  SWNTs, 193
  tissue engineering, 190
  types, 184
Cell fates, 5, 10, 14, 18, 21
Cell morphology, 57
Cell therapy, 123
Cellulose based nanofibers, 68
Cerium oxide nanoparticles, 131
Chitin, 96
Chitin/carbon nanotube, 164
Chitin nanofibrous (Chi-N) matrices, 51
Chitosan, 3, 15, 19–22, 24, 96, 157, 158
Chondral injury, 188
Chondrocytes, 7, 10, 186
Chondrogenesis, 186–187
Chondroitin sulfate proteoglycans (CSPG), 146, 147
Chronic wounds, 82, 89, 90
Collagen, 20, 93, 94, 153–155
Collagen fibrillary structure, 185
Collagens, 145
Conventional tissue engineering, 190
C-reactive protein (CRP), 208
Cryogels, 3, 17
Culture surface, 8
Cytocompatibility, 52
Cytokeratin staining, 53
Cytokines, 4
Cytoskeletal components, 12
Cytostaining methods, 8
Cytotoxic effects, 13

D
Dendritic cells (DCs), 123, 125, 131
Dental implants, 206
Dental nanorobots, 209
Dental resin nanocomposites, 202
Dentifrices, 199–200
Dermal fibroblasts, 86
Diabetic ulcers, 90
Diagnostic dentistry
  AFM, 199
  biofilm layer, 198
  biomarkers and nucleic acids, 199
  graphene, 198
  quantum dots, 198
  treatment, 198
Digital multimirror device (DMD), 55, 56
DNA origami, 18, 25

E
E7-BMP-2, 69, 70
ECM elements, 12
ECM-mimetic nanofibers, 9
ECM-mimetic scaffolds, 13
Elastin-based nanomaterials, 95
Electrospinning, 17, 18, 20, 148, 154, 192
  ADMSCs, 50
  cell migration and proliferation, 50
  cellular environment, 48
  characteristics, 49

components, 48
electrostatic and spinning, 48
electrostatic repulsion, 48
neural regeneration, 57–67
orthopedic tissue regeneration (*see* Orthopedic tissue regeneration)
parameters, 48, 49
preparation, 48
skin regeneration, 50–56
Electrospun chitosan nanofibers, 22
Electrospun nanofibers, 11, 15, 18
aligned form, 150
bio-nano research, 47
design and fabrication, 148
electrospinning, 148
functionalization, 48
non-aligned form, 152, 153
tissue regeneration, 148, 149
Electrospun polymeric nanofibers, 3
Electrospun polysaccharides, 3
Electrostatic repulsion, 48
Embryonic cells, 6
Embryonic stem cells, 9, 10, 14
Endodontics, 205
Environmental parameters, 18
Epidermis, 50
European Medicines Agency (EMA), 5
Extracellular matrix (ECM), 48, 82
biological signals, 4
biomacromolecules, 6
cartilage, 4
characteristics, 3
collagen/fibronectin, 4
complexity, 6
components, 3
composition, 2
fibrous *vs.* non-fibrous proteins, 2
fibrous/non-fibrous components, 4
flexibility/biodegradability, 4
functions, 2, 3
GAGs, 2, 5
growth factors, 5
laminins, 4
MMPs, 6
and molecular characteristics, 5
morphology, 4
MSCs (*see* Mesenchymal stem cells (MSCs))
non-fibrous proteins, 4
peptides, 5
pores, 3, 4
porous materials, 3
proteinaceous components, 5
reconstruction, 2
self-assembling peptide amphiphile nanofibers, 3
self-assembling synthetic amyloid-like peptide molecules, 3
short peptide sequence, 4
size of protein, 5
structure, 2
Extracellular matrix elements (ECM), 51

F

Fabrication process, 11
Fetal wounds heal, 91
Fibrillar proteins, 48
Fibrin, 94, 95, 155, 156
Fibrin glue, 21
Fibrin hydrogels, 155, 156
Fibrin nanoparticles (NPs), 94
Fibrinogen, 83
Fibroblast growth factor (FGF), 186
Fibroblasts, 20, 88
Fibronectin, 20, 21
Fibrosis, 86
Fibrous protein, 20
Fluorescence micrographs, 62
Fourier fast transfer (FFT), 58
Fullerenes, 12, 130

G

GFOGER, 4
Glycopeptide nanofibers, 24
Glycoproteins, 24
Glycosaminoglycans (GAGs), 48, 146, 147, 185
Gold and silica nanoparticles, 125
Gold nanoparticles, 14, 15, 102, 127
Granulation tissue formation
angiogenesis, 85–87
fibrosis, 86
Graphene, 14, 198
Graphene oxide, 165, 166
Graphene/zinc oxide nanocomposites (GZNC), 198

H

Hematopoietic stem cells, 6, 11
Hemostasis, 82, 83
Heparin/fibrin-based delivery system (HBDS), 70, 71
Heparin-mimetic PA nanofibers, 103
Hepatocytes, 10, 11
Hollow polymeric nerve guidance channels, 158

Human astrocytoma epithelial like cells, 58
Human induced pluripotent stem cells (hIPSCs), 152
Human osteoblasts (hOBs), 193
Hyaluronic acid (HA), 20, 98, 152
Hydrogels, 15, 22, 23, 191
Hydrogel system, 193
Hydrogen bonds, 22
Hydrophobic and electrostatic interactions, 22
Hydroxyapatite, 20
Hydroxyapatite nanoparticles, 203

I
Immune system, 120
Immunohistochemical analysis, 163
Immunomodulatory
 anti-allergy drugs, 121
 antigen, 120
 cancer vaccines, 124
 diagnostic systems, 121
 drugs, 120
 infections and toxins, 119
 nanomaterials, 122
 radiation and chemotherapy, 122
 types, 120
 vaccines, 120, 121
Immunostimulatory molecules, 129
Immunosuppression, 120, 131
In-body sensors, 13
Induced pluripotent stem cells (iPSCs), 10, 11
Inflammation, 83, 84
Influenza vaccines, 128
Intratumoral injection, 122
Iron oxide nanoparticles, 16, 130

K
Keratin, 95
Keratinocytes, 55
Keratocytes, 13

L
Lamina splendens, 185
Laminins, 4, 61, 145, 156
Layer-by-layer formation, 20
Ligaments, 20
Liposomal encapsulation, 132
Liposomes, 125, 129
Lymphocytes, 11

M
Macrophages, 84
Manganese-enhanced magnetic resonance imaging (MEMRI), 159
Matrix metalloproteases, 4
Matrix Metallo-Proteinase-8 (MMP-8), 208
Matrix metalloproteinases (MMPs), 188
Matrix-induced autologous chondrocyte implantation (MACI), 189
Mechanotransduction, 8
Megakaryocytes, 11
Mesenchymal stem cells (MSCs), 190
 adipocytes, 7
 adipose tissue derived, 7–9
 bone marrow derived, 7, 8
 chondrocytes, 7
 complex variables, 7
 controversial, 7
 ECM-mimetic nanofibers, 9
 fibroblasts, 7
 heterogeneity, 7, 9
 molecular mechanisms, 9
 osteoblasts, 7
 regenerative medicine applications, 6
 tissue origin, 7
 tissue regeneration, 7
 transplantation, 4
 treatments, 9
 umbilical cord-derived, 9
 umbilical vein, 7
 variety, 7
Mesoporous glass nanoparticles, 15
Mesoporous silica nanoparticles, 127
Metal nanoparticles, 18
 biocompatible and bioactive, 15
 biomedical applications, 14
 characteristics, 14
 drug delivery applications, 14
 gold, 14, 15, 127
 inorganic materials, 127
 iron oxide, 16, 127
 mesoporous glass, 15
 mesoporous silica nanoparticles, 128
 one-dimensional/two-dimensional, 14
 scaffold systems, 14
 silver, 15
 tissue regeneration, 14
 zinc oxide, 16
Metallic silver, 101
Metronidazole (MNA), 69
Microfracture method, 188
Microstereolithography, 56
Molecular biology, 2

Mouthwashes, 200
Multipotent stem cells, 6
Multiwalled carbon nanotubes, 164
Musculoskeletal injuries, 67
Musculoskeletal system, 183
Myofibroblasts, 86, 88

N

Nano-amorphous calcium phosphate (nACP), 202
Nanocarrier systems, 125
Nano light-curing glass ionomer, 202
Nanocellulose, 98
Nanoclusters, 14
Nanocomposites, 3, 201
Nanocubes, 14
Nanodentistry, 198
Nanofiber hydrogels, 192
Nanofibers, 13
Nanofiller particles, 201
Nanohydroxyapatite (nHA), 69
Nanomaterials, 191
  advantages, 12
  biological (*see* Biological nanomaterials)
  biomacromolecules, 12
  bottom-up approaches, 12
  bottom-up methods, 12
  definition, 11
  disadvantages, 12
  nonbiological (*see* Nonbiological nanomaterials)
  regenerative medicine (*see* Regenerative medicine)
  self-assembly, 12
  synthesis, 11
  top-town methods, 11
  types, 6
Nanoparticles, 132, 204
Nanoparticle vaccines, 125
Nanoporous anodic alumina, 206
Nanoribbon/nanoyarn, 13
Nanorobotic analgesics, 209
Nanorobotics systems, 209
Nanorods, 13, 14
Nanoscale materials, *see* Nanomaterials
Nanoshells, 14
Nanosized hydroxyapatite particles, 200
Nanospheres, 14
Nanostars, 14
Nanotechnology, 197
Natural biological materials
  alginate and chitosan, 19
  alginates, 21
  chitosan, 22
  collagen, 20
  donor material, 19
  fibronectin, 20, 21
  molecules, 19
  regenerative medicine applications, 19
Natural biomaterials
  alginate, 156, 157
  chitosan, 157, 158
  collagen, 153–155
  ECMs, 153
  fibrin, 155, 156
  laminin, 156
Natural biomimetic molecules, 3
Natural polymers, 191
  polysaccharide-based polymers
    alginate, 97
    chitosan, 96
  protein-based polymers
    collagen, 93, 94
    fibrin, 94, 95
Nerve guide conduits (NGCs), 65, 67
Netrins, 147
Neural ECM components
  collagens, 145
  GAGs, 146, 147
  laminins, 145
  netrins and slits, 147
  neural stem cells, 144
  reelin, 147
  tenascins, 146
Neural prostheses, 13
Neural regeneration, 149, 151, 157, 160–165, 167
  angled U-shaped patterns, 65
  BMSCs, 60
  cell viability and morphology, 58
  cells growth, 58
  controlled design of aligned and random nanofibers, 65, 66
  DRGs, 61, 62
  electrospinning setup, 59
  FFT, 58
  HA, 58
  human astrocytoma epithelial like cells, 58
  lower organisms, 57
  MA, 58
  NA, 58
  nanocomposite scaffolds, 63
  nanospheres, 63
  neural scaffold, 64
  neural tissue engineering, 60
  neurite outgrowth, 57, 61
  neurites, 62
  NGCs, 65, 67

Neural regeneration (*cont.*)
non-aligned (NA), 58
NSCs, 57
PCL, 62, 63
PCLM, 57
peripheral and central nervous system, 57
PLLA, 57
SEM, 58, 59
stacked nanofibers, 57, 58
3D AFs, 63, 65
3D tubular nerve guides, 63
tissue engineered, 63
2P electrospinning system, 63
uniaxially aligned nanofibers, 61
Neural stem cells (NSCs), 57
Neural tissue regeneration, 144
Neuronal lineage, 10
Neuroregeneration, 13
Non-aligned electrospun nanofibers, 152, 153
Nonbiological nanomaterials
biomedical applications, 12
carbon-based, 12–14
metal nanoparticles, 14–16
synthetic nonbiological polymers (*see* Synthetic nonbiological polymers)

O
Oil-Red-O staining, 8
One-dimensional carbon-based nanomaterials, 13
Oral Fluid Nano Sensor Test (OFNASET), 199
Organic-soluble chitosan (O-CS), 51
Orthopedic tissue regeneration
bone regeneration, 68–69
cartilage regeneration, 71, 73
musculoskeletal injuries, 67
tendon regeneration, 69–72
Osteoarthritis (OA), 184, 187
Osteoblastic MC3T3-E1 cells, 68
Osteoblasts, 7
Osteochondral injury, 189
Osteogenic cells, 11
Osteogenic differentiation, 8
Osteogenic proteins, 8

P
PCL-chitosan emulsion, 53
Peptide amphiphile nanofibers, 23, 24
Peptide amphiphiles (PAs), 103
aligned PA nanofibers, 162
GAGs, 162
hydrogels, 162
IKVAV epitope displaying nanofibers, 161, 162
macromolecular structures, 161
peptide sequences, 161
peripheral nerve regeneration, 163
self-assembly process, 161
Tenascin C, 163
3D culturing, 163
Peptide-based nanomaterials, 18
Peptide–carbon nanotube composites, 13
Peptide nanofibers, 11, 23
PAs, 161–163
self-assembling peptides, 159
Peptide nanospheres, 25
Perineural nets (PNNs), 147
Periodontology, 207
Peripheral nervous system (PNS), 145
Phase contrast, 52
Phase contrast light microscopy (PCLM), 57
Photolithography, 11
Platelet-derived growth factor BB (PDGF-BB), 70
Pluripotency, 9, 10
Pluripotent stem cells, 6
Poly (ethylene glycol)-poly(DL-lactide) (PELA), 51
Poly(3,4-ethylenedioxythiophene) (PEDOT), 158
Poly(3-hydroxybutyrate-co-3-hydroxyvalerate) (PHBV), 53, 55
Poly(glycerol sebacate) (PGS), 152
Poly(L/D-lactic acid) (PLDLA), 164
Poly(L-lactic acid) (PLLA), 57
Poly(L-lactide ε-caprolactone) and fibrinogen (PLCL/FBG), 50
Poly(L-lysine) (PLL), 61
Poly(methyl methacrylate) (PMMA), 150
Poly(γ-caprolactone), 10
Polycaprolactone (PCL), 100, 101
Polyethylene glycol (PEG), 99
Polyhydroxy butyrate (PHB), 51
Polylactic acid (PLA), 99, 100
Polylactic-co-glycolic acid (PLGA), 65, 100
Polylactide-cyclosporine A nanoparticles, 132
Poly-L-lactic acid, 10
Poly-L-lactide/hydroxyapatite (PLLA/HA), 68
Polymer fibers, 17
Polymeric materials, 18, 19
Polymeric nanofibers, 10
Polymeric nanoparticles, 126, 131
Polymethylmethacrylate (PMMA), 205
Polypyrrole/iodine coating, 56

Polysaccharide-based polymers
alginate, 97
chitosan, 96
HA, 98
Polyurethane (PU), 65
Polyvinylpyrrolidone fibers, 56
Porosity, 3
Pressure ulcers, 90
Preventive dentistry
dentifrices, 199–200
inorganic building blocks, 199
mouthwashes, 200
nano-toothbrush concept, 200
silica nanoparticles, 200
Processing parameters, 18
Proliferative chondrocytes, 186
Prostaglandins, 83
Prosthetic cements, 205
Prosthetic dentistry, 205
Protein-based polymers
collagen, 93, 94
elastin, 95
elastin-based nanomaterials, 95
fibrin, 94, 95
keratin, 95
silk fibroin, 95
Proteoglycans, 48
Proto-myofibroblasts, 86
Pulmonary exposure, 13

Q
Quantum dots, 131, 198
Quaternized polyethyleneimine (QPEI), 202

R
RADA peptides, 160
Random micro fibers (RMF), 57, 63
Random nanofibers (RNF), 57
Recombinant vaccines, 129
Reelin, 147
Re-epithelialization, 85
Regenerative dentistry
alloplastic bone grafts, 203
bone regeneration, 203–204
calcium phosphate and zinc oxide, 204
calcium sulfate, 203
graft material, 204
Regenerative medicine, 144
applications, 5
capabilities, 2
decellularization/acellularization, 2
mammals, 2
tissue regeneration, 1, 2
transplantation therapy, 47
Regenerative methods, 189
Regenerative therapies, 189
Regenerative wound healing, 90, 91
Rehabilitative dentistry
endodontics, 205
exopolymers, 207
Resin-modified glass ionomer cement (RMGIC), 202
Restorative dentistry
calcium peroxide nanoparticles, 203
nACP, 202
nanocomposites, 201
nanofiller particles, 201
nanorods, 202
nanotechnology, 201
restorative material, 201
Retinal applications, 13

S
Satellite cells, 11
Scar formation
tissue remodeling, 88, 89
Self-assembling natural biomaterials, 19
Self-assembling peptide amphiphile nanofibers, 3
Self-assembling peptide nanomaterials, 102–104
Self-assembling peptides, 23
laminin-derived IKVAV sequence, 159, 160
neural regeneration applications, 159
RADA peptides, 160
RGD/IKVAV, 160
spinal cord injury repair, 161
tissue regeneration studies, 159
traumatic brain injury models, 159
Self-assembling synthetic amyloid-like peptide molecules, 3
Self-assembling synthetic biological materials
biochemical rules, 22
bottom-up approach, 22
glycopeptide nanofibers, 24
intermolecular forces, 22
peptide amphiphile nanofibers, 23, 24
peptide nanofibers, 23
peptide nanospheres, 25
supramolecular structures, 22
top-down approach, 22
Silica-based nanoparticles, 127
Silk-fibroin (SF) nanofibers, 51, 95
Silk nanofibers, 15

Silver, 101
Silver nanoparticles (AgNPs), 15, 101, 102, 202, 207
Single-wall carbon nanotubes (SWNTs), 164–165, 193
Skeletal muscle cells, 11
Skin, 82
Skin regeneration
  bFGF, 51
  BOEA, 54, 55
  CECS/PVA, 51
  Chi-N matrices, 51
  core–sheath bFGF/PELA fibers, 51
  cytocompatibility, 52
  dermis, 50
  ECM, 51
  electrospinning technique, 51
  electrospun nanofibrous, 50
  emulsion solution, 53, 54
  epidermis, 50
  fabrication process, 55
  FESEM micrographs, 52
  gellan–PVA hydrogels, 52
  hypodermis, 50
  immunostaining, 53
  natural/synthetic materials, 52
  NHEF, 51
  NHEK, 51
  optical images, 54, 55
  PCL, 51
  PEGDA, 56
  phase contrast, 52
  PHBV polymeric solution, 55
  PLACL/SF/VE/Cur nanofibrous, 53
  SF nanofibers, 51
  stereolithography, 55
  surface characteristics, 51
  thermoregulation, 50
  3D silk fibroin nanofibers, 55
  3D structure, 53
  ultrafine fiber membranes, 51
  USSCs, 53
Skin wounds, 94, 100
Slits, 147
Soft tissue regeneration, 4
Solution parameters, 17
Solvent free methods, 18
South West Research Institute, 208
Stacked nanofibers, 57, 58
Stem cells
  classification, 6
  differentiation pathways, 3
  ECM, 6
  embryonic, 6, 9, 10
  epigenetic regulations, 6
  iPSCs, 10, 11
  multipotent, 6
  pluripotent, 6
  tissue regeneration, 3
  tissue specific, 11
  totipotent, 6
  transplantation, 3, 7
Stereolithography, 55
Supramolecular self-assembly, 102
Synthetic nanomaterials
  metallic NPs, 101, 102
  self-assembling peptide nanomaterials, 102–104
  synthetic polymers
    PCL, 100, 101
    PEG, 99
    PLA, 99, 100
    PLGA, 100
Synthetic nonbiological polymers
  advantages and disadvantages, 18, 19
  biomedical applications, 16
  preparation technique
    cryogels, 17
    electrospinning, 17, 18
    polymeric materials, 16
    porous structure, 16
    solvent free methods, 18
    3D printing, 18
    wet methods, 17
  regenerative medicine, 16
  tissue engineering, 16
Synthetic polymers, 126, 191
  PCL, 100, 101
  PEG, 99
  PLA, 99, 100
  PLGA, 100
  PU, 101

T
Tailored nanoparticles, 124
Targeted drug delivery, 12
Tenascin C (TN-C), 8, 146, 163
Tenascin R (TN-R), 146
Tenascins, 146
Tendon regeneration, 69–72
Tendons, 20
TheraCys®, 124
Theranostic agents, 5, 14
Therapeutic vaccines, 123
Thermoregulation, 50

3D AFs, 63, 65
3D printing, 18
3D silk fibroin nanofibers, 55
Tissue engineering, 2, 190
Tissue regeneration, 1–3, 16, 204
Tissue remodeling, 88, 89
Tissue specific stem cells, 11
Totipotent stem cells, 6
Transmission electron microscopy, 163
Triple helix structure, 20
Tumor microenvironment, 122
Two-pole (2P) electrospinning system, 63

U
Ultrafine fiber membranes, 51
Umbilical cord-derived MSCs, 9
Uniaxially aligned nanofibers, 61
Unrestricted somatic stem cells (USSCs), 53
Uronic acid, 21
U.S. Food and Drug Administration (FDA), 5

V
Van der Waals, 22
Vascular endothelial growth factor (VEGF), 84, 87
Vault nanoparticles, 128
Venous ulcers, 90
von Willebrand factor (vWF), 83

W
Wet methods, 17
Wound contraction, 88
Wound healing
  acute (*see* Acute wound healing)
  characteristics, postnatal and etal, 92
  chronic wounds, 89, 90
  contraction, 88
  natural polymers (*see* Natural polymers)
  regenerative, 90, 91
  scar formation, 88, 89 (*see also* Acute wound healing)
  synthetic nanomaterials (*see* Synthetic nanomaterials)
  tissue remodeling, 88, 89
Wound/scar tissue, 20
Wounds
  classification, 82
  definition, 81

Z
Zinc oxide nanoparticles, 16

MIX
Papier aus verantwortungsvollen Quellen
Paper from responsible sources
FSC® C105338

If you have any concerns about our products,
you can contact us on
ProductSafety@springernature.com

In case Publisher is established outside the EU,
the EU authorized representative is:
Springer Nature Customer Service Center GmbH
Europaplatz 3, 69115 Heidelberg, Germany

Printed by Libri Plureos GmbH
in Hamburg, Germany